2nd International Conference on Mechanical Manufacturing and Industrial Engineering 2019

IOP Conference Series: Materials Science and Engineering
Volume 784

Beijing, China
14 - 16 December 2019

ISBN: 978-1-7138-2668-2
ISSN: 1757-8981

Printed from e-media with permission by:

Curran Associates, Inc.
57 Morehouse Lane
Red Hook, NY 12571

Some format issues inherent in the e-media version may also appear in this print version.

Institute of Physics
Dirac House, Temple Back
Bristol BS1 6BE UK

Phone: 44 1 17 929 7481
Fax: 44 1 17 920 0979

techtracking@iop.org

Additional copies of this publication are available from:

Curran Associates, Inc.
57 Morehouse Lane
Red Hook, NY 12571 USA
Phone: 845-758-0400
Fax: 845-758-2633
Email: curran@proceedings.com
Web: www.proceedings.com

TABLE OF CONTENTS

DESIGN AND APPLICATION OF ENGINEERING MATERIALS

MODERN MECHANICAL THEORY AND DESIGN

MECHANICAL DESIGN MANUFACTURING AND AUTOMATION

IOP Conf. Series: Materials Science and Engineering **784** (2020) 011001 doi:10.1088/1757-899X/784/1/011001

Preface

It is our great pleasure to welcome you to the 2nd International Conference on Mechanical Manufacturing and Industrial Engineering (MMIE2019) held in Beijing, China during December 14-16, 2019 with the workshop, 2019 The 2nd International Conference on Advanced Composite Materials (ICACM 2019). The conference is dedicated to issue related to the areas of advanced composite materials, mechanical manufacturing and industrial engineering.

Prof. Alan Lau from Swinburne University of Technology, Australia, Prof. Yong Suk Yang from Pusan National University, Korea, Prof. Lihui Lang from Beihang University, China and Prof. Tingkai Zhao from Northwestern Polytechnical University, China were invited to deliver keynote speeches and provide latest findings in their respective areas of expertise. It was a golden opportunity for the students, researchers and engineers to interact with the experts and specialists to get their advice or consultation.

Due to the constraint of the capacity and suitability of the topics, 34 papers were accepted as full papers for publication and presentation from 75 submissions. We would like to express our gratitude and appreciation to the authors for your contributions. Many thanks go as well to all of the reviewers who helped us maintain the high quality of manuscripts included in the Proceedings published by *IOP Conference Series: Materials Science and Engineering*. We also express our sincere thanks to the members of the organizing team for their hard work. We wish that all the authors and delegates find MMIE2019 interesting, exciting and inspiring. We also wish that our guests will enjoy your stay in Beijing, China at this beautiful time of the year.

Looking forward to see you again in Tokyo, Japan during August 26-28, 2020!

Prof. Alan Lau
Swinburne University of Technology, Australia,
On behalf of Conference Committee

Conference Committees

Conference Committee Chairs

Prof. Lihui Lang, Beihang University, China

Prof. Alan Lau, Swinburne University of Technology, Australia

Prof. Yong Suk Yang, Pusan National University, Korea

Program Committee Chair

Prof. Young Hoon Rim, Semyung University, Korea

International Technical Committees

Shen-Yung Lin, National Formosa University, Taiwan

Arnold C. Alguno, MSU-Iligan Institute of Technology, Philippines

Abdullah Chik, Unversiti Malaysia Perlis, Malaysia

Wan-Yang Gao, Shanghai Jiao Tong University, China

Rey Y. Capangpangan, Caraga State University, Philippines

Radhika Pathak, Priyadarshini Institute of Architecture and Design studies, India

Ibetombi Soibam, NIT Manipur, India

R.Yuvakkumar, Alagappa University, India

Omid Khalaj, University of West Bohemia, Czech Republic

Mohamad Kahar Ab Wahab, Unversiti Malaysia Perlis, Malaysia

Hendra Suherman, Universitas Bung Hatta, Indonesia

Hao Wang, Northeastern University, China

Omid Reza Baghchesaraei, Western Sydney University, Australia

Bo Xu, Shawcor Ltd., 25 Bethridge Rd. Toronto, ON, CANADA

Chye Lih Tan, Universiti Malaysia Perlis, Malaysia

Yuanyuan CUI, National Institute of Metrology, China

Budisantoso Wirjodirdjo, Institut Teknologi Sepuluh Nopember, Indonesia

Rungchat Chompu-inwai, Chiang Mai University, Thailand

Danardono Agus Sumarsono, University of Indonesia, Indonesia

IOP Conf. Series: Materials Science and Engineering **784** (2020) 011001 doi:10.1088/1757-899X/784/1/011001

Statement of Peer Review

All papers published in this volume of *IOP Conference Series: Materials Science and Engineering* have been peer reviewed through processes administered by the proceedings Editors. Reviews were conducted by expert referees to the professional and scientific standards expected of a proceedings journal published by IOP Publishing.

Effects of refractory nitride reinforcements on the microstructure, densification and microhardness of Ti6Al4V-based binary composites produced by spark plasma sintering

J O Abe[1,2, a], A P I Popoola[1], O M Popoola[2] and E Ajenifuja[1,2]

[1] Department of Chemical, Metallurgical and Materials Engineering, Tshwane University of Technology, P.M.B. X680, Pretoria, South Africa
[2] Centre for Energy and Electric Power, Tshwane University of Technology, Pretoria, South Africa

[a]*johnabe.fm@gmail.com*

Abstract. The effects of different refractory nitrides: aluminium nitride (AlN), titanium nitride (TiN) and hexagonal boron nitride (*h*-BN) reinforcements on the microstructure, densification and microhardness characteristics of spark plasma sintered Ti6Al4V-based binary composites were studied. The microstructure was examined by optical microscopy, densification was evaluated according to the method of Archimedes and the microhardness of the sintered composites was measured by Vickers microhardness tester. The SPS technique was successfully used to consolidate Ti6Al4V powder and 3 wt.% nanoparticles of AlN, TiN and *h*-BN respectively. The binary composites produced attained nearly full theoretical densification (98.22 to 99.54%) due to enough diffusional mass transport ensuing in solidly bonded compact of the matrix and the refractory nitride reinforcements. Ti6Al4V-3*h*-BN composite gave the optimal combination of relative densification and microhardness values of 99.54% and 7.03 GPa which exceeds 200% that of the monolithic alloy and about 48% superior to both composites with TiN and AlN reinforcements.

1. Introduction

Titanium and its alloys find wide use in airframes, jet engine and automotive structures owing to their exceptional combination of properties: moderate density, high structural properties (e.g. strength, stiffness, toughness and fatigue resistance), creep strength, and excellent resistance to corrosion and oxidation [1-2]. Of the many types of titanium alloys with different compositions in use today, the most important is Ti6Al4V. The alloy is comprised of both α-Ti and β-Ti, and therefore, regarded as $\alpha+\beta$-Ti alloy [1,3]. Despite the undeniable benefits, extensive applications of Ti6Al4V alloy in the automotive and aerospace industries has been restricted because of low hardness and poor tribological properties, characterized by a high friction coefficient, weak self-lubricating property and severe adhesive wear, and insufficient high-temperature oxidation resistance [4-6]. The last five decades have witnessed new developments in titanium matrix composites (TMCs) which possess high strength, good wear resistance, high temperature stability owing to the added reinforcement. Superior desirable set of mechanical properties are obtained when fine and thermally stable ceramic particulates are dispersed in a TMC [5,7-8].

In recent times, studies have shown tremendous advantages of applications of refractory nitrides reinforcements in the development of materials meant for use in the automotive and aerospace industries. These nitrides are of a strong structural and chemical nature especially in applications involving high

temperature and/or corrosive environments where most metals are no longer suitable. Refractory nitrides consist of the nitrides of Ti, V, Zr, Nb, Hf, Ta, B, Al and Si [7-8]. Ray et al. [8] fabricated Al-based composites with 10 and 30 wt.% titanium nitride (TiN) by using pressureless sintering and hot-pressing techniques. The presence of titanium nitride (TiN) particles at grain boundaries played an important role in improving the densification, and results in improved mechanical properties. de Araugo et al. [9] reinforced Al with aluminium nitride (AlN) and silicon nitride (Si_3N_4) separately by mechanical alloying, cold uniaxial compaction and vacuum sintering. The results showed better Vickers hardness as mass fraction of reinforcements were increased. Yang et al. [10] and Matizamhuka [11] reported the frequent challenges manufacturers encounter in the synthesis of nanosized materials by conventional sintering methods. However, at the instance of the development of spark plasma sintering (SPS) technology, such recurrences are no more menacing. For instance, two dissimilar materials of Al and BN which cannot be prepared in a composite form via conventional powder metallurgical method as reported by Du et al. [12] were successfully achieved by Firestein et al. [13] where Al-based composite with BN nanoparticles reinforcement was synthesized via spark plasma sintering method. Kgoete et al. [14] also produced Ti6Al4V- Si_3N_4 composite by using spark plasma sintering process. In all these studies, incredible improvement in the mechanical properties of the MMCs were achieved. Besides, SPS offers rapid densification, with minimal grain growth, enhanced microstructure combined with improved mechanical properties and more within reduced sintering times in contrast to the conventional hot-pressing technology [11,15].
In this work, SPS was utilized to synthesize Ti6Al4V-based composites reinforced with nanoparticles of AlN, TiN and *h*-BN. The effect of the refractory nitride reinforcements on the microstructure, densification and microhardness of the composites were also investigated.

2. Materials and methods
The raw materials in this work were commercially pure Ti6Al4V powder (>99%, 45-90 μm spherical particle sizes by TLS Technik, Germany) used as the matrix and AlN, TiN and *h*-BN (>99%, 100-200 nm by Hongwu International Group Ltd., China) as the reinforcements respectively. Three sets of Ti6Al4V-based powders of the same weight with each consisting 3 wt.% of each reinforcement and a control of pure Ti6Al4V were measured with a high precision scale and mixed in a tubular mixer rotating at a speed of 100 rpm for 10 hours to achieve homogeneous mixture. Some steel balls of approximately ϕ8mm were incorporated into the powder container at the ball-to-powder weight ratio of 2:5 during the mixing for better distribution of the reinforcement in the matrix.
The admixed powders were measured into a graphite die of ϕ30mm to produce 10mm thickness of the composite and consolidated by using spark plasma sintering machine (SPS FCT Systeme GmbH, Germany) under argon atmosphere with optimized sintering parameters (1000 °C temperature, 30 MPa pressure, 100 °C/min heating rate and 10 min dwell time) obtained from an initial optimization experiment designed by Taguchi method. The density measurements of the sintered samples were carried out by using Archimedes' principle where the average value of six repeated measurements was taken for each sample and recorded as the effective density and the theoretical density was calculated according to the rule of mixture. Subsequently, the relative densification of each sample was calculated as the ratio of the effective density and theoretical density multiplied by 100% [14,16].
Later, the samples were first grinded by using 120, 400, 500, 800, 1200, 2400, 4000 grit papers and then polished by using diamax powder (6 μm, 3 μm and 1 μm) in solution to achieve mirror-like polish. Afterwards, Kroll's reagent was used to etch the samples, and the surfaces were observed under an optical microscope (Olympus BX 51 TRF, Japan).
Finally, microindentation of the sintered samples was done by using Vickers microhardness tester (FUTURE-TECH FM-800 from Japan) with a dura scan diamond indenter and test load of 300gf for 15 seconds with a spacing of 0.5. The average value of five indentations was taken for each sample and recorded as the effective microhardness value.

3. Results and discussion
The details of the results of the experiment obtained for relative densification and microhardness measurement is shown in table 1 and figure 1 shows the graph of the comparison of the relative

densification and microhardness values of Ti6Al4V alloy, Ti6Al4V-3AlN, Ti6Al4V-3TiN and Ti6Al4V-3*h*-BN.

Table 1: Densification and microhardness measurement details

Sample	Effective density (g/cm^3)	Theoretical density (g/cm^3)	Relative densification (%)	Microhardness (GPa)
Ti6Al4V alloy	4.42	4.50	98.22	3.25
Ti6Al4V-3AlN	4.41	4.45	99.10	4.79
Ti6Al4V-3TiN	4.44	4.52	98.23	4.75
Ti6Al4V-3*h*-BN	4.36	4.38	99.54	7.03

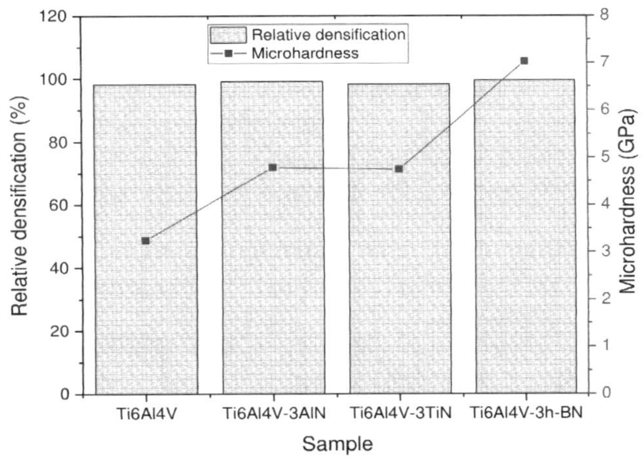

Figure 1. Graph showing the comparison of the relative densification and microhardness values of Ti6Al4V alloy, Ti6Al4V-3AlN, Ti6Al4V-3TiN and Ti6Al4V-3*h*-BN

Figure 2. Optical micrographs depicting the microstructure and representative diamond-shaped microindentation on the surface of: a. Ti6Al4V alloy b. Ti6Al4V-3AlN c. Ti6Al4V-3TiN d. Ti6Al4V-3*h*-BN

The optical micrographs of the samples sintered at 1000 °C of temperature, 30 MPa of pressure, 100 °C/min of heating rate and 10 min of dwell time are given in figure 2 (a-d) in order to show the distribution of the reinforcement nanoparticles of AlN, TiN and *h*-BN and their effects on the densification and hardness property of the matrix. The relative densification of the matrix calculated from an effective density of 4.42 g/cm^3 and theoretical density of 4.50 g/cm^3 and Vickers microhardness were 98.22% and 3.25 GPa respectively. The micrograph in figure 2a shows a dominance of α-Ti lamellar (which is responsible for toughness and creep resistance) and little evidence of prior β-Ti grain boundaries (which is responsible for high strength and fatigue resistance) in the Ti6Al4V-rich matrix can be said to be the reason for the low microhardness value of the alloy [1,3] as shown by the large-sized diamond indentation. The relative densification results (98.22 to 99.54%) obtained as presented in table 1 and figure 1 indicate that the SPS consolidation technique applied to synthesize the Ti6Al4V-based composites was effective and produced nearly full theoretical densification with *h*-BN reinforcement (99.54%). Figure 2b presents the micrograph of a decent distribution of coarsened nanoparticles of AlN in the Ti6Al4V-rich matrix with a slightly increased microhardness value of 4.79 GPa than the TiN reinforcement as denoted by the relatively moderate-sized diamond indentation. This increase can be traced to the presence of hard AlN phase which imparted more hardness and acted as further nucleation points for strong β-Ti phase [17]. Similarly, the moderate microhardness value of 4.75 GPa of Ti6Al4V-3TiN as depicted in figure 2c can be ascribed to the formation of dendritic structure demonstrating a significantly coarsened TiN and a relatively refined homogeneous dispersion at the grain boundaries of the Ti6Al4V-rich matrix. However, Ti6Al4V-3*h*-BN binary composite gave the optimal combination of relative densification and microhardness values of 99.54% and 7.03 GPa which exceeds 200% that of the matrix (3.25 GPa) and about 48% superior to both composites with TiN and AlN reinforcements as ably characterized by the smallest diamond indentation in figure 2d. This phenomenon can be attributed to adequate mixing of the powders where the nanoparticles of *h*-BN filled up pore spaces within the titanium alloy matrix which are the primary causes of poor densification and the effectiveness of the consolidation method which enables development of dense products [18]. Besides, there is an indication of better closure of pores and refined grains which could be the reason for the complete densification [16]. In general, it is found that the hardness property is proportional to the relative densification which is promoted by enough diffusional mass transport ensuing in solidly bonded compact of the matrix and the refractory nitride reinforcements [19]. Likewise, the considerable improvement in the hardness property of the matrix suggests that SPS aids the synthesis of refractory nitride nanoparticles of AlN, TiN and *h*-BN reinforced titanium matrix composites with superior mechanical characteristics useful in the automotive and aerospace industries.

4. Conclusion

In this work, SPS was usefully applied to synthesize Ti6Al4V-based binary composites with three different refractory nitride nanoparticles of AlN, TiN and *h*-BN with desirable characteristics. The effects of these reinforcements on the microstructure, densification and microhardness of the sintered composites were studied. The following conclusions are therefore made:

1. The consolidation of Ti6Al4V powder and nanoparticles of 3 wt.% of AlN, TiN and *h*-BN by SPS produced nearly full theoretical densification (98.22 to 99.54%) of the sintered composites which gives credence to the effectiveness of the consolidation technique enabled by enough diffusional mass transport ensuing in solidly bonded compact of the matrix and the refractory nitride reinforcements.
2. Ti6Al4V-3*h*-BN binary composite gave the optimal combination of relative densification and microhardness values of 99.54% and 7.03 GPa which exceeds 200% that of the monolithic alloy (3.25 GPa) and about 48% superior to both composites with TiN and AlN reinforcements.

5. Acknowledgement

The authors acknowledge the support received from the Department of Chemical, Metallurgical and Materials Engineering and Centre for Energy and Electric Power (CEEP), Tshwane University of Technology (TUT), Pretoria, South Africa.

6. References

[1] Mouritz, A. P. (2012). *Introduction to aerospace materials.* Philadelphia New Delhi: Woodhead Publishing Limited.

[2] ASM International. (2000). Introduction to Selection of Titanium Alloys. In ASM International, *Titanium: A Technical Guide, 2nd Edition* (pp. 5-11).

[3] Leyens, C. & Peters, M., 2003. *Titanium and Titanium Alloys: Fundamentals and Applications.* 1st ed. Weinheim: WILEY-VCH Verlag GmbH & Co. KGaA.

[4] Chen, J., Guo, C. & Zhuo, J., 2012. Microstructure and tribological properties of laser cladding Fe-based coating on pure Ti substrate. *Trans. Nonferrous Met. Soc. China 22,* Volume 22, p. 2171−2178.

[5] Kim, I. Y., Choi, B. J., Kim, Y. J., & Lee, Y. Z. (2011). Friction and wear behaviour of titanium matrix (TiB+TiC) composites. *Wear 271*, 1962-1965.

[6] Ochonogor, O. F., Akinlabi, E. T., Nyembweb, K. D., Pityana, S., & Shongwe, M. B. (2017). Particle varariation and effect on the microstructure and microhardness of Ti6Al4V hybrid metal matric system. *Internatinal conference on sustainanble materials processing and manufacturing, SMPM* (pp. 616-621). Kruger National Park: Procedia manufacturing.

[7] Ray, A. K., Venkateswarlu, K., Chaudhury, S. K., Das, S. K., & Kumar, B. R. (2002). Fabrication of TiN reinforced aluminium metal matrix composites through a powder metallurgical route. *Materials Science and Engineering A338*, 160-165.

[8] Pierson, H. O., 1996. *Handbook of Refractory Carbides and Nitrides: Properties, Characteristics, Processing and Applications.* 1st ed. Norwich: Noyes Publications.

[9] de Araujo, E. R. et al., 2012. Preparation of Metal Matrix Aluminum Alloys Composites Reinforced by Silicon Nitride and Aluminum Nitride Through Powder Metallurgy Techniques. *Materials Science Forum* , Volume 727-728, pp. 259-262.

[10] Yang, H; Fang, H; Yu, H; Chen, Y; Wang, L; Jiang, W; Wu, Y; Li, J. (2019). Low temperature self-densification of high strength bulk hexagonal boron nitride. *Nature Communications 10*, 1-9.

[11] Matizamhuka, W. R. (2016). Spark plasma sintering (SPS) – an advanced sintering technique for structural nanocomposite materials. *The Journal of the South African Institute of Mining and Metallurgy Volume 16*, 1171-1180.

[12] Du, Y., Li, S., Zhang, K., & K, L. (1997). BN/Al composite formation by high-energy ball milling. *Scripta Materialia, Vol. 36, No. 1*, 7-14.

[13] Firestein, K. L., Steinman, A. E., Golovin, I. S., Cifre, J., Obraztsova, E. A., Matveev, A. T., Golberg, D. (2015). Fabrication, characterization, and mechanical properties of spark plasma sintered Al–BN nanoparticle composites. *Materials Science & Engineering A 642* , 104–112.

[14] Kgoete, F. M., Popoola, A. P., Fayomi, O. S., & Adebiyi, I. D. (2018). Influence of Si3N4 on Ti-6Al-4V via spark plasma sintering: Microstructure, corrosion and thermal stability. *Journal of Alloys and Compounds 763* , 322-328.

[15] Cao, Y., Zeng, F., Lu, J., Liu, B., Liu, Y., & Li, Y. (2015). In Situ Synthesis of TiB/Ti6Al4V Composites Reinforced with Nano TiB through SPS. *Materials Transactions, Vol. 56, No. 2* , 259-263.

[16] Ujah, C. O., Popoola, A. P., Popoola, O. M., & Aigbodion, V. S. (2018). Optimisation of spark plasma sintering parameters of Al-CNTs-Nb nano-composite using Taguchi Design of Experiment. *The International Journal of Advanced Manufacturing Technology*, 1-11.

[17] Falodun, O. E. et al., 2017. Synthesis of Ti-6Al-4V alloy with nano-TiN microstructure via spark plasma sintering technique. *IOP Conf. Series: Materials Science and Engineering* , 272(012029), pp. 1-6.

[18] Ocak, B. C., Yavas, B., Akin, I., Sahin, F., & Goller, G. (2018). Spark plasma sintered ZrC-TiC-GNP composites: Solid solution formation and mechanical properties. *Ceramics*

[19] Cavaliere, P., 2019. *Spark Plasma Sintering of Materials: Advances in Processing and Applications.* 1st ed. Switzerland: Springer Nature.

IOP Conf. Series: Materials Science and Engineering **784** (2020) 012002 doi:10.1088/1757-899X/784/1/012002

Investigation into Innovative Fabrication of Fiber Metal Laminated Pipe and Application

Lang Zijian[1,a] ,Zhang Meng[2,b],Li Xiaoxing[2,c]

[1]Beijing University of Civil Engineering and Architecture,No. 15 Yongyuan Road, DaXing District, Beijing, P.R. China
[2]Beihang University, No. 37 Xueyuan Road, Haidian District, Beijing, P.R. China

[a]lzjleo@163.com

[b]Stu_zhangmeng@buaa.edu.cn
[c]li.xiaoxing@buaa.edu.cn

Abstract. Fiber reinforced metal laminated tube (FMLT) is a kind of multi-layer super-hybrid material which is cured at a fixed pressure and temperature after alternately laying metal laminated tube and fiber composite material. The development of GLARE composite pipe hydroforming technology and performance testing is of great significance to the development of aviation industry and automotive industry in terms of lightweight and safety.

1. Introduction

For a long time, aircraft and automobile manufacturers have been looking for a new material with high specific strength, good fatigue performance and corrosion resistance. The birth of GLARE composite pipe has solved many problems, and now scientists have been researching it successively. Based on the research of fibre metal laminates, Tao Jie used Yan Huigeng's graphic method to determine the matching pipes for fibre metal reinforced pipes. Through Lame's formula, Tao Jie gives the minimum internal pressure required for expansion by assuming that the expansion nozzle is in a plane stress state. KUL CAN obtains the optimum pressure curve through dynaform during the expansion process, and gives that the best forming condition is that the feed speed of the axis is 8mm/s. Based on Dai Qiwei's KUL CAN, the preparation and formability of Ti/CF/PEEK/Ti composite layer tube were studied. It was pointed out that the critical bulging value of the composite crown was 7.9 MPa, and the strength of the inner layer titanium tube was lower than that of the outer layer titanium tube.The deep drawing experiments of the Ti/CF/PEEK/Ti GLARE composite pipe were carried out, and the deep drawing performance was obtained. In addition, KUL CAN also carried out the deep drawing performance of the Ti/CF/PEEK/Ti GLARE composite pipe. The energy absorption experiment of LARE tube under axial compression was carried out. The main purpose of this study is to further study the hydraulic forming technology of GLARE composite pipe and to further study the physical objects.Formatting the title, authors and affiliations

2. Current Situation of Development

2.1. Research Background and Significance

Lightweight is the development trend of aerospace, high-speed railway, automobile and other transport machinery manufacturing industry. Thin-walled, integral and lightweight structure is an important measure to realize lightweight products. The first ten-year action plan of the Chinese government to implement the "Made in China 2025" strategy clearly emphasizes and points out that green manufacturing, aerospace equipment and new energy vehicles are key areas for development. In addition, COP15, the World Climate Conference held in Copenhagen, Denmark in 2009, advocated a green low-carbon lifestyle and enterprise production and consumption. For green manufacturing, one of the important measures is lightweight components. At the same time, various countries and regions are stricter and stricter in terms of environmental protection, safety and corrosion resistance of the automobile industry, which greatly promotes the automobile manufacturers all over the world to actively develop some environmentally friendly automobile products. According to statistics, the automobile manufactured by lightweight design has reduced about 2% compared with the previous steel manufacturing. 5%[1].

Under this background, tube hydroforming and its derivative technology have gained wide attention and development in the field of aviation, aerospace and automobile manufacturing all over the world. This technology is one of the key technologies to realize lightweight structural manufacturing proposed in the late 1970s. It was first used in large-scale engineering practice by Germany in the 1990s. In the field of automobile manufacturing, its industrial chain covers liquid-filled forming equipment, liquid-filled forming pipes, liquid-filled forming parts and related inspection and testing equipment. In Japan, Europe, America, Korea and other countries and regions, a relatively independent and complete supplier system has been formed. The parts produced are designed for chassis, body and exhaust system of automobiles. And many other fields.

Pipe hydroforming technology has also made extensive development in aerospace industry. In the field of aircraft, rocket and engine manufacturing, the requirement of lightweight is more urgent. The integral complex thin-walled structural parts suitable for tube hydroforming technology are widely used in the above fields, so this technology has attracted much attention. In 2004, the Wright Foundation of the United States made the hydroforming technology one of the key developing forming technologies. In view of the current situation in China, the technology has also been funded by the National Natural Science Foundation, the National Science and Technology Support Plan and the National Key Basic Research Development Plan. The development prospects are broad [2].

In addition to the development of new technologies, the search for new materials is also an inevitable requirement for the development of aircraft with better performance. Compared with other materials, the plasticity of traditional metal materials is better and easier to be processed and moulded, but the corrosion resistance is relatively poor. Fiber reinforced resin matrix composites have high specific modulus, specific strength, corrosion resistance and fatigue properties, but have poor impact damage resistance and ductility, and are vulnerable to humidity. The water aging is affected by the environment. Considering that these two materials have their own advantages, a new type of composite material with colloidal connection structure, fiber reinforced metal laminates, emerged as the times require. “Figure 1” shows the application of GLARE composite laminated tubes in various fields.

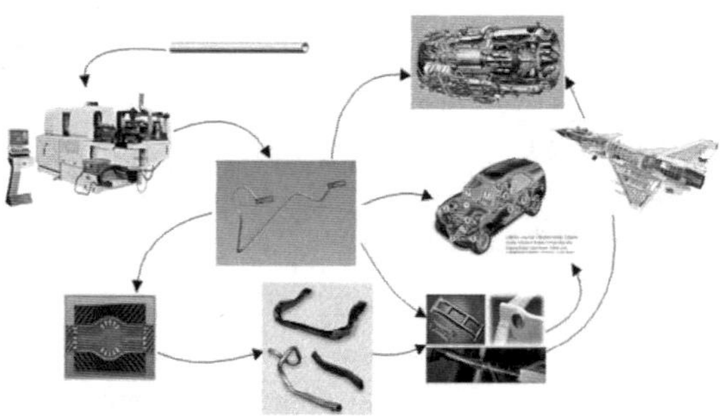

Figure 1. Application of GLARE Composite Laminated Tubes in Various Fields

3. Development status

Fiber reinforced metal laminates can be divided into the following generations according to the material of fiber reinforced matrix: aramid reinforced aluminium laminates (ARALL), glass reinforced aluminium laminates (GLARE) and glass reinforced aluminium laminates (GLARE) in the first generation and in the third generation. Carbon fiber-titanium alloy laminates (CARE) and graphite fiber-titanium alloy laminates (TiGr) are the fourth generation.

Fiber reinforced metal laminated tube (FRMT) is a kind of interlaminar superhybrid material which is cured at a fixed pressure and temperature after alternately laying metal laminated tube and fiber composite material.

Among them, the composite laminated tube formed by similar manufacturing process of second generation GLARE fiber reinforced metal laminates has good impact resistance, which can be used for aircraft crashworthiness structure, aircraft landing buffer structure, automobile body crashworthiness device, air transmission material protection and so on; at the same time, because of its excellent corrosion resistance. It can be widely used in chemical industry. The development of GLARE laminated tube hydroforming technology and performance testing is of great significance to the development of aviation industry and automotive industry in terms of lightweight and safety.

In this paper, the formability and liquid filling formability of the fibre metal laminated tube are studied as an example. The structure is shown in the following "figure 2".

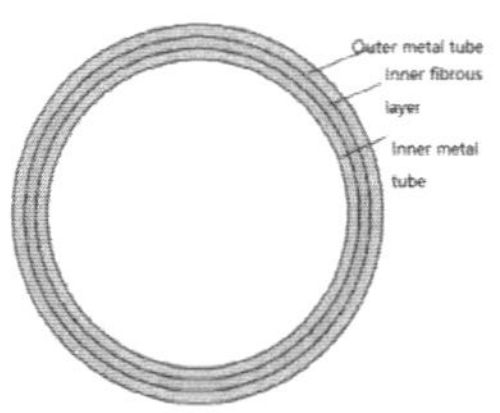

Figure 2. Structural sketch of GLARE layered tube

The manufacturing technology of Grail tube is mainly liquid-filled forming, which has developed rapidly in recent years.

After the pre-treatment (deburring, elbowing and oiling) of the tube, the tube is usually placed in a pre-set cavity die, and the tube is deformed by controlling the axial feed (lateral thrust) and internal pressure. Finally, the technology of the given shape is obtained[3]. "Figure 3" shows the process schematic diagram of hydroforming pipe.

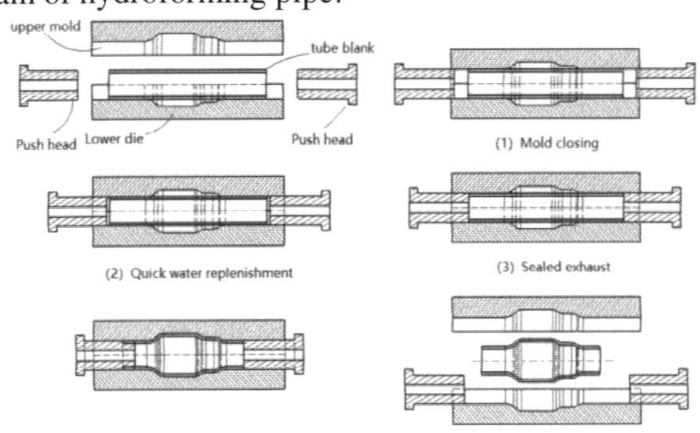

Figure 3. Process schematic diagram of hydroforming pipe

Because the material is formed under the action of real-time controllable uniform fluid surface force, and the axial thrust force can continuously push the appropriate material into the deformation area, thus greatly improving the forming limit of the tube hydroforming and the precision of the part clamping. Because water (or emulsified liquid) is used as force transfer medium in industrial production and test, and hydraulic proportional servo control technology is applied, forming pressure and lateral thrust can also be precisely controlled by real-time servo, and real-time pressure holding treatment is carried out according to specific conditions, thus the quality of formed parts can be accurately controlled. Pipe hydroforming technology is suitable for the production of any plastic thin-walled metal materials, including stainless steel, carbon steel, copper alloy, aluminum alloy, magnesium alloy and titanium alloy. Pipe filling parts are widely used in advanced manufacturing fields such as automobiles, aerospace, military nuclear power and so on. Twist beams, towing arms, front and back frames, energy absorbing boxes, etc. in automobile manufacturing are typical application cases, such as fuel conduits, heat dissipating vanes, tee tubes, small radius elbows, etc.

3.1. Development Status of Hydraulic Expansion Joint

Generally, the methods of manufacturing laminated pipes include hot extrusion, explosive cladding (forming), coil welding (welding), centrifugal casting, cold drawing and hydraulic bulging, etc[4]. Compared with other technologies for manufacturing composite laminated pipes, hydraulic expansion has the advantages of low cost and good formability. The manufacturing process of common composite pipes is listed in “Table 1”.

Table 1. Common composite pipe manufacturing process

Forming method	Forming characteristics	Combination method
Hot extrusion	It is suitable for alloy metal composite with low plasticity and poor workability; small deformation resistance and high surface roughness of composite pipe.	Mechanical combination
Explosive forming	It can realize the compounding between various metals and has high efficiency.The thickness of the covering metal can be large or small, and the interface is tightly combined; the explosion site is dangerous and the technical requirements are high.	Mechanical combination
Welding method	It can be used to manufacture composite steel pipes for natural gas transportation with a diameter of more than 300mm; the production process is complicated and the production cost is high.	Metallurgical combination
Diffusion bonding	The material combination is more diversified, the quality is good, the cost is slightly higher; the production process is complicated, the equipment investment is large, and the production efficiency is low.	Metallurgical combination
Drawing method	The production process is simple and the cost is low; the interface is non-diffusion connection, the temperature is high and easy to stratify, and the application temperature is low.	Mechanical combination

The expansion joint of the composite pipe refers to the plastic deformation after the inner pipe is convex, and the outer pipe is elastically deformed.When the internal pressure is unloaded, the rebound amount of the outer tube is greater than the rebound amount of the inner tube.Finally, the residual contact stress at the interface stroke causes the two layers of tubes to mechanically bond, as shown in the following figure.As a plastic forming method, the expansion-molded composite layer tube can greatly improve the material utilization efficiency and has high forming precision.In addition, the hydraulic expansion joint pipe is used, the expansion joint force is uniform, and can be calculated according to the set parameters and mechanical properties of the two metal pipes, the wall thickness distribution is uniform, and the inner surface quality of the pipe is high[5]. "Figure 4" is a expansion joint schematic.

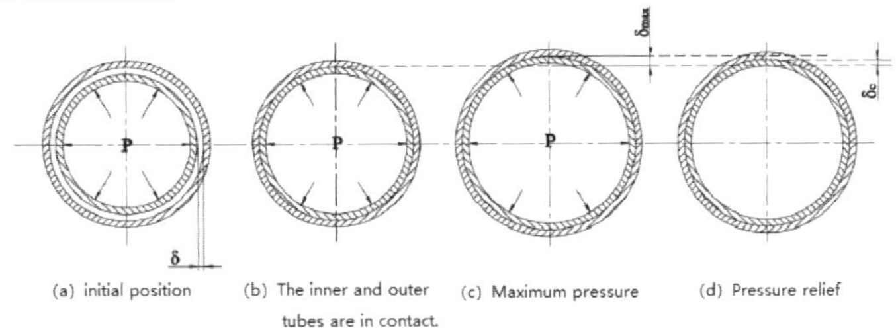

Figure 4. Expansion joint schematic

3.2. Research status of fiber reinforced metal layer tube at home and abroad

3.2.1. Status of foreign research

In the early 1970s, Fokker discovered the effect of metal thickness on the cementitious laminate in the bonding structure between the laminates. It was unexpectedly found that the fracture toughness of the thin plate can be improved by cementation and the fatigue crack propagation can be suppressed.[6].Based on the above phenomenon, the University of Delft in the Netherlands developed in cooperation with Fokker to add a reinforcing base (single aramid fiber) to an aluminum alloy metal laminate to obtain a composite-fiber reinforced metal laminate (ARALL) having a novel structure.Since then, Delft University has systematically studied fatigue performance, durability, processing properties, fracture toughness, impact damage, structural design, etc., and laid a precedent for fiber reinforced metal laminate research.[7].

In 1987, the second generation of fiber reinforced metal laminates, which is the main technical idea used in this experiment, was the glass fiber-aluminum laminate (glare).Compared with the first generation of fiber reinforced metal laminates, glare laminates have the advantages of higher fatigue properties, notch strength and compressive strength.Airbus became the first person to eat crabs. The company first applied glare laminates and its technology to the a340 and a330 fuselage sections. After more than 100,000 flight tests, no damage was found. Boeing will then Glare laminates are used in the cabin floor of the 777 and 757. The first large-scale application of the glare laminate is in the upper fuselage skin, fairing, fairing, vertical and horizontal tail, upper fuselage siding and upper double cabin A380 side panel."Figure 5" shows the vault of the barrel portion of the fuselage made of glare.According to statistics, the upper body of the a380 has a total area of about 470 square meters and a total of 27 glare laminates[8].The longest part is 11 meters and the weight is reduced by about 800 kilograms.

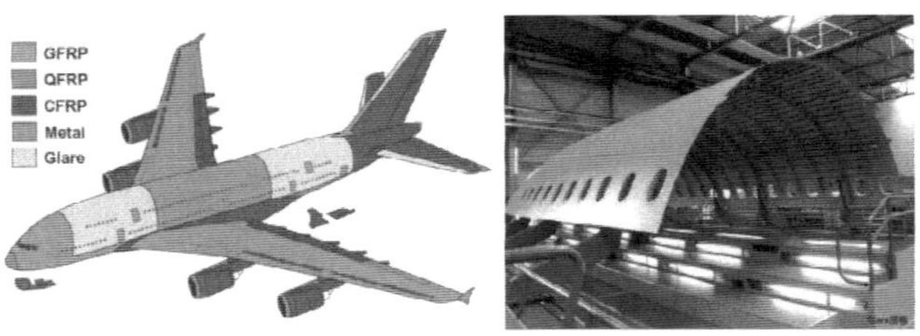

Figure 5. A380 fuselage dome made with glare laminate

The third-generation fiber reinforced metal laminate is a carbon fiber reinforced aluminum alloy laminate, but due to the large difference between the galvanic sequence of carbon and aluminum, it is prone to electrical corrosion that is common in life.To solve this problem, a solution for filling a barrier layer between the carbon fiber layer and the aluminum alloy laminate has appeared, but the use of the barrier layer increases the difficulty in preparing the carall laminate.For these various reasons, carbon fiber reinforced aluminum alloy laminates have not been able to achieve large-scale applications in actual production and life.

The main research of glare laminate mechanics focuses on both shear properties and anisotropy.

The glare laminates are directional and exhibit anisotropy along the direction of the different fiber arrangements.In general, due to the high modulus of elasticity of the fibers, the tensile strength of the unidirectionally arranged glare laminate in the fiber arrangement direction is much higher than that of the aluminum alloy sheet, but the lateral tensile strength is low.If the fiber layer is selected as the prepreg, the horizontal and vertical drawing strengths are guaranteed.

Hale et al. first studied the related effects of the orientation direction of glass fiber prepreg on the tensile properties of GLARE laminates.Donnellan et al. studied the effect of temperature on the tensile properties of ARALL laminates by operating at varying ambient temperatures.The results show that low temperature and normal temperature have no effect on the tensile properties of various types of ARALL laminates. However, when the temperature is raised to 120 °C, the tensile strength of ARALL-1, -2, -3 ARALL laminates will be The above phenomenon[9] occurs for the ARALL-4 type to reach 170 °C or higher.

Liu Cheng et al. used the short beam method to study the shear behavior between the plate and the plate of the galle laminate. The results show that when the span-thickness ratio is 8, the sample undergoes pure shear failure, ie near the neutral layer. Interlayer peeling delamination.The apparent interlaminar shear strength values obtained by this method can characterize the layers of the composite laminate.Shear performance.When the span-thickness ratio is 5, the shear test is the squeeze-shear failure mode, and when the span-thickness ratio reaches 10, the failure mode is the bending effect[10]. “Table 2” lists the glare laminate shear performance.

Table 2. glare laminate shear performance

Shear properties	Test Direction	GLARE 4–5/4	GLARE 4–3/2	GLARE 4–2/1	Al 2024–T3
yield shear stress/MPa	Lengthways	101	105	119	207
	Transverse	98	107	117	207
shear modulus/GPa	Lengthways	13.67	15.26	15.77	28
	Transverse	13.84	14.33	17.50	28

In addition to the study of basic mechanical properties, the damage of the galle laminate and its mechanics theory are also hot topics.Volume fraction theory is a classical theory[11] that characterizes the tensile properties of galre laminates.The theoretical formula is as follows:

$$MVF = \sum_{p_{metal}}^{t} t_{metal} / t_{lam}$$

$$\sigma_{t,lam} = MVF \cdot \sigma_{t,met} + a \cdot (1 - MVF) \cdot \sigma_{t,FRP}$$

$$E_{lam} = MVF \cdot E_{met} + a \cdot (1 - MVF) \cdot E_{FRP}$$

$$\sigma_{0.2,lam} = [MVF + a \cdot (1 - MVF) \cdot \frac{E_{FRP}}{E_{met}}] \cdot \sigma_{0.2,met}$$

among them, t_{lam} Is the total thickness of the laminate; t_{metal} Is the thickness of a single layer of metal; p_{metal} Is the number of metal layers; $\sigma_{t,lam}$ 、 $\sigma_{t,metal}$ 、 $\sigma_{t,FRP}$ Corresponding to the tensile strength of the laminate, the metal layer and the fiber layer respectively; E_{lam} 、 E_{met} 、 E_{FRP} Corresponding to the elastic modulus of the laminate, the metal layer and the fibrous layer; $E_{0.2,lam}$ 、 $E_{0.2,met}$ Is the yield strength of the laminate and the metal; a is the fiber volume fraction of the laminate in the tensile direction.

Delfe University studied the uniaxial tensile properties of GLARE1-to-GLARE-6 and other GLARE. The results show that when 0.45<MVF<0.85, MVF theory can give the prediction of tensile strength[12].

The glare laminate also follows the industry-recognized classical laminate theory. Although the theory ignores the plastic deformation of the aluminum alloy sheet after yielding, it can reflect its main mechanical properties through some data and according to the theory.Perform relevant mechanical analysis and calculation of common clt theory before the yield behavior of the aluminum alloy, and calculate the residual stress generated by the glare laminate during the curing process, as shown in FIG.During the cooling of the glare laminate, the coefficient of thermal expansion of the aluminum alloy layer and the fibrous layer during cooling will result in residual stress between the materials, and the mechanical properties of the glare laminate will be significantly affected[13]. “Figure 6” uses clt theory to explain the effect of participating stress on glare laminates.

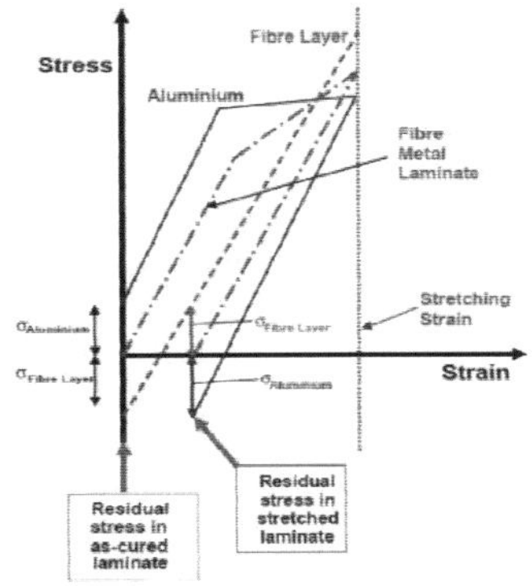

Figure 6. uses clt theory to explain the effect of participating stress on glare laminates

3.2.2. *Status of domestic research*

Based on the research of fiber reinforced metal laminates, Tao Jie used Yan Huigen's graphic method to determine the matching of the fiber metal reinforced tube.Through the Lame formula, Tao Jie gives the minimum internal pressure required for expansion by assuming that the expansion joint is in a plane stress state.KUL CAN obtains the optimum pressure curve during the expansion process through dynaform, as shown in the figure below, and also gives the axis feed rate of 8mm/s which is most favorable for forming.Based on the theory, Dai Qijun studied the preparation and forming properties of Ti/CF/PEEK/Ti composite layer tubes. It is pointed out that the critical value of the composite crown is 7.9 MPa, and the inner titanium tube strength is required to be lower than The outer layer of titanium tube and the tube drawing of Ti/CF/PEEK/Ti tube were subjected to the deep drawing test, and the deep drawing performance was obtained. In addition, KUL CAN also carried out the axial compression energy absorption experiment on the GLARE layer tube. Its compression curve[14]. “Figure 7” shows the internal pressure loading curve.

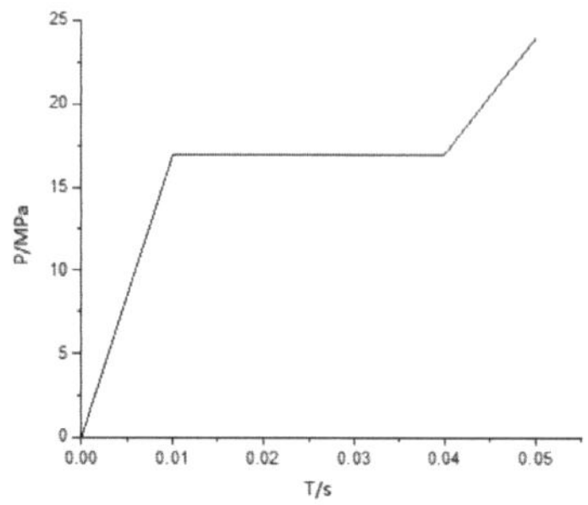

Figure 7. Internal pressure loading curve

3.3. Interlayer "viscous" effect test

Because semi-curing has the characteristics of large tangential deformation and complicated deformation process compared with traditional cementation, in addition to considering the fluid hydroforming process in semi-cured forming in this subject, the interface has always had a normal pressure value with time.In order to better describe the viscous effect of the metal and prepreg interface in the glare laminate and the tube semi-cured, the physical simulation test shown in "Figure 8" below is designed.

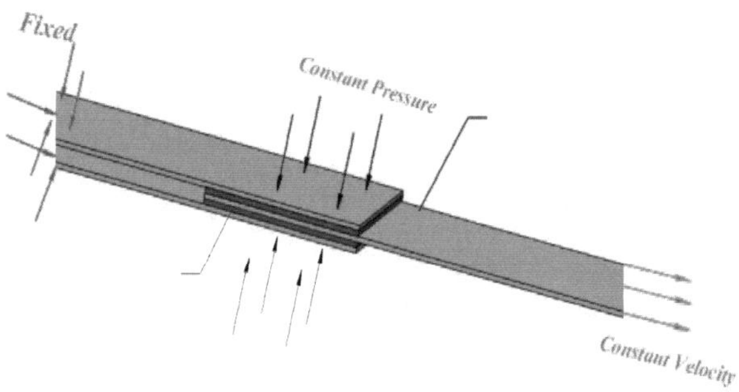

Figure 8. "viscous" effect test

As shown in the figure above, the 3+2 GLARE layer form is used.The thickness of the aluminum plate is 3mm, the material is aluminum alloy 2024-T3, and its performance is shown in "Table 3" below; the prepreg is glass fiber WP9011(25*25), and its performance is shown in "Table 4" below:

Table 3. Mechanical properties of aluminum alloy 2024-t3

Rolling direction	Yield stress/MPa	tensile strength/MPa	Elongation /%	Thick anisotropy index	Hardening index n
0	298	425	19.2	0.78	0.195
45	290	414	16.8	0.82	0.184
90	295	421	18.3	1.10	0.191

Table 4. Mechanical properties of glass fiber prepreg wp9011

Tensile Strength/MPa	Tensile modulus/GPa	Compressive strength/MPa	Flexural modulus/GPa	Shear strength/MPa
400	18	450	19	55

As shown in “Figure 8”, there is a 25*25mm pressure zone in the middle of the test piece, and there is a constant pressure in the normal direction to ensure that the prepreg is tightly bonded to the metal.Due to the large thickness of the aluminum plate, the amount of tensile deformation can be neglected in this experiment.Under the action of the axial force, the intermediate layer aluminum plate continuously slides in the axial direction.By performing the above experiments under different pressures (five in each group, averaging), the tensile force and sliding displacement

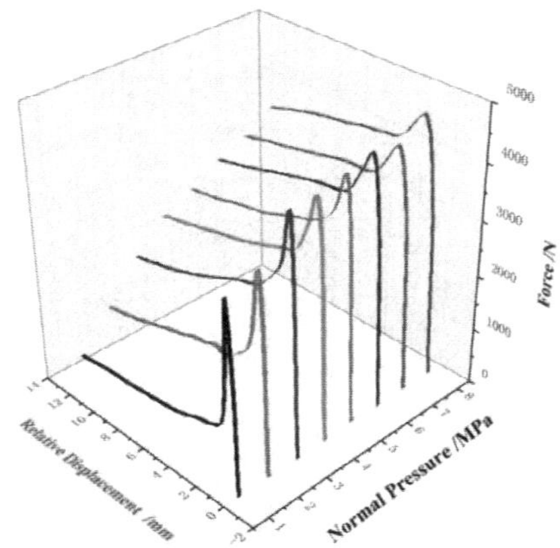

maps are obtained as shown in “Figure 9” below.

Figure 9. Tensile-displacement curve at different pressures

It can be seen from “Figure 9” that the curve distribution law is basically the same under different pressures, that is, the slope gradually becomes smaller after linear increase, and then exponentially decays after reaching the peak value, and finally shows a linear trend; in addition, the curve peak value and the final linear segment value Positively related to stress.

For any pressure, the relationship between tensile force and displacement can be expressed by the following “figure 10”:

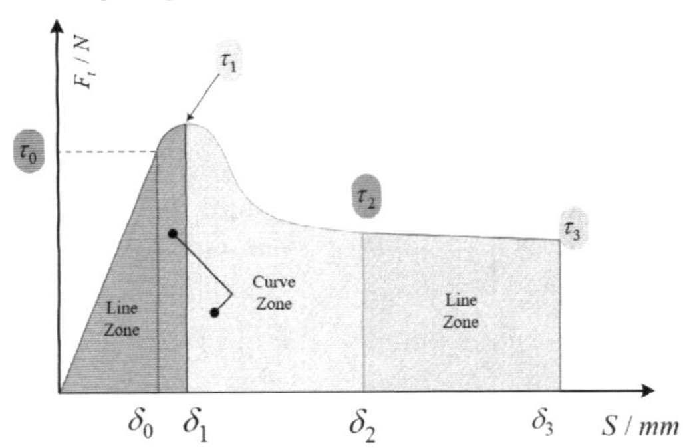

Figure 10. Typical tensile-displacement curve

As shown in the above figure, the typical tensile-displacement curve can be divided into four parts: linear area 1, curve area 2, curve area 3, and linear area 4.The following four areas are analyzed:

(1) Linear area 1:

For the linear region 1, the main deformation mode of the cementing unit is the shear deformation in the elastic region.In this area, the cementing unit remains stable without major deformation.

(2) Curve area 2:

As the elastic shear deformation of the cementing unit becomes larger and larger, the deformation of the cementing unit exceeds the elasticity, and the deformation is permanent deformation, and the deformation amount is large.In addition, the tensile force and displacement are not linear.

(3) Curve area 3:

As the deformation gradually increases, the tensile-displacement curve tends to decline.At this time, the amount of deformation is extremely large, and itis not restored after unloading.

(4) Until the deformation to the linear region 4, the stretch-displacement curve region is stable and tends to decrease linearly

It can be seen that when the curve reaches its apex, the interlaminar viscous effect degrades, which is conducive to subsequent forming.

4. development prospects

4.1. Technology development prospects

Composite material refers to a material composed of two or more different substances and combined in different ways. It can overcome the defects of a single material and utilize the advantages of various materials to expand the application range of materials.Due to its light weight, high strength, convenient processing, excellent elasticity, chemical resistance and weather resistance, composite materials gradually replace wood and metal alloys, and are widely used in automobiles, aerospace, construction, electrical and electronic, fitness equipment and so on.In other areas, it has grown rapidly in recent years.

As a new type of composite material, the chemical fiber reinforced metal layer tube is still in the laboratory development stage. Due to its good impact resistance, it can be used in aircraft crashworthy structures, aircraft landing buffer systems, and automobile body anti-collision devices. Airborne material protection, etc.; at the same time, due to its excellent corrosion resistance, it can be widely used in the chemical industry.

4.2. Development prospects of the transportation market

The development potential of composite materials in the transportation market is even more important.Although composite materials are commonly used in the field of automation, composites account for only 1% of the weight of ordinary vehicles.Not to mention the proportion of fiber-reinforced metal sheet materials, it can be seen that the market development potential of fiber-reinforced metal materials is enormous.The driving force of regulations and technology has made people's optimism about the growth of composite materials become reasonable, but the adoption of composite materials is still difficult, and the market conditions in the next few years will challenge the industry.Composite materials will be used in aerospace and other industries that are in urgent need of light weight, but will still occupy a certain share in the general market.

To be like other materials, composites are the choice of people, and only the overall cost, weight and performance can be compared to the extraordinary value.Although in automotive and other fields where lightweighting is not demanding, the reason for the application of composite materials is not only its lightweight, but its main disadvantage is that the processing cost is higher than that of ordinary manufacturing materials.At this point, the decline in sales of light vehicles and competition from other materials has become an obstacle to composite applications. "Figure 11" shows the use of Ford gt's wide range of carbon fiber materials

Figure 11. Use of Ford gt's wide range of carbon fiber materials

The results of the survey show that a 10% reduction in vehicle volume will result in a 7% increase in fuel economy.In addition, lightweight vehicles require less power to accelerate, so even with smaller, low-burning engines, these vehicles still have the potential to drive.Therefore, original equipment manufacturers are investing in lightweight materials to produce economical but at the same time enjoyable vehicles.

Composite materials - here is the fiber reinforced metal tube, which will be used more in automotive structural design due to its outstanding performance compared to other materials.

4.3. Development prospects of the aerospace market

At present, the aerospace industry is one of the main application fields of carbon fiber, which is mainly due to the high strength and light weight properties of carbon fiber.Compared with aluminum or steel, carbon fiber can reduce weight by 20% to 40%. "Figure 12" shows the carbon fiber aviation interior.

Figure 12. Carbon fiber aviation interior

In the aerospace industry, it is mainly used for aircraft structural materials (about 40% of aircraft weight), so the use of carbon fiber is comprehensive. It can reduce the weight of the aircraft by 6% to 12%, which significantly reduces the fuel cost of the aircraft.In the aerospace industry, composite materials - carbon fiber was first used in the manufacture of satellite antennas and satellite supports, and because of its heat and fatigue resistance, carbon fiber has also been widely used in solid rocket engine casings and nozzles.Some aircraft tubular structures, such as the fuselage, oil pipelines, landing gear, etc., use this product to not only improve the strength, but also reduce the quality, is a good choice.

5. Conclusions

1.A new preparation method of metal fibre tube has been studied and successfully formed. The results show that the tube has higher specific strength, specific modulus, fatigue property and corrosion resistance than common composite materials and uniform metal components.
2.Hydroforming is a good way.For hollow variable cross-section structural parts, the traditional manufacturing process is to stamp two halves and then weld them into a whole. Hydraulic forming can form hollow structural parts with variable cross-section at one time. Compared with stamping welding process, hydroforming technology and process have the following main advantages:
(1). Reduce quality and save materials. For hollow shaft parts, the weight can be reduced by 40%-50%.
(2). Reduce the number of parts and moulds and reduce the cost of moulds. Hydraulic forming parts usually need only one set of dies, while stamping parts mostly need multiple sets of dies.
(3). It can reduce the welding quantity of follow-up mechanical processing and assembly.
(4). Improve strength and stiffness, especially fatigue strength.
(5). Reduce production cost. According to the statistical analysis of the parts already used in hydroforming, the cost of production of hydroforming parts is 15%-20% lower than that of stamping parts on average, and the cost of die is 20%-30% lower than that of stamping parts.
3. After hydroforming, the interface property can be improved greatly. Hydraulic expansion of GLARE composite pipes refers to a mechanical connection mode which realizes expansion by residual stress caused by different deformation of inner and outer pipes before solidification. The basic principle is that the two ends of the pipe are sealed by the left and right sealing devices, and the high-pressure liquid enters the inner pipe through the sealing device, and the plastic deformation of the inner pipe occurs under the action of the fluid surface force; when the internal pressure is unloaded, the rebound of the outer pipe is greater than that of the inner pipe, and finally the residual contact stress at the interface stroke makes the two-layer pipe. Mechanical bonding occurs. As a plastic forming method, the bulging forming of composite tube can greatly improve the material utilization efficiency and the forming accuracy. In addition, the hydraulic expansion joint compound pipe has the advantages of uniform expansion force, uniform wall thickness distribution and high surface quality, which can be calculated according to the set parameters and mechanical properties of the two metal pipes.

References

[1]Zhang Xuguang. Analytical-numerical simulation of numerical control elbow springback [d]. Xi'an: Northwestern Polytechnical University, (2003)
[2]Jia Meihui, Tang Chengtong. Research on springback prediction model of stainless steel pipe bending forming[J]. Journal of Beijing Institute of Technology.(2012), 32(9): 910-914
[3]Endow J, Murofa T. Flattening of a circular tube section during uniform bending [A]. The latest technical translation of the world plastic processing
[4] Liu Yong, Hou Yuansheng, Wang Yi, et al. Application of bimetallic composite pipeline in Yaha condensate gas field [j]. Oil and Gas Field Surface Engineering, (2009): Home.
[5] Ling Xingzhong. Inner composite bimetal tube manufacturing technology [j]. Welded pipe, (2001), 24 (2): 43-46.
[6]Yan Huigeng, Shi Tingrui. Liquid bag hydraulic expansion technology of heat exchanger and its application [J]. Chemical Machinery, (1998), 25(6): 339-341.
[7] Xue Jia, Wang Wenxue, Xu Yuanming, et al. Residual Thermal Stress Analysis of Carbon Fiber Reinforced Aluminum Alloy Laminates[J]. Aircraft Design, (2008), 28(5): 22~26.
[8] Chen Shaojie. Composite technology and large aircraft[j]. Acta Aeronautica Sinica, (2008), 29(03): 605~610.
[9] Wang Ping. One of the civil aircraft design reference models Boeing 787-8 double-width wide-body medium-range passenger aircraft [j]. Civil Aircraft Design and Research,(2010), (2): 37~37.
[10] Liu Cheng. Study on failure behavior and mechanism of interlaminar shear in glare laminates [d]. Nanjing University of Aeronautics and Astronautics, (2017).

[11] Ferrante L, Sarasini F, Tirillò J, et al. Low velocity impact response of basalt-aluminium fibre metal laminates[J]. Materials & Design, (2016), 98: 98~107.
[12] Wu H F, Wu L L, Slagter W J, et al. Use of rule of mixtures and metal volume fraction for mechanical property predictions of fibre-reinforced aluminium laminates[J]. Journal of Materials Science, (1994), 29(17): 4583~4591.
[13]KUL CAN. Preparation and Properties of Glass Fiber-Aluminum Alloy Laminated Pipes[D].Nanjing·Nanjing Aerospace, (2016).
[14]Dai Qiwei.Preparation Process of Ti/CF/PEEK/Ti Composite Pipes And performance research [D]. Nanjing. Nanjing Aerospace, (2016)

Effect of temperature variations on interfacial debonding of FRP-plated beams: A coupled mix-mode cohesive zone modeling

Julian Wilson[1,2], Wan-Yang Gao[1] and Jian Yang[1]

[1]State key laboratory of ocean engineering, school of naval architecture, ocean and civil engineering, Shanghai Jiao Tong University, Shanghai 200240, China.

[2]Corresponding author's email address: julianwilson9@sjtu.edu.cn

Abstract. The use of fiber-reinforced polymer (FRP) plates for the strengthening of structural members involves their exposure to service temperature variations (i.e., thermal loadings). Such temperature variations may lead to thermal stress and the interfacial debonding of the FRP-plated beam as a result of the thermal mismatch between the FRP plate and the substrate beam. There is lacking research related to the effect of temperature variations on the interfacial debonding of FRP-plated beams. This paper presents the first-ever analytical study on the behavior of the FRP-plated beam under combined thermal and mechanical loadings. The interfacial stress and slip evolutions as well as the debonding process of the FRP-plated beam are estimated based on a coupled mix-mode cohesive zone modeling. In addition, a finite element (FE) approach is proposed to verify the analytical solution in which the analytical results are compared with the FE model predictions, indicating that a good agreement was achieved between the analytical and FE results.

Keywords: Fiber reinforced polymer (FRP); Interfacial stress; Cohesive zone modeling; Mixed mode; Temperature variations; Plate end debonding.

1. Introduction

Fiber-reinforced polymer (FRP) composites have been increasingly used for the repairing and strengthening of infrastructure elements. The main advantages of the FRP composites include their high strength-to-weight ratio, excellent corrosion resistance, and ease of application [1-2]. One of the most common strengthening methods involves the flexural strengthening of deficient structural members with externally bonded FRP laminates or sheets [3-4] (hereafter referred to as plates for simplicity of presentation). The FRP-plated beams in service are likely to experience significant temperature variations (i.e., thermal loadings) due to the seasonal ambient temperature changes. Such temperature variations may induce the thermal stresses at the FRP-to-substrate interface due to the thermal mismatch between the FRP plate and the substrate (e.g., steel, concrete, or timber), which may further influence the debonding process of the FRP-plated beam under combined thermal and mechanical loadings [5-6]. However, limited research studies are available on the effect of temperature variations on the debonding failure of FRP-to-substrate interfaces [7-10]. Among the limited studies, most of them were focused on the bond behavior of the FRP-to-substrate bonded joints under combined thermal and mechanical loadings [7-9]. This paper presents the first-ever analytical study on the effect of temperature variations on the interfacial debonding of FRP-plated beams. In the existing literature, the cohesive zone model was usually used for predicting the interfacial stresses and debonding failure of FRP-plated beams under mechanical loading solely [11-12]. This paper follows the research initiated in [12] and moves forwards the understanding of the behavior and debonding process of the FRP-plated beam under combined thermal and mechanical loadings. Details of the closed-form analytical solution using a coupled mixed-mode cohesive zone modeling approach are provided in the following sections.

2. Research significance

Although some analytical studies in the literature were conducted on the interfacial debonding of FRP-plated beams under mechanical loading only [11-14], there is lacking research on the effect of temperature variations on the debonding process of such beams. This paper aims to provide an analytical solution for the first time on the debonding process of the FRP plated beams under combined mechanical and thermal loadings. Within the analytical solution, the interfacial thermal stress induced by the differences in the thermal expansion coefficients between the FRP plate and the substrate beam is considered. Such an analytical solution is expected to be useful to obtain an in-depth understanding of the thermal stress and the debonding failure of FRP-plated beams when they are subjected to seasonal ambient temperature changes.

3. Theoretical background and analytical solution

The analytical solution and the nomenclature parameters for a FRP-plated beam subjected to three-point bending as presented in this paper, is shown in figure 1. As the analysis is related to the plate end debonding process, all the materials involved at the interface are assumed as linear elastic following the assumption adopted by De Lorenzis *et al.* [12]. The cohesive zone (CZ) model defined for the bond interface is illustrated in figure. 2, in which the coupled damage CZ model proposed by previous studies [12-13, 15] is adopted. Bilinear bond-slip models were implemented as they best suitable for capturing the nonlinear characteristics of the bond interface [7-8, 16-18].

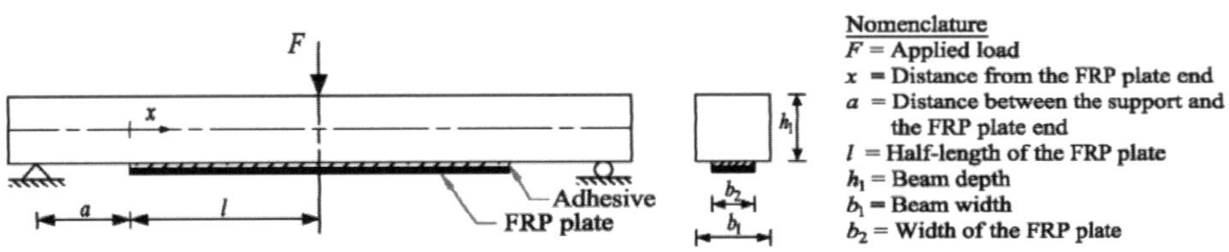

Figure 1. Simply supported FRP-plated beam under three-point bending load.

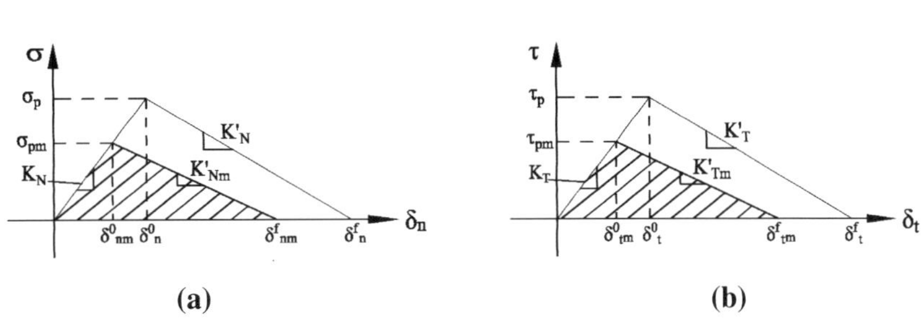

Figure 2. Mixed-mode cohesive zone laws. (a) Mode I. (b) Mode II.

The bond-slip relationships considered for the normal and shear directions of the bond interface under a coupled mix-mode loading can be defined as follows:

$$\sigma = \begin{cases} K_N \delta_N & \delta_M \leq \delta_M^0 \\ (1\text{-}d)\, K_N \delta_N & \delta_M^0 < \delta_M < \delta_M^f \\ K_N \langle -\delta_N \rangle & \delta_M > \delta_M^f \end{cases} \quad (1)$$

$$\tau = \begin{cases} K_T\, \delta_T & \delta_M \leq \delta_M^0 \\ (1\text{-}d) K_T \delta_T & \delta_M^0 < \delta_M < \delta_M^f \\ 0 & \delta_M > \delta_M^f \end{cases} \quad (2)$$

The strains of the bottom fiber of the beam and of the top fiber of the plate can be obtained from the equilibrium equations:

$$\varepsilon_1 = \frac{du_1}{dx} = \frac{y_1}{E_1 I_1} M_1(x) - \frac{1}{E_1 A_1} N_1(x) + \alpha_{T1} \Delta T \quad (3)$$

$$\varepsilon_2 = \frac{du_2}{dx} = -\frac{y_2}{E_2 I_2} M_2(x) + \frac{1}{E_2 A_2} N_2(x) + \alpha_{T2} \Delta T \quad (4)$$

The strain solutions presented in equation (3) and equation (4) are different from those in done by De Lorenzis *et al.* [12] as they include the thermal strains due to the thermal mismatch between the FRP and the substrate beam. Due to the space limitation, the detailed process on how to obtain the solution for the three stages (namely elastic stage, elastic-softening stage, and elastic-softening-debonding stage) is not given in the conference paper; the results from the analytical solution are presented herein to illustrate the effect of thermal loading on the plate end debonding of FRP-plated beams.

***3.1.** Interfacial stress and slip components for the elastic stage (E-stage).*
The shear stress and slip components can be obtained as follows:

$$\tau_e(x) = B_1 \cosh\lambda x + B_2 \sinh\lambda x + m_1 \frac{F}{2} \quad (5)$$

$$\delta_{Te}(x) = \frac{1}{K_T}\left(B_1 \cosh\lambda x + B_2 \sinh\lambda x + m_1 \frac{F}{2}\right) \quad (6)$$

where the constants in the above equations are determined as follows:

$$B_1 = \frac{1}{\lambda}\left[\frac{m_2 a}{2} F - K_T(\alpha_{T2} - \alpha_{T1})\Delta T\right] \tanh\lambda L - m_1 \frac{F}{2\cosh\lambda L} \quad (7)$$

$$B_2 = \frac{1}{\lambda}\left[-\frac{m_2 a}{2} F + K_T(\alpha_{T2} - \alpha_{T1})\Delta T\right] \quad (8)$$

Similarly, for the normal stress and slip components, the equations are

$$\sigma_e(x) = (C_1 \cos\beta x + C_2 \sin\beta x)\exp(-\beta x) - \eta_1 \lambda (B_1 \sinh\lambda x + B_2 \cosh\lambda x) \quad (9)$$

$$\delta_{Ne}(x) = \frac{1}{K_N}\left[(C_1 \cos\beta x + C_2 \sin\beta x)\exp(-\beta x) - \eta_1 \lambda (B_1 \sinh\lambda x + B_2 \cosh\lambda x)\right] \quad (10)$$

where the constants in the above equations are determined as follows:

$$C_1 = \frac{K_N F}{4\beta^3 E_1 I_1}(1 + \beta a) + \frac{\eta_1 \lambda^3}{2\beta^2}\left(\frac{\lambda B_1}{\beta} + B_2\right) - \frac{\eta_3}{2\beta^3}\left(B_1 + \frac{m_1 F}{2}\right) + \frac{K_N}{2\beta^2}(\alpha_{T2} - \alpha_{T1})\Delta T \quad (11)$$

$$C_2 = -\frac{K_N}{2\beta^2}\left[\frac{Fa}{2E_1 I_1} + (\alpha_{T2} - \alpha_{T1})\Delta T\right] - \frac{\eta_1}{2\beta^2} B_2 \lambda^3 \quad (12)$$

where $\tau_e(x)$, $\sigma_e(x)$, $\delta_{Te}(x)$, and $\delta_{Ne}(x)$ relates to the shear and normal stresses in addition to their respective relative slip within the elastic range. K_T and K_N are the elastic interfacial stiffness in the shear and normal direction respectively, while G_a and E_a are the shear and elastic modulus of the adhesive.

3.2. Determination of the load corresponding to the onset of softening.

When considering the mixed mode effects, the onset of softening occurs upon the fulfilment of the quadratic stress failure criterion which can be described in equation (13). Assuming that the maximum stresses (at $x = 0$) are reached and therefore causes the onset of softening, the maximum values for shear and normal stresses would produce equation (14). After substituting the relevant constants and terms, equation (14) could be simplified to equation (15), where K_1 to K_6 are all constants and the force corresponding to the onset of softening could be easily computed.

$$\left(\frac{\tau}{\tau_p}\right)^2 + \left(\frac{\sigma}{\sigma_p}\right)^2 = 1 \tag{13}$$

$$\left(\frac{B_1 + m_1 \frac{F}{2}}{\tau_p}\right)^2 + \left(\frac{C_1 - \eta_1 \lambda B_2}{\sigma_p}\right)^2 = 1 \tag{14}$$

$$\frac{K_1 F^2 + K_2 F + K_3^2}{\tau_p^2} + \frac{K_4 F^2 + K_5 F + K_6}{\sigma_p^2} = 1 \tag{15}$$

3.3. Interfacial stress and slip components for the elastic-softening stage (E-S stage).

If the applied load is greater than the load to cause the onset of softening, $\delta_{Te,max}$ is larger than the tangential relative slip at the onset of mixed-mode softening (δ_{tm}^0) and/ or $\delta_{Ne,\,max}$ is larger than the normal relative slip at the onset of mixed-mode softening (δ_{nm}^0), then the equations for the elastic stage would not be valid due to the length of softening, denoted as ($\bar{x}$). Therefore, the equations given below are for the analytical solution of the E-S stage.

Step 1. By defining the displacement-based mode-mixity ratio [equation (16)] according to De Lorenzis *et al.* [12], a value for $\delta_{Te,max}$ is firstly selected, and the corresponding $\delta_{Ne,max}$ can be calculated by equation (16). Also, since $\delta_{Te,max}$ is known, the value can be applied to its definition as observed in equation 17.

$$\gamma_{e,min} = \gamma_e(x) = \gamma_e(0) = \frac{\delta_{Te,max}}{\delta_{Ne,max}} = \frac{K_N\left(B_1 + m_1 \frac{F}{2}\right)}{K_T(C_1 - \eta_1 \lambda B_2)} \tag{16}$$

$$\delta_{Te,max}(x) = \frac{1}{K_T}(B_1 + m_1 \frac{F}{2}) \tag{17}$$

Step 2. Rearranging equation (17) and by substituting B_1 into the equation, the expression for the force component can be found. This component is then substituted into the shear stress equation for the elastic stage [equation (5)] to produce equation (18). After repeating these steps to obtain the normal stress force component ($F_{\sigma e}$), this expression is then also substituted into the normal stress equation for the elastic stage [Equation (9)].

$$\text{Shear stress force component, } F_{\tau e} = \frac{2\left[K_T \delta_{Te,max}(x) + \frac{K_T \tanh\lambda L}{\lambda}(\alpha_{T2} - \alpha_{T1})\Delta T\right]}{\left(\frac{m_2 a \tanh\lambda L}{\lambda} - \frac{m_1}{\cosh\lambda L} + m_1\right)} \tag{18}$$

Step 3. After having all known terms in equations (5) and (9), the shear and normal stress equations would be inserted into equation (13) would now take the form of equation (19) and the length of softening ($\bar{x}$) could now be computed.

$$\left(\frac{\tau_e}{\tau_p}\right)^2 + \left(\frac{\sigma_e}{\sigma_p}\right)^2 = 1 \tag{19}$$

Step 4. Having $\bar{x}$, δ_T, δ_N, total mixed-mode relative displacements (δ_m), mode-mixity ratio (γ), total mixed-mode relative displacement at the onset of softening (δ_m^0), mixed-mode critical total displacement (δ_m^f) and the damage evolution (d) can be all calculated as functions of x. Further information on these equations can be found in the previous study conducted by De Lorenzis *et al.* [12] to analyze the FRP-plated beam under mechanical loading solely.

3.4. Determination of the ultimate load.

The approximate ultimate load which includes the thermal stress effects is given in the following equation. For the load which corresponds to the length of softening, equation (20) can be used in its current form. For obtaining the ultimate load, equation (20) should be maximized with respect to $\bar{x}$.

$$F=\frac{2\left[\tau_p(\tan\lambda'\bar{x}+r\coth[\lambda(1-\bar{x})])+K'_T(\alpha_{T2}-\alpha_{T1})\Delta T\left(\frac{\sin^2\lambda'\bar{x}}{\cos\lambda'\bar{x}}+\frac{\cos\lambda'\bar{x}}{\lambda'}\right)\right]}{\frac{m_1\sin\lambda'\bar{x}}{\cos\lambda'\bar{x}}+\frac{m'_2\,a}{\lambda'\cos\lambda'\bar{x}}+m_1 r\left(\frac{\cosh[\lambda(1-\bar{x})]-1}{\sinh[\lambda(1-\bar{x})]}\right)} \tag{20}$$

4. Finite element (FE) model

A 3D FE model using the general purpose software ABAQUS was proposed to verify the analytical solution as stated above. A CFRP-plated steel beam was taken as an example using the geometry, material and interface data which were taken from the previous study [12]. The specimen is a 127×76×13 steel beam, which had a 1.2m long and a clear span of 1.1m. A 3-mm-thick carbon FRP (CFRP) plate was bonded to the bottom of the steel beam with an epoxy bonding adhesive (1mm thick). More details of the geometrical dimensions of the steel beam are listed in Table 1, while the mechanical and bond properties are depicted in Table 2. As previously stated, the cohesive zone model was implemented within the FE model and could be observed in figure 3. In the FE model, the steel beam and the CFRP plate were modeled using shell elements. The bond interface between them was modelled using a cohesive interface element. The coefficient of thermal expansion for the CFRP plate and steel beam was taken as 0/°C [7] and 12×10^{-6} /°C [8], respectively.

Table 1. Details of the geometry for FRP-strengthened steel beam

b_2 (mm)	y_1 (mm)	y_2 (mm)	A_1 (mm^2)	A_2 (mm^2)	I_{1v} (mm^4)	I_2 (mm^4)	t_a (mm)	l (mm)	a (mm)
76	63.5	1.5	1602	228	4588602	171	1	150	400

Table 2. Details which consists of the material properties and interfacial data

E_1 (N/mm^2)	τ_p (N/mm^2)	δ_t^0 (mm)	δ_t^f (mm)	G_{IIC} (N/mm)	α_1 (/°C)	E_2 (N/mm^2)	σ_p (N/mm^2)	δ_n^0 (mm)	δ_n^f (mm)	α_2 [/°C]
205000	26.73	0.0526	0.1197	1.60	12×10^{-6}	212000	29.70	0.0037	0.0044	0

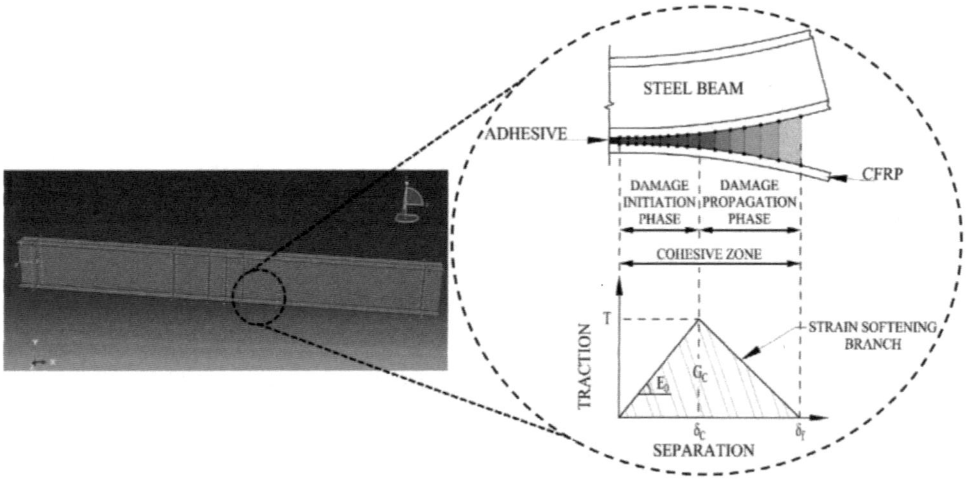

Figure 3. The FE model.

5. Comparison with FE results

Since the stresses vary across the interface, the FE results examined include the interfacial stresses at a distance of 5 mm from the side and at the mid-width of the beam, respectively (the curves are labeled as "side" and "mid-width" in figure 4). A very close agreement between the analytical solutions and the FE results can be observed in Figures 4a and 4c when the FRP-plated beam under mechanical loading only (i.e., ΔT=0 °C). As the temperature variation of 50 °C is taken into account, the results presented in figures 4b and 4d are still satisfactory in addition to the lengths of softening being close to each other. In figure 4a, the applied load is less than the softening load (i.e., the loading corresponding to the onset of softening) and hence the interface is within the elastic stage. However, due to the presence of thermal loading, the stresses shown in figure 4b indicate that the interface enters into the elastic-softening stage at the same applied load. Figures 4c and 4d show the stress responses of the interface (i.e., the elastic-softening stage) under a much higher applied load. It is clearly seen that the thermal stresses can not only influence the interface stress responses but also result in the reductions in the softening load.

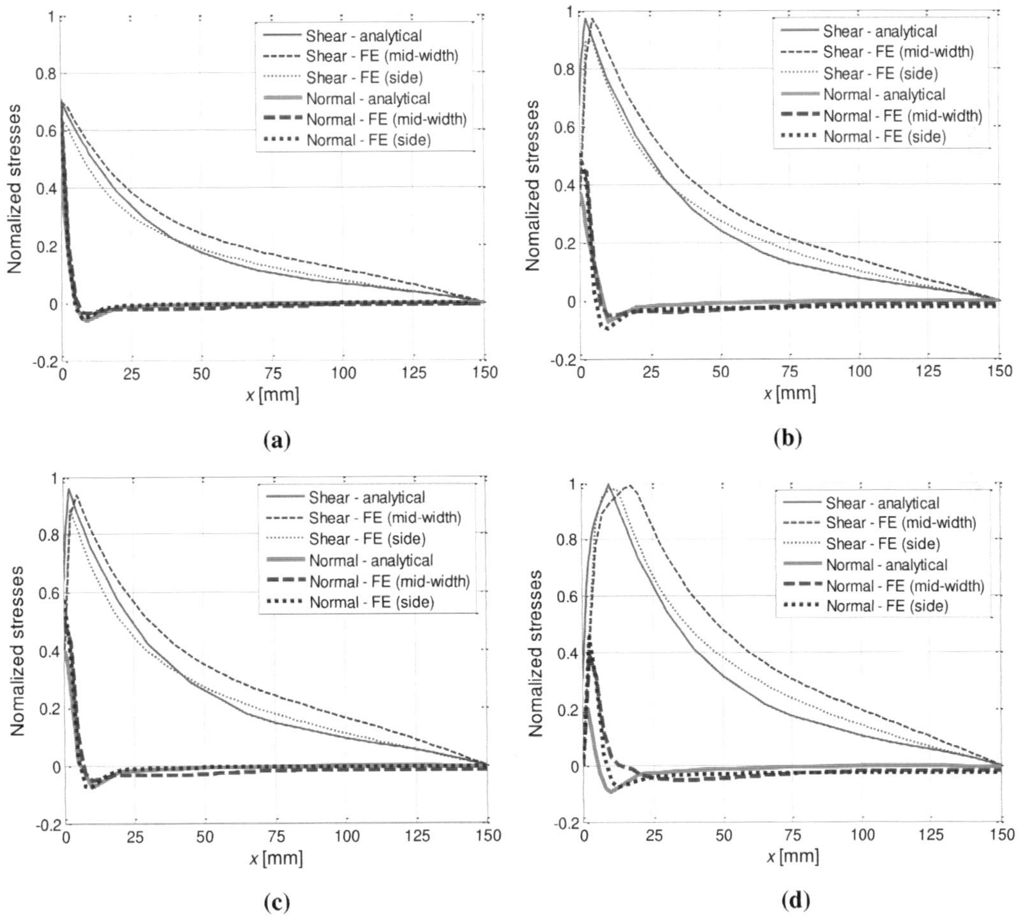

Figure 4. Normalized interfacial stresses τ/τp and σ/σp along the CFRP length. (a) F = 85kN, ΔT = 0°C. (b) F = 85kN, ΔT = 50°C. (c) F = 121kN, ΔT = 0°C. (d) F = 121kN, ΔT = 50°C.

6. Parametric Study

In this section, the verified analytical solution is used to conduct a parametric study by examining the thickness and length of the FRP plate on the softening load of the FRP-plated beam under combined thermal and mechanical loadings. Figure 5 presents the effects of FRP thickness and length on the softening load as the temperature changes. From this parametric study, it is seen that a thinner and longer

FRP plate has experienced more load changes with the temperature variations. In addition, the softening load has changed from around 140-170 kN at a minus 50 °C variation to around 60-70 kN at a positive 50 °C variation.

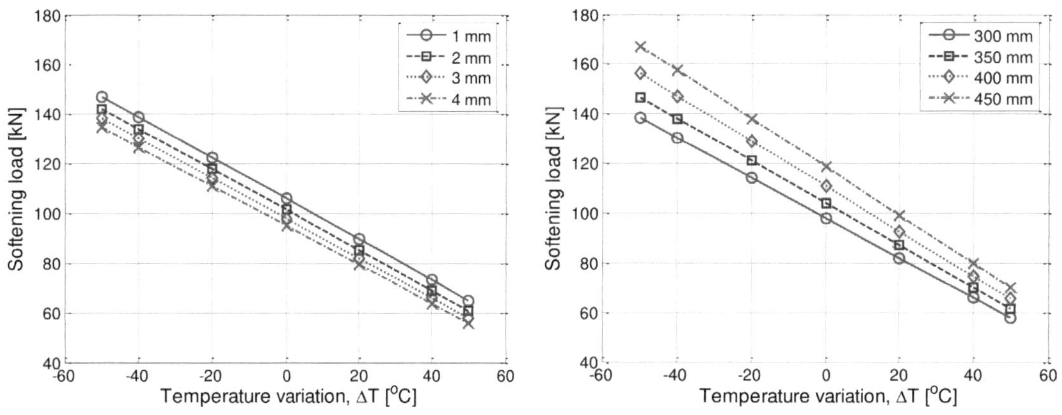

Figure 5. Effect of temperature variations on the softening load. (a) Effect of CFRP plate thickness. (b) Effect of CFRP plate length.

7. Concluding remarks

This paper presented an analytical solution for predicting the interfacial stress and slip evolutions as well as the debonding process of the FRP-plated beam under combined thermal and mechanical loadings. In the analytical model, a coupled mix-mode cohesive zone model with bilinear traction-separation relationships was used to describe the bond behavior of the FRP-to-substrate interface. The thermal stress induced by the different thermal expansion coefficients of the FRP plate and the substrate beam was the primary focus of the present study. Closed-form expressions for the interfacial stresses, the interfacial slips, the load corresponding to the onset of softening and the ultimate load were determined by the analytical solution. The analytical results indicated that the thermal loading had significant effects on the interfacial stress and slip evolutions as well as the loading behavior of the FRP-plated beam. Besides, an FE model was proposed to verify the analytical solution. The comparisons between the analytical results and the FE model predictions indicated that a good agreement was achieved between them, which further demonstrated the accuracy and reliability of the analytical solution. In addition, a parametric study was conducted to investigate the effects of two main design parameters (i.e., thickness and length of the CFRP plate) have on the performance of the FRP-plated beams under combined thermal and mechanical loadings. The results showed that the thermal loading had a significant effect on the softening loads of the FRP-plated beams. More research work is needed to propose a predictive model for the load capacity of the FRP-plated beam under combined thermal and mechanical loadings, which will be expressed as a function of various design parameters.

8. References

[1] Bank L C 2006 *Composites for Construction: Structural Design with FRP materials.* (USA: John Wiley & Sons Inc.)

[2] Hollaway L C and Teng J G 2008 *Strengthening and Rehabilitation of Civil Infrastructure Using Fibre-Reinforced Polymer (FRP) Composites.* (UK: Woodhead Publishing)

[3] Teng J G, Chen J F, Smith S T and Lam L 2003 Behaviour and strength of FRP-strengthened RC structures: a state-of-the-art review. *Proc. of the Insti. of Civil Engineers - Structures and Buildings* **156** 51-62.

[4] Dai J G, Gao W Y and Teng J G 2015 Finite Element Modeling of Insulated FRP-Strengthened RC Beams Exposed to Fire. *J. Comp. Const.* **19** 04014046.

[5] Denton S R 2001 Analysis of stresses developed in the FRP plated beams due to thermal effects. *J. FRP Composites in Civil Engineering* **1** 527-36.

[6] Benyoucef S, Tounsi A, Yeghnem R and Bachir M 2014 An Analysis of Interfacial Stresses in Steel Beams Bonded With a Thin Composite Plate Under Thermomechanical Loading. *J. Mech. Comp. Mat.* **49** 641-50.

[7] Gao W Y, Teng J G and Dai J G 2012 Effect of Temperature Variation on the Full-Range Behavior of FRP-to-Concrete Bonded Joints. *J. Comp. Const.* **16** 671-83.

[8] Gao W Y, Dai J G and Teng J G 2015 Analysis of Mode II debonding behavior of fiber-reinforced polymer-to-substrate bonded joints subjected to combined thermal and mechanical loading. *J. Eng. Fract. Mech.* **136** 241-64.

[9] Biscaia H C 2019 The influence of temperature variations on adhesively bonded structures: A non-linear theoretical perspective. *Int. J. of Non-Linear Mechanics* **113** 67-85.

[10] Rabinovitch O 2010 Impact of thermal loads on interfacial debonding in FRP strengthened beams. *Int. J. of Solids and Structures* **47** 3234-44.

[11] De Lorenzis L and Zavarise G 2009 Cohesive zone modeling of interfacial stresses in plated beams. *Int. J. of Solids and Structures* **49** 4181-91.

[12] De Lorenzis L, Fernando D and Teng J-G 2013 Coupled mixed mode cohesive zone modeling of interfacial debonding in simply supported plated beams. *Int. J. Solids and Structures* **50** 2477-94.

[13] Zhang L and Teng J G 2010 Simple General Solution for Interfacial Stresses. *J. Comp. Const.* **14** 434-42.

[14] Deng J, Lee M M K and Moy S S J 2004 Stress analysis of steel beams reinforced with a bonded CFRP plate. *J. Comp. Struc.* **65** 205-15.

[15] Camanho P P, Davila C G and De Moura M F 2003 Numerical simulation of mixed-mode progressive dalamination in composite materials. *J. Comp. Mat.* **37** 1415-38.

[16] Wang H-T and Wu G 2018 Bond-slip models for CFRP plates externally bonded to steel substrates. *J. Comp. Struc.* **184** 1204-14.

[17] Dehghani E, Daneshjoo F, Aghakouchak A A and Khaji N 2012 A new bond-slip model for adhesive in CFRP–steel composite systems. *J. Eng. Struc.* **34** 447-54.

[18] Wahab M M A, Ashcroft I A, Crocombe A D and Shaw S J 2001 Prediction of fatigue thresholds in adhesively bonded joints using damage mechanics and fracture mechanics. *J. Adhesion Sci. and Tech.* **15** 763-81.

Study on application of adhesives for Carbon Fiber Composites connection in High-speed Rail

Jiangyi Yan[1,2] and Qian Deng[3]

[1] Department of Mechanical Engineering, Tsinghua University, Beijing, 100084, China
[2] Beijing Key Lab of Precision/Ultra-precision Manufacturing Equipments and Control, Beijing, 100084, China
[3] Department of Materials Science and Engineering, Shenzhen International Graduate School of Tsinghua University, Shenzhen, 518000,China

yanjy18@mails.tsinghua.edu.cn

Abstract. At present, the high-speed train mainly made up of carbon fiber composites adopts a new flame-retardant multi-material system. It is necessary to select adhesives for the Adhesive bonding structures of different materials in order to meet the overall technical and performance requirements of high-speed trains. Through this research, mechanical performance data of several structural adhesives and sealant adhesives are obtained when they are bonded to the high-speed trains with carbon fiber composite material as the main body. Finally, the optimum adhesives for bonding different material systems in the high-speed trains are determined under different operational aging environments, which also provides data support for the formulation of connection specifications for carbon fiber reinforced composites in the field of high-speed railway. In addition, a multi-dimensional data reduction method is proposed for the analysis of sealant performance data, which is very helpful for the future analysis of big data of adhesives.

1. Introduction

Integral High-speed trains are facing many problems, such as harsh service environment, complex coupling effect, improvement of operation speed and service quality, and lightweight is the key to solve these problems. As one of the latest fiber reinforced composites, Carbon Fiber Reinforced Composites (CFRP) have become the best choice for high-speed trains to solve the problem of lightweight due to their excellent comprehensive performance of lightweight, high strength and high weather resistance, and their mature application in aviation, aerospace, ship, automobile and sports medical fields[1-5]. However, CFRP is anisotropic: its interlaminar strength is low and ductility is small. Consequently, the connection is the weak link of performance[6-7]. At present, the bonding mode of CFRP is mainly bonded by different kinds of adhesives. Adhesive bonding refers to bonding two or more components through adhesives, it can avoid the stress concentration caused by hole-making, and it has the advantages of high bonding efficiency and light structure[8-10]. As a result, it has become a common connection mode in CFRP.

At present, the high-speed train mainly made up of carbon fiber composites adopts a new flame retardant multi-material system. Based on the diversity of its structure, materials and service conditions, it is necessary to select a variety of adhesives to meet the performance requirements of different parts of high-speed trains.

2. Experimental

Generally speaking, the selection of adhesives for high-speed trains made up of carbon fiber composites is mainly divided into two aspects. One is to select the best structural adhesives for bonding different materials in the high-speed train, and the other is to select the best sealant for specific materials in the high-speed train.

2.1 Tensile shear property test of structural adhesives

The tensile shear property test of structural adhesives is mainly to select the best structural adhesive for the adhesion between two different materials in the high-speed train. It is mainly to select by testing and comparing the tensile shear strength of the adhesive samples with different adhesives under the standard tensile shear aging environment[11]. The test standard that Adhesives-Determination of tensile lap-shear strength of rigid-to-rigid bonded assemblies (GB/T 7124-2008 of China, idt ISO 4587:2003) is followed to test the lap-shear, and the whole shear tests were tested at room temperature(23℃). Among them, the bonding between the two materials adopts the way of single lap. The specific lapping mode is shown in Figure 1:

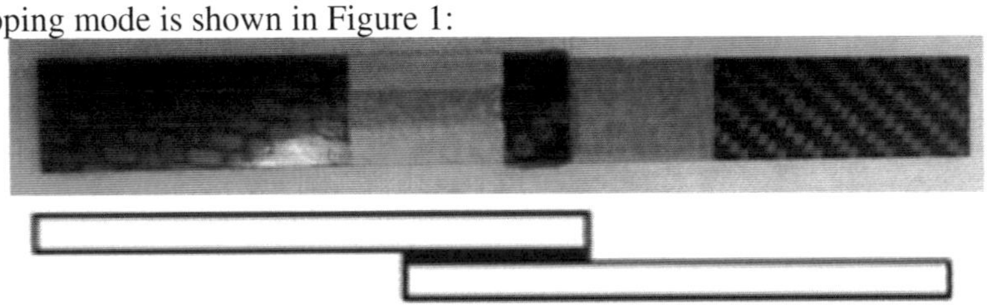

Figure 1. Single lap bonding between Carbon Fiber Composites

The structural adhesives selected in this study are mainly as follows:

(1) Polyurethane structural adhesives: UK1351 (1351), TS845(845) and 8100, suitable for assembling aluminum plates, steel plates and plastic part;

(2) Modified polyurethane structural adhesive: 3381, mainly used for bonding polypropylene(PP), polyvinyl chloride(PVC), polyethylene(PE), engineering plastics(ABS) and other plastics;

(3) Acrylic structural adhesive: 880, suitable for bonding plastics, glass and rubber.

There are four kinds of plates that need to be bonded with adhesives as a whole: stainless steel, CFRP, PVC and bakelite.

The types and sizes of adhesive samples with different adhesives are as follows:

(1) Adhesive 1351/845/8100:

①Bakelite 25×100×2(mm) + Stainless steel 25×100×1(mm)

②CFRP 25×100×2(mm) + PVC 25×100×2(mm)

③CFRP 25×100×2(mm) + CFRP 25×100×2(mm)

(2) Adhesive 880/3381: ①CFRP 25×100×2(mm) + PE 25×100×1(mm)

The standard test environment for aging is ISO9142D3 High and Low Temperature Cycle Curve with 14 cycles (seven days in total). The temperature change curve of a specific cycle is shown as Figure 2:

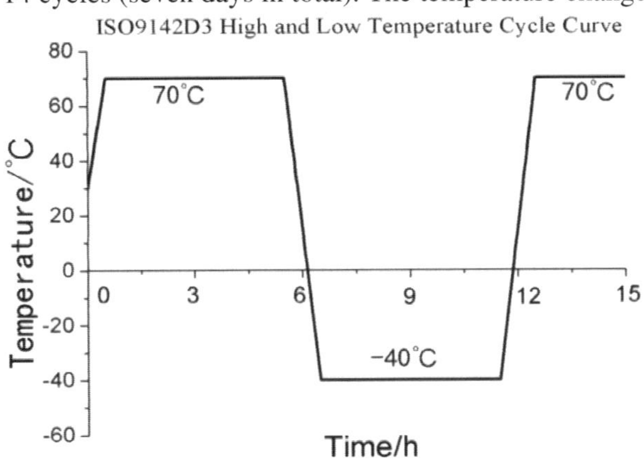

Figure 2. ISO9142D3 High and Low Temperature Cycle Curve

IOP Conf. Series: Materials Science and Engineering **784** (2020) 012004 doi:10.1088/1757-899X/784/1/012004

The test process of structural adhesive tensile shear property test mainly includes plate sample grinding, alcohol cleaning, drying, adhesive tape, curing, tensile shear aging environment treatment and tensile shear strength test.

2.2 The stripping performance test of sealant

The stripping performance test of sealant is mainly to select the best sealant adhesive for a specific material in the high-speed train made up of CFRP. It is mainly to select by testing and comparing the stripping performance of different adhesives in different standard aging environment. The test standard that Structural adhesives-Testing of adhesively bonded joints-Grub peel test (DIN54457-2007) is followed to perform 180° stripping test.

The sealants selected in this study are mainly silane modified polymers, which mainly include the following three types: Terostat MS 939 FR(939), MS1939(1939) and ISR 70-03 SSKF (7003). And there are three kinds of materials that need to be tested for adhesive stripping performance: stainless steel, CFRP, PVC.

For each type of sealant, the adhesive sample type and size are as follows:

(1)Stainless steel 150 x 300 x 2(mm); (2)CFRP150 x 300 x 2(mm); (3)PVC150 x 300 x 2(mm)

The standard aging treatment test environment for the stripping performance test of sealed adhesives mainly includes four types: (1) solidification for168 hours; (2) solidification for 168 hours + soaking water for 168 hours; (3) solidification for 168 hours + soaking water for 168 hours + 80℃ high temperature for 24h; (4) solidification for 168 hours + soaking water for 168 hours + 80℃ high temperature for 24 hours + high temperature and humidity 168 hours.

In general, the stripping performance test process of sealant mainly includes plate sample grinding, alcohol cleaning, drying, coating adhesive in surface, curing, different aging environment treatment and stripping performance test[12-13]. Figure 3 is a sealant bonded to the surface of CFRP. The stripping performance test of sealant is mainly through 180° stripping experiment to measure the ratio of the stripping cohesive failure of each adhesive, that is, the ratio of the adhesive cohesive failure surface to the total coating area when the bond sample is strippinged off. The higher the value is, the better the adhesion performance is. Figure 4 is the stripping performance test of stainless steel for different adhesives, and the stripping cohesion failure ratio on the left is 98%, and the right is 10%.

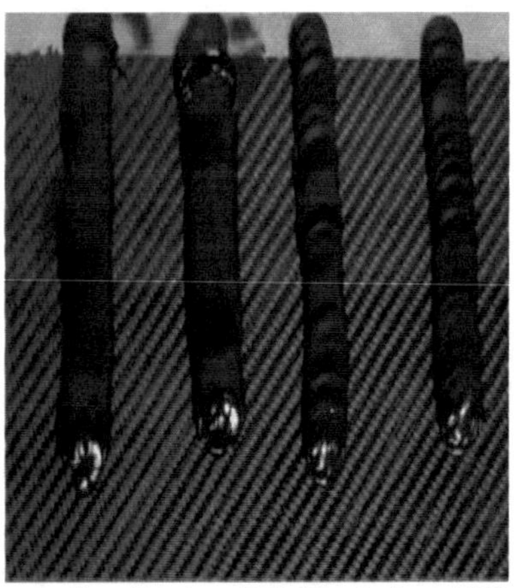

Figure 3. adhesive 939 bonded to CFRP

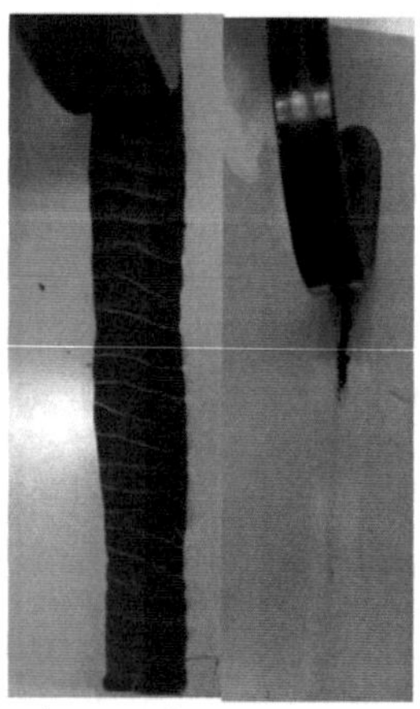

Figure 4. the stripping performance test of stainless steel

2.3 Experimental Equipment

The main experimental equipments needed in this research are as follows: Aging Testing Machine, Programmable constant temperature and humidity testing machine, Commercial compressed air gun, Static screw mixer, Beaker, glass rod, sandpaper, tape, etc.

3. Results and discussions

3.1 Tensile shear properties of structural adhesives

By testing the tensile shear properties of adhesive specimens with different structures, the tensile shear strength of different adhesives treated under standard tensile shear aging environment is finally obtained, and the error bar for the test data is ± 0.1Mpa, as shown in Table 1 below:

Table 1. tensile shear strength after aging treatment of different adhesives bonding plates(MPa)

Aging condition	ISO9142D3 high and low temperature cycle										
Material combination type	Bakelite+Stainless steel			CFRP+PVC			CFRP+CFRP			CFRP+PE	
Adhesive type	8100	845	1351	8100	845	1351	8100	845	1351	880	3381
Tensile shear strength/MPa	1.2	0.7	0.9	5.7	5.8	∞	13.7	7.4	17.0	3.1	0.9

Structural adhesives are shown in Figure 5. Obviously, the following conclusions can be drawn:

(1)For bonding stainless steel and bakelite, the order of tensile shear strength of different adhesives from high to low is adhesive 8100>1351>845, so 8100 can be preferred from the performance of tensile shear properties.

(2)Adhesive 1351 has the strongest tensile shear strength when bonding CFRP and PVC, because the PVC sheet has been damaged before the adhesive bonding site is damaged during the tensile shear performance test. The tensile shear performance of adhesive 8100 and 845 are not much different, and their tensile shear strength is around 5.7MPa.

(3)The tensile shear strength is adhesive 1351>8100>845 when CFRP is bonded to each other from high to low, and the tensile shear strength of adhesive1351 reaches 17MPa, indicating that the tensile shear performance of adhesive 1351 for bonding CFRP is very good, so adhesive 1351 is preferred.

(4)When bonding CFRP with PE, the tensile shear performance of adhesive 880 is better than 3381, so it is preferred.

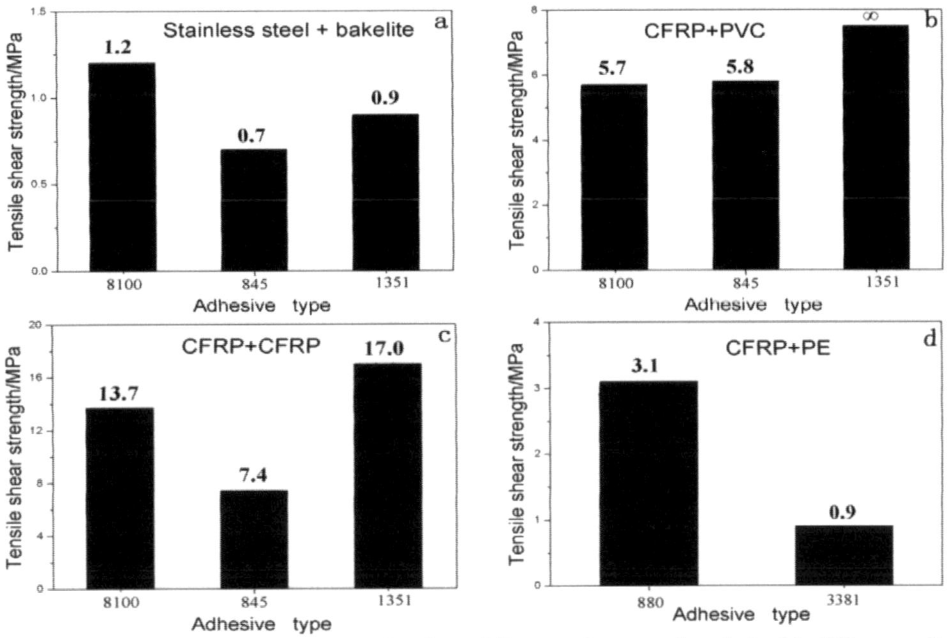

Figure 5. Bar chart of tensile shear strength when different plates are bonded with different adhesives, (a)Stainless steel+bakelite, (b)CFRP+CFRP, (c)CFRP+PVC, (d)CFRP+PE.

3.2 Stripping properties of sealants

By testing the tensile shear properties of adhesive specimens with different structures, the tensile shear strength of different adhesives treated under standard tensile shear aging environment is finally obtained, and the error bar for the test data is ± 0.1Mpa, as shown in Table 1 below:

By testing the stripping performance of the specimens bonded with different sealants, the stripping cohesive failure radio of different adhesives after successively superimposed treatment in the standard stripping aging environment are finally obtained, and the error bar for the test data for the stripping test is ± 3%, as shown in Table 2:

Table 2. The stripping cohesion failure radio of different adhesives under different aging environments

Aging conditions (Sequential superposition)	Matiral	Stainless steel			CFRP			PVC		
	Adhesive type	939	1939	7003	939	1939	7003	939	1939	7003
solidification 168h		85%	98%	80%	85%	98%	75%	95%	98%	98%
soaking water 168h		10%	60%	80%	75%	98%	80%	70%	95%	98%
80℃ high temperature 24h		90%	98%	80%	60%	95%	90%	40%	90%	95%
high temperature and humidity 168h		95%	95%	90%	90%	98%	98%	60%	95%	97%

The above experimental test results are summarized to obtain the change curves of the stripping cohesion failure radio of different adhesives after the treatment of successively superimposed aging environment, as shown in Figure 6. It is obvious that the following conclusions can be drawn:

(1)For stainless steel, adhesive 939 is highly sensitive to water when bonding, its stripping cohesion failure ratio after soaking water environment is only 10%, so the adhesion stripping performance is very poor. Adhesive 7003 and 1939 have better bonding properties overall. In addition, water will have a little impact on the stripping performance of adhesive 1939, and adhesive 7003 is more stable.

(2)For CFRP, the stripping performance of adhesive 1939 is more stable and safe, and the ratio of stripping cohesive failure under all four successively superposition aging environment treatment is higher than 90%. However, stripping performances of adhesive 7003 and 939 fluctuate greatly in different aging environments, and adhesive 939 performs poor in 80℃ high temperature environment.

(3)For PVC, the overall stripping performance of adhesive 939 is significantly lower than that of 1939 and 7003, there are safety risks in bonding. In addition, the stripping performance of adhesive 1939 and 7003 under all four successively superposition aging environment treatment is higher than 90%, the bonding is relatively safe and can be used. And, the stripping performance of adhesive 7003 is better than that of 1939.

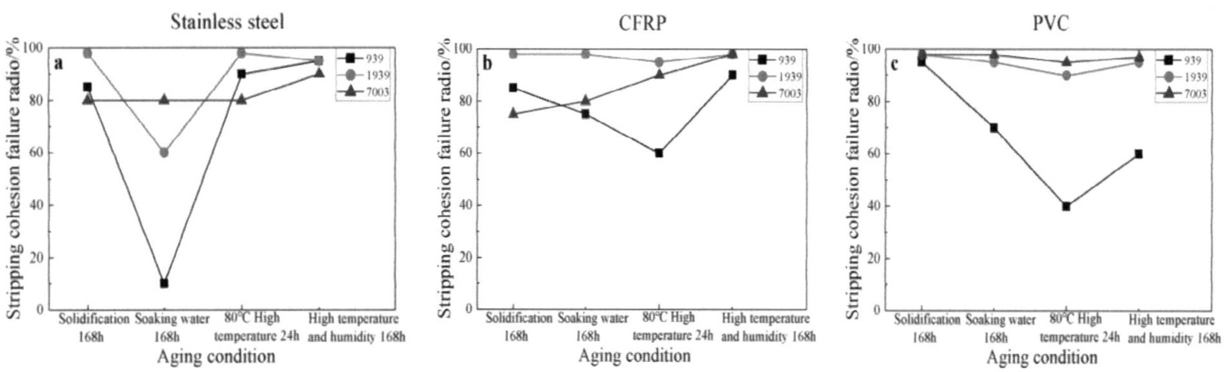

Figure 6. The stripping cohesion failure radio curves of different adhesive bonding materials after the treatment of four successively superimposed aging environment(a)Stainless steel, (b)CFRP, (c)PVC.

3.3 Establishment of comprehensive stripping performance index of sealant

According to the stripping performance test of the above sealant, for each adhesive, it can be selected only when it shows better stripping performance under four different aging experimental environments. It has become the core problem of choosing adhesives for the high-speed train made up of CFRP that how to select the best adhesive quickly and accurately considering the stripping performance under the four different aging environments, especially when the adhesive in four different conditions show the jagged stripping performance (performances in some aging environments are good, also some are bad) , this problem will become more important.

Considering the different stripping performance of different adhesives in four aging environments, a comprehensive evaluation index can be formulated to select sealants, which can solve the above problem well. And, according to the standard performance requirements of sealants, the comprehensive index needs to meet the following conditions:

(1)The comprehensive stripping index must be the same as the stripping cohesion failure ratio when the strippinging cohesion failure ratio in four different aging environments is the same.

(2)Only when the stripping cohesion failure ratio is good in all four aging environment can the comprehensive index be good. As long as the stripping cohesion failure ratio is low in one of the circumstances, the comprehensive index will eventually be low and close to the lowest stripping cohesion failure ratio. The requirement of the comprehensive index also indicates that the average value is not satisfied;

(3)According to the demand of different adhesives for different environments, a comprehensive stripping performance index focusing on one or more environments can be formulated.

Based on the above requirements, the following function is finally selected to calculate the comprehensive stripping performance index, and its specific expression is shown as in eqution (1):

$$H_{com} = \left(\frac{1}{\sum_{i=1}^{n} m_i} \times \left(\frac{m_1}{x_1} + \frac{m_2}{x_2} + \cdots + \frac{m_n}{x_n}\right)\right)^{-1} \quad (1)$$

Among them, H_{com} is the formulated comprehensive stripping performance index, m_i is the weight coefficient of each environment, x_n is the stripping cohesion failure radio in different aging environments, and n is the number of aging test environments needed. For the above functions, the comprehensive stripping performance index focusing on different environments can be worked out by adjusting the size of each weight.

Since the consideration of the four environments in this experimental study is equivalent, that is, the weight coefficients are the same, which can be assumed to be 1. In addition, there are only four different aging environments, that is, n=4. So, the final comprehensive stripping performance index is calculated as follows eqution (2):

$$H_{com} = \left(\frac{1}{4} \times \left(\frac{1}{x_1} + \frac{1}{x_2} + \frac{1}{x_3} + \frac{1}{x_4}\right)\right)^{-1} \quad (2)$$

In the above formula, x1,x2,x3 and x4 is the actual stripping cohesion failure radio in four different aging environments of this experimental study. Bring them into the above formula, we could get the final comprehensive stripping performance index as shown in Table 3.

Table 3. The comprehensive stripping performance index of different adhesives under four different aging environments

Matiral	Stainless steel			CFRP			PVC		
Adhesives	939	1939	7003	939	1939	7003	939	1939	7003
comprehensive stripping performance(%)	29.7%	84.6%	82.3%	74.7%	97.2%	84.7%	60.2%	94.2%	97%

The above data are summarized to obtain the bar chart of the comprehensive stripping performance indexes of different adhesives, as shown in Figure 7, and the following conclusions can be drawn:

(1)For stainless steel, the comprehensive stripping performance of different sealants from high to low is adhesive 1939>7003>939, of which adhesive 1939 has the highest comprehensive stripping performance, so adhesive 1939 is the best choice.

(2)For CFRP, the comprehensive stripping performance of different sealants from high to low is adhesive 1939>7003>939, and the comprehesive stripping index of adhesive 1939 is more than 90%, the bonding is very safe and adhesive 1939 can be selected.

(3)For PVC, the comprehensive stripping performance of different sealants from high to low is adhesive 7003>1939>939, the comprehesive stripping index of adhesive 7003 and 1939 are higher than 90%, so they both can be selected.

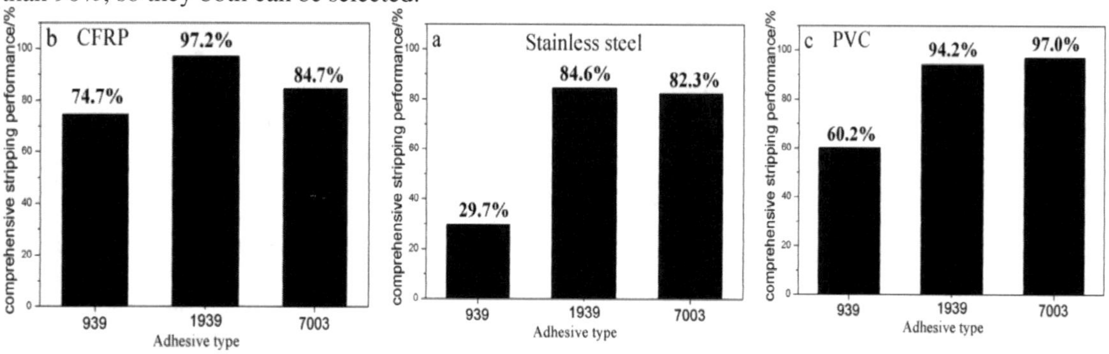

Figure 7. Bar chart of the comprehensive stripping performance indexes of different adhesives(a)Stainless steel, (b)CFRP, (c)PVC.

It can be seen that the above conclusions are consistent with those obtained in 3.2 analysis, which also indicates the effectiveness of the calculation method to the comprehensive stripping performance index.

4. Conclusions

Through the overall project research, the following conclusions are drawn for the selection of structural and sealant adhesives for carbon fiber composites connection in High-speed Rail:

(1)Structural adhesives

1) Adhesive 1351 is preferred when CFRP is bonded to each other;

2) Adhesive 1351 is preferred when bonding PVC with CFRP;

3) Adhesive 8100 is preferred when bonding stainless steel with bakelite;

4) Adhesive 880 is preferred when bonding PE with CFRP.

(2)Sealant adhesives

1) Adhesive 1939 is preferred for CFRP;

2) Adhesive 7003 is preferred for PVC ;

3) Adhesive 1939 is preferred for stainless steel.

In addition, in the stripping performance analysis for sealants, a new comprehensive stripping performance index is introduced for the stripping performance of the same adhesives in various environments, the essence is to reduce the dimension of multidimensional data and improve the efficiency of data processing, especially helpful for the big data analysis of the accumulation of different adhesive properties in the future. However, it is only a preliminary calculation method, which can be continuously improved according to the actual demand and the amount of data accumulated to make it more perfect and comprehensive.

Acknowledgment

1.This work was supported by the National Natural Science Foundation of China (51875312).

2.This work is supported by Tianjin research institute for advanced equipment of Tsinghua University in China.

3.This work is supported by National high speed train assembly engineering research center of CRRC Qingdao Sifang Co., Ltd in China.

References

[1] Ray BC. Temperature Effect During Humid Ageing on Interface of Glass and Carbon Fibers Reinforced Epoxy Composites[J]. Journal of Colloid and Interface Science, 2006,1:111-117.

[2] Alaa A M, Riadh A M, Zhao X L. Effect of CFRP properties, on the bond characteristics between steel and CFRP laminate under quasi-static loading [J]. Construction and Building Materials, 2015, 98(15):489-501.
[3] Sabrina F, Riadh A M, Zhao X L. Experimental and finite element analysis of a double strap joint between steel plates and normal modulus CFRP. Composite Structures, 2006.75(1-4):156-162.
[4] Mair R I. Advanced composite structures research in Australia[J]. Composite structures, 2002, 57(1): 3-10.
[5] Liu S F, Cheng X Q, Xu Y H, et al. Study on impact performances of scarf-repaired carbon fiber reinforced polymer laminates[J]. Journal of Reinforced Plastics and Composites, 2015, 34(1):60-71.
[6] Dattaguru B, Everett R. Geometrically-Nonlinear Analysis of Adhesively Bonded Joints[J]. 1982:16-25.
[7] Renton W J, Vinson J R. The efficient design of adhesive bonded joints[J]. The Journal of Adhesion, 1975, 7(3): 175-193.
[8] Lewis T.B., Nielsen L.E. Dynamic mechanical properties of particulate-filled composites[J]. Journal of Applied Polymer Science, 1970, 14(6): 1449-1471.
[9] Yu J.H., Huang X.Y., Wang L.C., et al. Preparation of hyperbranched aromatic polyamide grafted nanoparticles for thermal properties reinforcement of epoxy composites[J]. Polymer Chemistry, 2011, 2(6): 1380-1388.
[10] Gu J.W., Zhang Q.Y., Dang J., et al. Thermal conductivity epoxy resin composites filled with boron nitride[J]. Polymers for Advanced Technologies, 2012, 23(6): 1025-1028.
[11] Phillips H J, Shenoi R A. Damage tolerance of laminated tee joints in FRP structures[J]. Composites Part A: Applied Science and Manufacturing, 1998, 29(4): 465-478.
[12] Takahashi I, Ito Y, Takeda S, et al. Impact damage detection on scarf-repaired composites using Lamb wave sensing[C]. 16th international conference on composite materials, Japan, 2007.
[13] J. Epaarachchi, H. Ku and K. Gohel. A Simplified Empirical Model for Prediction of Mechanical Properties of Random Short Fiber/Vinylester Composites[J]. Journal of Composite Materials, 2009(44):779-788.

IOP Conf. Series: Materials Science and Engineering **784** (2020) 012005 doi:10.1088/1757-899X/784/1/012005

Prediction of flexural strength of natural pozzolana and limestone blended concrete using machine learning based models

Xiao-Yong Wang

Department of architectural engineering, Kangwon National University, Korea

wxbrave@kangwon.ac.kr

Abstract. Natural pozzolana and limestone ternary blended concrete is a new type of limestone calcined clay cement (LC3) concrete. Natural pozzolana and limestone can benefit each other and contribute to the performance improvement of concrete. Flexural strength is an important engineering property of concrete. This study predicts the development of flexural strength of ternary blended concrete using machine learning based models, such as artificial neural networks (ANN) and gene expression programming (GEP). Concrete mixtures and curing ages are used as input parameters of ANN and GEP. Flexural strength is output parameter of ANN and GEP. The correlation coefficients of GEP and ANN model are 0.98 and 0.99, respectively. GEP and ANN can make reliable evaluations of flexural strengths of ternary blended concrete.

1. Introduction

Natural pozzolana and limestone blended concrete belongs to the family of LC3 concrete because the alumina phase in pozzolana can react with limestone and form carbo-aluminate. The pozzolanic reaction of natural pozzolana can increase the late age strength of concrete. The nucleation effect of limestone can increase the early age strength of concrete. In addition, the additions of natural pozzolana and limestone can increase the chloride ingression durability, acid resistance, and reduce the CO2 emissions [1]. On the other hand, flexural strength is an important engineering property of concrete. Flexural strength is useful for pavement design, crack control, and prestress applications of reinforced concrete structures. The prediction of flexural strength is valuable for engineering practices of natural pozzolana and limestone blended concrete.

Many machine learning-based models, such as ANN and GEP, have been proposed for predicting properties of concrete. Altun et al. [2] evaluated the compressive strength of steel fiber reinforced concrete using ANN and found ANN can make better predictions than multiple linear regression. Bingol et al. [3] predicted the compressive strength of lightweight concrete after high temperatures. The prediction results were in accordance with experimental results. Atici [4] found ANN was useful for evaluating the nonlinear functional relationships. Mahdinia et al. [5] predicted the strength of concrete using GEP models. The testing ages, cement class, and sand to cement ratios were considered in Mahdinia et al.'s study. Ozcan et al. [6] found that GEP could accurately evaluate the splitting strength of concrete with various sizes and mixtures. Jiao et al. [7] predicted the shear debonding strength between high-performance fiber reinforced concrete and normal concrete.

Although lots of studies have been made about properties evaluation with machine learning-based models, the evaluation of flexural strength is relatively limited. Especially, the prediction of flexural strength of natural pozzolana-limestone ternary blended concrete is scarce. In this study, we evaluate the flexural strength of ternary blended concrete based on ANN and GEP model. The performance of ANN and GEP prediction is compared and discussed.

2. ANN and GEP model

2.1. ANN model

Artificial Neural Network (ANN), is a mathematical model that, based on the basic principle of neural network in biology, after understanding and abstracting the structure of human brain and the response mechanism of external stimulus, simulates the processing mechanism of complex information by the neural system of human brain on the theoretical basis of network topology knowledge. ANN is actually a complex network with a large number of simple components connected to each other, with a high degree of nonlinear, can carry out complex logic operations and nonlinear relations to achieve the system [8-11].

Experimental results from Ghrici et al. [1] are used for building ANN model. Ghrici et al. [1] measured flexural strength of ternary blended concrete with ten mixtures at four testing ages 2, 7, 28, and 90 days. The maximum replacement ratios of limestone and natural pozzolana were 20% and 30%, respectively. Water to binder ratio was .5. The total experimental results are divided into three sets, i.e., training set, validation set, and test set. The training set, validation set, and test set have 70%, 15%, and 15% experimental results, respectively. The ANN model is shown in Figure 1. The input parameters are mass ratios of cement, limestone, natural pozzolana, and ages. The hidden layer consists of 4 neurons. The initial values of the hidden layer are random numbers. Figure 2 shows the training process. Prediction error of the training set decreases with the evolution of the training process. Table 1 shows the results of the weight matrix and bias vector of ANN model. Because the input layer has four parameters, and the hidden layer has five neurons, the weight matrix has four columns and five rows. The bias vector has one column and five rows. Figure 3 shows the correlation coefficients of the training set, validation set, and testing set are 0.995, 0.979, and 0.998, respectively. The global correlation coefficient between output and target is 0.991. Summarily, ANN can accurately evaluate the flexural strength of ternary blended concrete.

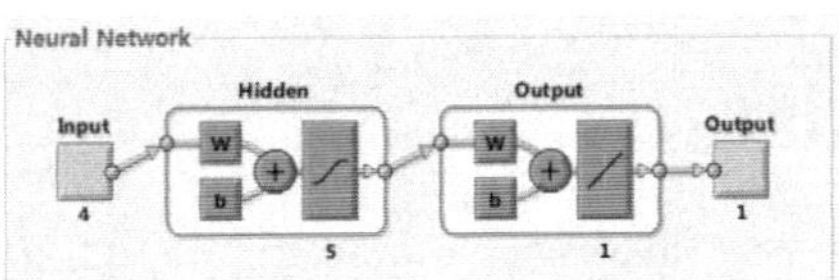

Figure 1. ANN model.

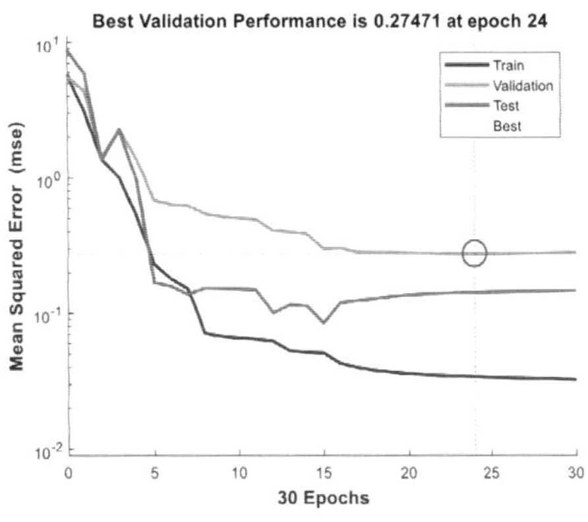

Figure 2. training process of ANN.

Table 1. Weight matrix and bias vector of ANN model

Weight matrix				Bias vector
-0.381	-0.419	-2.998	-0.960	2.521
-1.074	-0.399	-1.177	-1.022	1.766
0.147	1.298	-0.768	-3.023	-2.923
-1.977	4.288	1.060	-0.938	-1.843
0.596	0.395	0.162	3.523	3.687

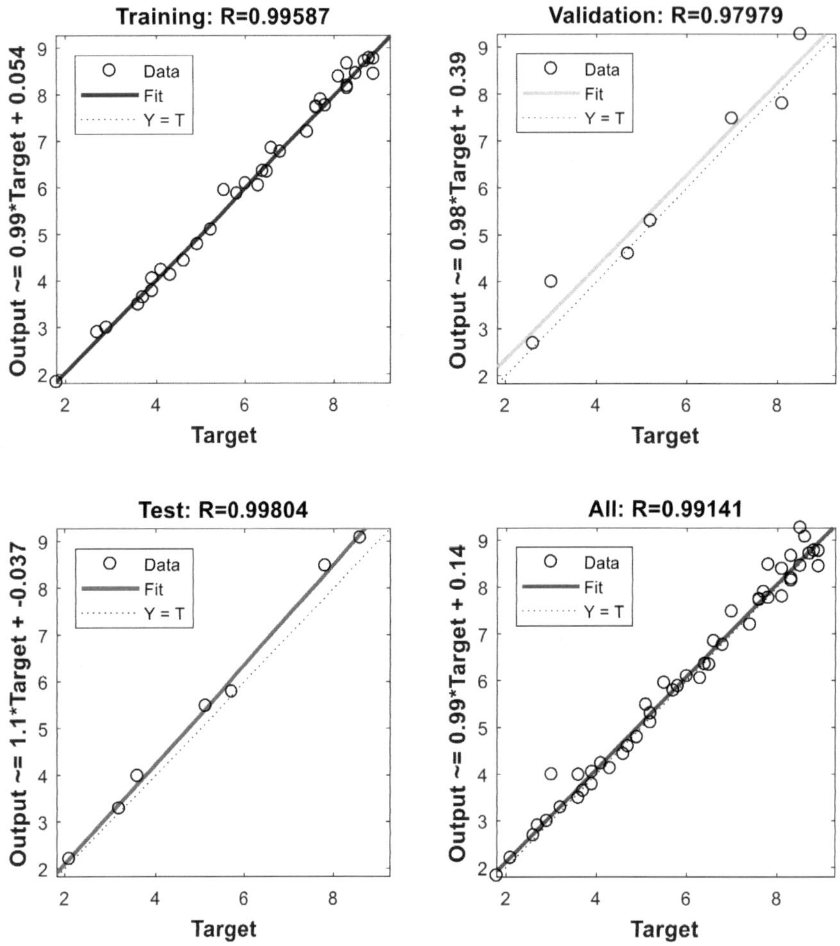

Figure 3. prediction results of ANN.

2.2. GEP model

Gene expression programming (GEP) is a new member of the family of genetic computing based on the inheritance and development of genetic algorithms and genetic programming. Gene-expression Programming differs markedly from traditional Genetic Algorithms (GA) and Genetic Programming (GP) in terms of the form of individual representations, processes, and results. Genetic programming is developed on the basis of genetic algorithm, which is superior to genetic algorithm in performance. Compared with genetic programming, gene-expression programming has more advantages in symbol and expression processing.

Based on experimental results of Ghrici et al. [1], we can obtain the expression tree of GEP model. The expression tree is shown in Figure 4. The prediction equation is shown as follows:

$$\begin{aligned} strength = &(1.0/(((((min(-10.451, 8.714)+d(3))/2.0)-min(d(0),8.714))/(1.0/(min(d(0),d(3))))))) \\ &+max((gep3Rt((d(3)+-6.108))+(tanh(4.485)^2)), gep3Rt((d(1)-d(1)))) \\ &+(((d(0)+-7.197)/(2.541*8.288))/max((d(3)/d(0)),(-7.197/-6.987))) \end{aligned} \quad (1)$$

where d(0), d(1), and d(3) means mass ratios of cement, natural pozzolana, and ages, respectively. The function Gep3Rt is cubic root function. On the other hand, although the number of the input parameter is 4, the number of used variables in regression equation is 3. It is because the sum of mass ratios of cement, natural pozzolana, and limestone equals to 100%. Based on the mass ratio of cement and natural pozzolana, we can calculate the mass ratio of limestone. In other words, the mass ratios of cement, natural pozzolana, and limestone are not independent variables. GEP method can automatically select independent variables and make prediction equations using these independent variables. On the other hand, ANN default uses all the input parameters for making prediction models. ANN can not automatically select the input parameters. Figure 5 shows the prediction results generally agree with experimental results. The correlation coefficient between output and target is 0.98. Although the prediction accuracy of GEP is slightly lower than that of ANN, the accuracy of GEP is acceptable for engineering applications of civil engineering.

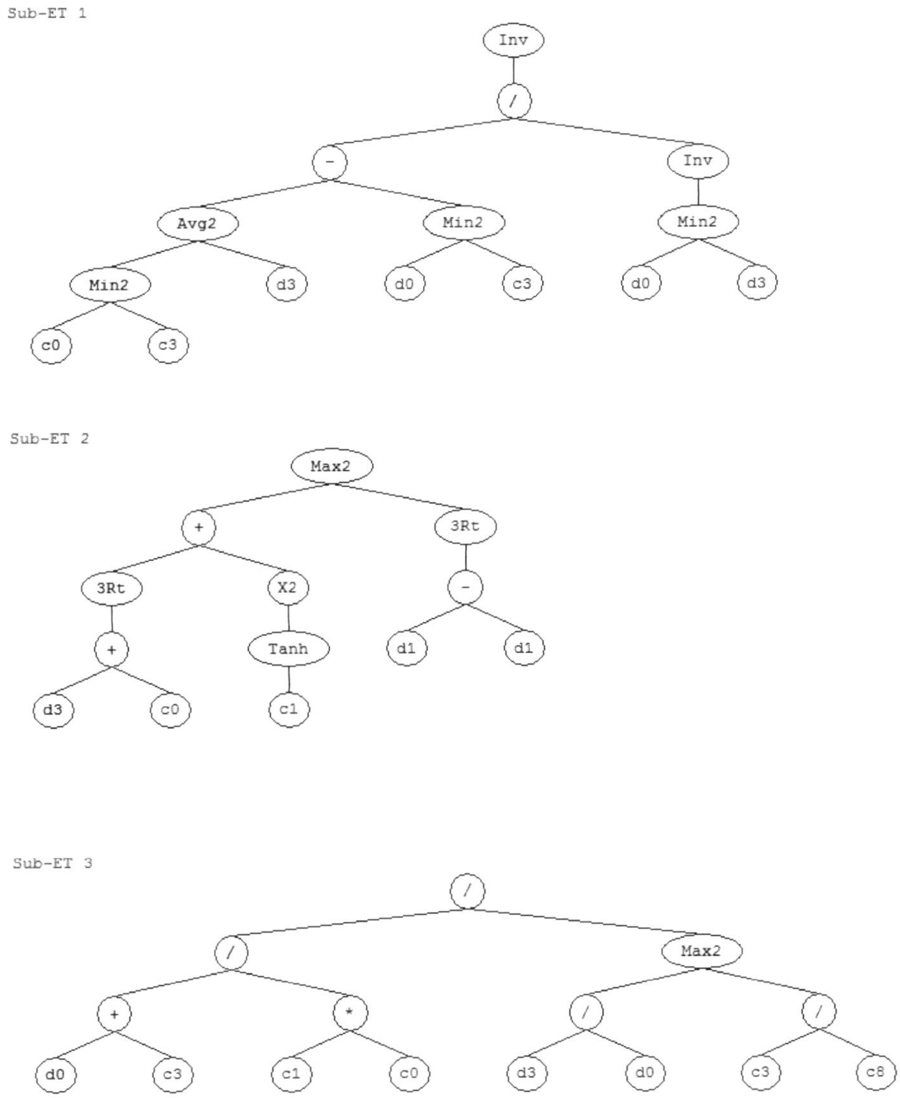

Figure 4. GEP model.

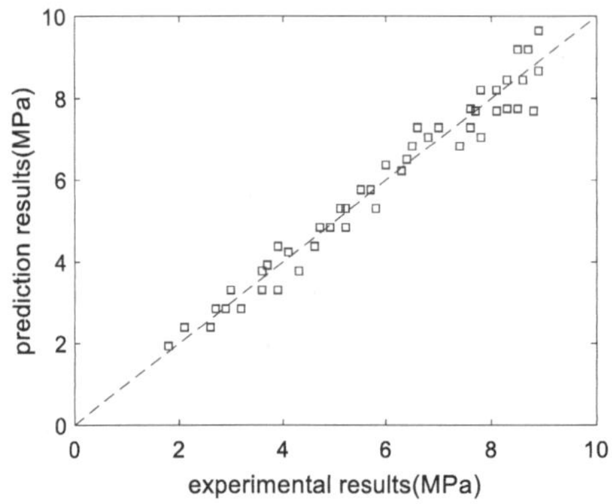

Figure 5. prediction results of GEP model.

2.3. Discussions of ANN and GEP

In this paper, neural network and gene expression programming are used to predict the strength of composite concrete. Neural network and gene expression programming present a strong predictive power. On the other hand, the neural network belongs to a black box system and is not convenient for engineering practice. The results of gene expression programming are presented in the form of formulas, which are convenient for practical application of engineering. Gene expression programming belongs to the global optimization algorithm, which overcomes the local optimal problem in neural network regression.

3. Conclusions

Artificial neural networks (ANN) and gene expression programming (GEP) are used to predict the flexural strength of LC3 concrete with various mixtures and ages.

Concrete mixtures and ages are used as input parameters of models. ANN model consists of an input layer, a hidden layer with four neurons, and an output layer. The total experimental results are divided into three sets, i.e., training set, validation set, and test set. Based on ANN, the flexural strength of LC3 concrete is estimated. The correlation coefficients of the training set, validation set, and testing set are 0.995, 0.979, and 0.998, respectively. The global correlation coefficient between output and target is 0.991.

GEP model calculates the flexural strength as the sum of three expression trees. GEP method can automatically select independent input variables and make prediction equations using these independent variables. ANN default uses all the input parameters for making predictions. ANN can not automatically select the input parameters. The correlation coefficient between output and target is 0.98. Although the prediction accuracy of GEP is slightly lower than that of ANN, the accuracy of GEP is acceptable for engineering applications of civil engineering.

Acknowledgment

This research was supported by the Basic Science Research Program through the National Research Foundation of Korea (NRF) funded by the Ministry of Science, ICT, and Future Planning (No. 2015R1A5A1037548) and an NRF grant (NRF-2020R1A2C4002093).

References

[1] Ghrici M, Kenai S, Said-Mansour M. Mechanical properties and durability of mortar and concrete containing natural pozzolana and limestone blended cements. *Cement & Concrete Composites* **29** (2007) 542-549

[2] Altun F, Kisi O, Aydin K. Predicting the compressive strength of steel fiber added lightweight concrete using neural network. *Computational Materials Science* **42** (2008) 259-265

[3] Bingol AF, Tortum A, Gul R. Neural networks analysis of compressive strength of lightweight

concrete after high temperatures. *Materials and Design* **52** (2013) 258-264
[4] Atici U. Prediction of the strength of mineral admixture concrete using multivariable regression analysis and an artificial neural network. *Expert Systems with Applications* **38** (2011) 9609-9618
[5] Mahdinia S, Naddaf HE, Shadnia R. Effect of cement strength class on the prediction of compressive strength of cement mortar using GEP method. *Construction and Building Materials* **198** (2019) 27-41
[6] Ozcan F. Gene expression programming-based formulations for splitting tensile strength of concrete. *Construction and Building Materials* **26** (2012) 404-410
[7] Jiao PC, Roy M, Barri K, Zhu RH, Ray I, Alavi AH. High-performance fiber reinforced concrete as a repairing material to normal concrete structures: Experiments, numerical simulations and a machine learning-based prediction model. *Construction and Building Materials* **223** (2019) 1167-1181
[8] www.mathworks.com
[9] Wang XY, Luan Y. Modeling of Hydration, Strength Development, and Optimum Combinations of Cement-Slag-Limestone Ternary Concrete. *International Journal of Concrete Structures and Materials* **12** (2018) 1-13
[10] Lim CH, Yoon YS, Kim JH. Genetic algorithm in mix proportioning of high performance concrete. *Cement and Concrete research* **34** (2004) 409-420
[11] Papadakis VG, Antiohos S, Tsimas S. Supplementary cementing materials in concrete Part II: A fundamental estimation of the efficiency factor, *Cement and Concrete research* **32** (2002) 1533-1538

Numerical simulation of mesoscopic bubble defect and optimization of process parameters in fiber wetting process of high voltage post insulator resin

B Y Yan[1], Z Zhang[1], W Yang[1], W Chu[2], L Yin[1], K Wang[1] and X Y Dai[1]

[1]State Key Laboratory of Advanced Power Transmission Technology, State Grid Global Energy Interconnection Research Institute Co.LTD, Beijing 102211, China
[2]Mechatronic School, Xi'an University of Architecture and Technology, Xi'an 710055, China

yanbingyue.good@163.com

Abstract. Based on the method of fluid volume function, the two-phase flow model of mesoscopic resin and air between unit cell fiber bundles was established. The causes of mesoscopic bubble defects were discussed and the controllable process parameters were analyzed. A near-real mono-cell mesoscopic geometry model of strut insulator was constructed and meshed. The solution of the resin and air two-phase flow model was completed by adopting the finite volume method and using the Fluent software. The numerical simulation of the mesoscopic resin air two-phase flow and the air bubble defect was applied. The relationship between process parameters and air bubble defect was studied, and process parameters were optimized. Through the contrastive analysis of the capillary number, the correctness of the numerical simulation and parameter optimization was theoretically verified. Further, the effectiveness of the method in this paper was verified through the experiment.

1 Introduction

High-voltage strut insulator adopts vacuum assisted resin impregnated fiber forming process, which is widely used in the manufacture of composite materials. However, the determination of its process parameters still relies mainly on a large number of tests and experiences. This design method is of high cost and low efficiency. With the development of numerical simulation technology, the method of. simulating the resin in the mold to infiltrate the fiber, predicting the defects that may occur during the infiltration process and optimizing the process parameters [1, 2], can reduce the cost, improve the design efficiency, and provide theoretical guidance for the actual infiltration process design. In the study of numerical simulation, Vanegas [3] established the resin flow model in the Resin Transfer molding (Resin Transfer Moulding, RTM) mold filling process based on Darcy's law . They simulated the RTM mold filling process, and verified the validity of the algorithm by experiment. According to Stokes and Brinkman equations, Verleye and Wei yajun established a two-scale RTM resin flow model for fiber unit cell. Resin simulation and calculation were carried out for mold filling process and permeation process, and the validity of the model was verified by experimental results. De Valve C et al. [4] simulated the entrapment process of bubbles in the fiber bundle in RTM process by VOF method, and analyzed the relationship between the bubble content and size, Reynolds number and capillary number. Yang bo analyzed the flow of resin in the two-scale pores of preformed body and the formation mechanism of bubbles, established a prediction model for bubble formation in the mold filling process, and verified the correctness of the prediction model through experiments. Jin tianguo established the control equation of single-cell resin and air flow in the process of RTM mold filling by using the VOF two-phase flow method, and realized the simulation and prediction of air inclusion process, and compared with the experimental values in previous research results, verified the correctness of bubble prediction model. In the application of numerical simulation, researchers have carried out numerical simulation of resin charging and resin infiltrated fibers on actual products [5]. Also, molding process was optimized by the simulation, such as sprue gate and vent port layout, etc., and used to guide the practical production.

The above numerical simulation research had greatly promoted the development of resin mold filling and resin infiltration fiber technology, but these researches mainly focus on the fields of machinery and ships, and it is still rare in the field of electrical insulation, especially in the field of high voltage insulation. In order to improve the optimization efficiency of resin infiltrating fiber molding process of high-voltage insulation parts, the numerical simulation of meso-bubble defects and the optimization of controllable process parameters was carried out for the vacuum-assisted resin-infiltrating fiber process of high-voltage pillar insulators in this paper.

2 Construction of mono-cellular two-phase flow model of post insulator resin impregnated fiber

In the process of fiber infiltration by vacuum assisted of high-voltage strut insulator, the resin flows in the mold cavity where the fiber is laid, which can be regarded as the seepage process of the viscous Newtonian fluid in the porous medium. Due to the complex structure of fiber preforming parts, it is difficult to describe the flow process of resin in the fiber at one time. Therefore, the whole fiber is considered to be composed of multiple single cells, and the resin flow in the preforming body is studied by analyzing the resin flow in the fiber unite cell according to the periodic characteristics of fiber preforming parts. Among them, monocellular mesoscopic flow, that is, resin flow in mesoscopic pores between fiber bundles, is prone to produce large bubbles, which has a fatal impact on the electrical insulation performance of strut insulators. Numerical simulation method was used to simulate the flow of resin, bubble generation and the effect of process parameters on bubble defects in the infiltration process, further optimizing the process parameters and eliminating bubbles, which can improve the process design efficiency and reduce the design and experiment cost.

Bubble formation process is a typical two-phase flow process of gas-liquid (air and resin). In order to effectively simulate the infiltration of mesoscopic resin flow, the generation of large bubbles and the influence of process parameters on bubble defects, it is necessary to establish a correlation model for tracking the interface of the two-phase flow for resin and air to correctly reflect the unsaturated flow of the resin. Among the methods to build the model, the VOF method is currently the most widely used multi-phase fluid interface tracking technology [6]. Therefore, in this paper we build a single-cell mesoscopic resin two-phase flow interface tracking model based on the VOF method for high-voltage strut insulators, as follows:

The VOF method determines the free surface through the resin and mesh volume ratio function F, and tracks the change of the resin. When F=1, the resin fills the unit; When F=0, there is no resin in the unit; When 0≤F≤1, the unit is a resin-air interface unit. In the unicellular mesoscopic region, that is, the inter fiber bundle region, it can be considered that there is only pure fluid flows, and the corresponding momentum equation of two-phase flow for resin and air can be expressed by navier-stokes equation and surface tension term:

$$\frac{\partial(\rho_m v)}{\partial t}+\nabla\cdot(\rho_m v\times v)=\rho_m g-\nabla P+\nabla\cdot\left[\mu_m\left(\nabla v+\nabla v^T\right)\right]+F_{st}\,. \quad (1)$$

Where ρ_m is the density of the unit, P is pressure, v is velocity of resin at time t, g is acceleration of gravity, μ_m is viscosity of the unit, and F_{st} is surface tension term of the unit. ρ_m and μ_m are as follows:

$$\rho_m = F\rho_r+(1-F)\rho_a=0\,. \quad (2)$$

$$\mu_m = F\mu_r + (1-F)\mu_a = 0. \tag{3}$$

Where ρ_r and ρ_a a are the density of the resin and air, respectively, and μ_r and μ_a are the viscosity of the resin and air, respectively. The transport equations of two-phase flow continuity equation and volume ratio function F are shown in Eq. (4) and (5) respectively.

$$\frac{\partial \rho_m}{\partial t} + \nabla \cdot (\rho_m v) = 0 . \tag{4}$$

$$\frac{\partial F}{\partial t} + \nabla \cdot (\mu_m F) = 0 . \tag{5}$$

According to literature [7], the element surface tension F_{st} is reflected in the form of pressure difference on both sides of the resin-air interface. According to the divergence theorem, F_{st} can be expressed as volume force, specific as follows:

$$F_{st} = \sigma_{ra} \frac{\rho_m \kappa \nabla F}{\frac{1}{2}(\rho_r + \rho_a)}. \tag{6}$$

Where, σ_{ra} is the tension coefficient at the air interface of the resin, κ is the curvature of the air interface of the resin, and the ∇ F is the gradient of the volume fraction of the resin unit.

3 An overview of mesoscopic bubble formation and analysis of controllable process parameters

3.1 An overview of the origin of meso-large bubbles

According to the literature [4,7], the mesoscopic flow of resin between fiber bundles is mainly driven by dynamic pressure due to the large mesoscopic space between fiber bundles (millimeter-level) and small capillary effect. While the micro voids in the fiber bundles are small (micron scale), and the flow of resin in the fiber bundles mainly depends on the capillary force. When the flow rate is low, the microcosmic flow of the resin is ahead of the mesoscopic flow. When the front the microcosmic flow meets the braiding knot of the fiber or the transverse fiber, the resin flows laterally, thus forming a larger microscopic bubble between the fiber bundles.

3.2 Analysis of controllable process parameters

Resin type, fiber type, fiber laying method, fiber layer number, resin feeding port, discharging port, feeding mode and discharging mode have been determined in the process of high-voltage post insulator resin infiltration fiber. The pressure applied the feed port during infiltration is small and negligible compared to vacuum negative pressure. Therefore, the adjustable process parameters are vacuum degree and resin viscosity. Insulation rod resin wetting fiber process requires more than 97% vacuum, so only the resin viscosity can be adjusted Resin viscosity is not a directly controllable process parameter, but viscosity is related to temperature. The corresponding relationship between the resin viscosity and temperature is shown in Figure 1. Meanwhile, the change of temperature will cause the change of resin surface tension and the corresponding relationship between the surface tension and temperature is shown in Figure 2. Therefore, the wetting temperature can be adjusted to change the viscosity and surface tension of the resin, thereby the infiltration process can be indirectly controlled to minimize the negative effect of the bubble.

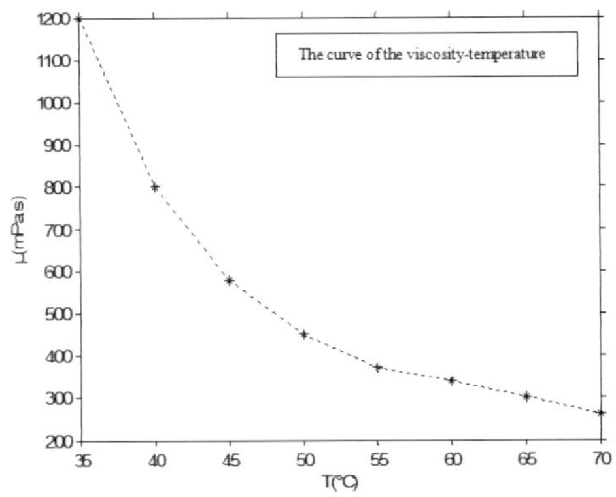

Figure 1. The curve of the viscosity-temperature of a certain type of resin.

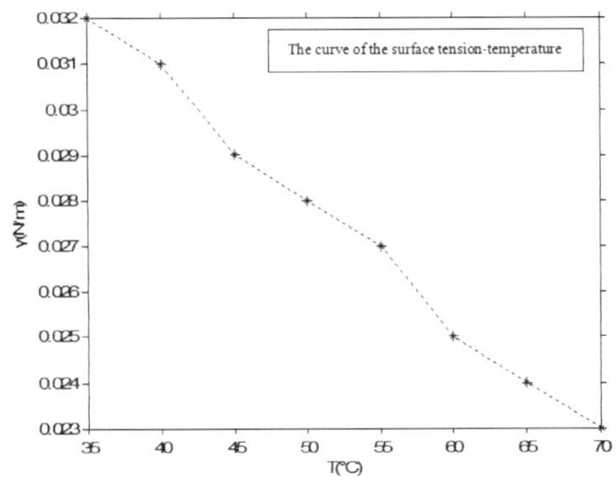

Figure 2. The curve of the surface tension-temperature of a certain type of resin.

4 Construction and mesh generation of cellular geometry model

The corresponding geometric model was built according to the real cellular structure, which was used as the structural matrix for simulation. The mesh division of the structural matrix was used as the basis for solving the resin air two-phase flow model. The cellular mesoscopic geometric model and mesh division process were as follows.

4.1Construction of geometric model of single cell

In order to reduce the intensity of simulation calculation, the two-dimensional section of the model was adopted, and the two-dimensional section geometric model of the intercellular fiber bundles was constructed. Transverse fiber bundle width was 0.9 mm; longitudinal dimension bundle width was 1.8 mm; and the distance between transverse and longitudinal fiber bundles was 0.2 mm.

4.2 Mesh generation

The model is divided into small simple geometric unit, which provides a basis for solving the resin air two-phase flow model. The mesh size is 0.02 mm, and the total mesh number is 25,000.

5 Discretization and solution of governing equations

In this simulation, the discrete model of resin air two-phase flow model was established by FVM method. FVM divides the calculation area into a series of control volumes, which are interrelated and isolated. All of them contain a grid node as the representative. Through the method of control volume integration, the two-phase flow model in the form of conservation is transformed into a discrete model. This model is widely used in numerical simulation of incompressible flow, because of the conservation of integral in the whole computational domain. The basic idea of FVM is simple and understandable. The core of FVM is regional discretization. In essence, it replaces continuous space by a finite number of discrete points.

In the numerical solutions of incompressible fluid flow field, the pressure correction method is the main method, and the semi-implicit method of pressure coupled equations is the most widely used pressure correction method. It is the most popular algorithm in various CFD software. The method is adopted to solve the discrete model.

6 Microscopic bubble defect simulation and process parameter optimization

The mono-cellular infiltration process of vacuum-assisted resin infiltrating fiber of high-voltage strut insulator was numerically simulated and the controllable process parameters were analyzed by Fluent software. Which is based on the mono-cellular geometric model of strut insulator and mesh division results, and dispersion model of resin and air two-phase flow.

Boundary conditions: the left side of the cellular mesoscopic view was the inlet, and the inlet boundary conditions were the velocity inlet; The right side of the cell is the outlet, and the boundary condition of the outlet is the pressure outlet. Considering the fact of multilayer fiber coating, the upper and lower boundary of a single cell is set as symmetrical boundary. The axial permeability of fiber bundles of a certain type of fiber is $2.246\times10^{-9}m^2$ and the radial permeability is $4.776\times10^{-10}m^2$. The resin viscosity was set at 300mPa·s. According to the viscosity - temperature and surface tension - temperature relationship in section 3, the temperature was 65°C and the surface tension was 0.024N/m. The flow rate was set at 0.472mm/s. The simulation results of mesospheric infiltration between fiber bundles were shown in Figure 3. The blue area was resin, and the red area was air. According to the literature [6], the capillary number less than 0.007 is prone to produce mesoscopic large bubbles. However, between 0.007 and 0.008, resin infiltration of fibers has the best effect. The capillary value Ca=0.0059 was obtained according to the capillary number calculation method [7], indicating that the simulation results of microscopic infiltration between fiber fascicles were consistent with the theory.

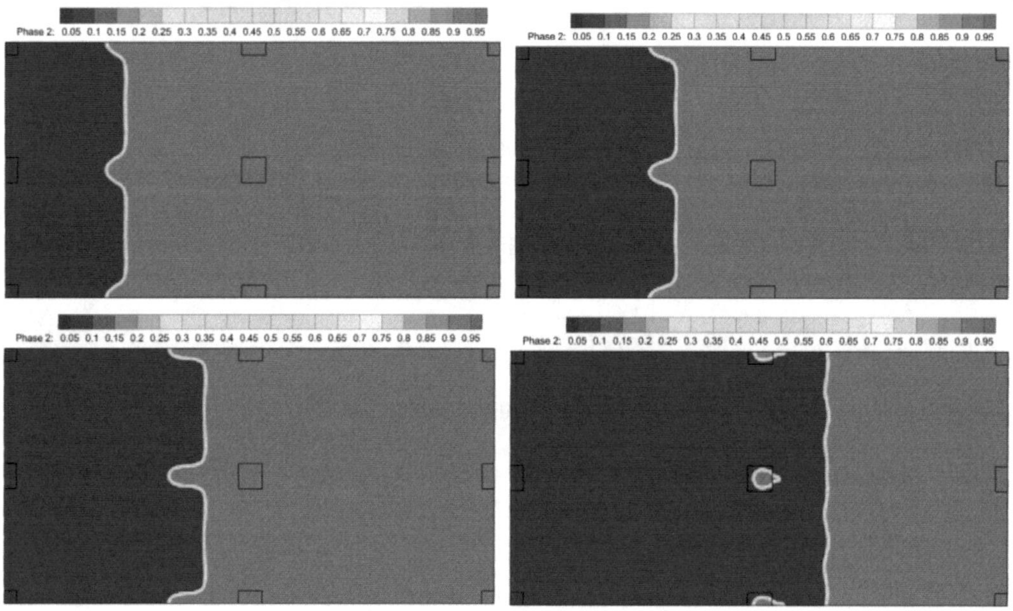

Figure 3. Cellular mesoscopic bubble wrap simulation.

On the basis of the above simulation, this paper carried out a series of simulations on the process of a certain type of resin infiltrating a certain type of fiber at different temperatures. The results showed that there was no microscopic bubble defect when the temperature was 50°C, and the resin viscosity was 450mPa.s. The surface tension was 0.028N/m, and the flow rate was 0.472mm/s. The corresponding capillary number is calculated based on the process conditions of series simulation, which are shown in table 1. As can be seen from the table, when the temperature is 50°C, the capillary value Ca is between 0.007 and 0.008, which theoretically verifies the effectiveness of the simulation.

Table 1. Capillary values under different simulation conditions

No.	Temperature (°C)	Viscosity (mPa.s)	Surface tension(N/m)	Flow velocity(mm/s)	Capillary number
1	50	450	0.028	0.472	0.0076
2	55	370	0.027	0.472	0.0065
3	60	340	0.025	0.472	0.0064
4	65	300	0.024	0.472	0.0059
5	70	260	0.023	0.472	0.0053

On the basis of the above series of simulation, the effectiveness of simulation is further verified by experiments. The temperature process parameters in No. 1 and 4 in table 1 and the flow rate is 0.472mm/s. The corresponding prop insulator test pieces are obtained by vacuum assisted resin fiber infiltration process. Sample pieces are cut for the test pieces respectively and siphon experiments are conducted. It can be seen from the figure 4 that the large bubble defect was produced at 65°C, while the product had no defect at 50°C. This result is consistent with the numerical simulation results, which further verifies the effectiveness of the numerical simulation, which indicates that the numerical simulation method can provide theoretical guidance for the actual infiltration process design.

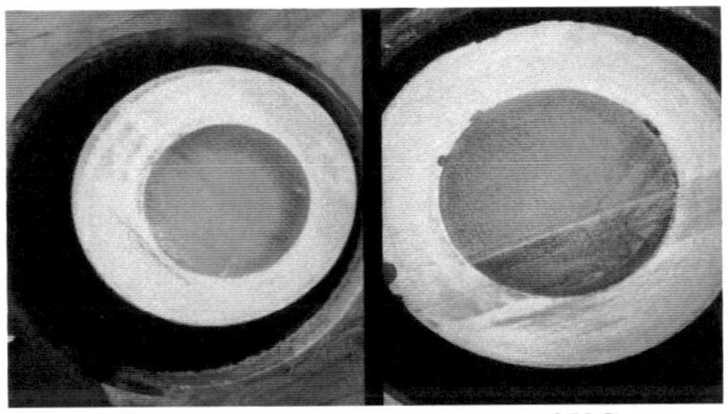

50°C 65°C

Figure 4. The siphon test results of the molded strut insulator samples at different temperatures.

7 Conclusion

In this paper, the numerical simulation method of bubble defect of vacuum assisted resin wetted fiber of high voltage strut insulator is studied. Specific conclusions include: the resin and air two-phase flow model of vacuum assisted resin infiltration fiber of high-voltage post insulator was established based on VOF method, taking the single-cell mesoscopic view of high-voltage post insulator as the object. The causes of mesoscopic bubble defects were summarized, and the controllable process parameters were analyzed. Finally, the wetting temperature was determined as the main control parameter. Closing to the real high-voltage insulating rod unit cell mesoscopic geometrical model is built. With the help of the Fluent software and by FVM method, single cell in the process of mesoscopic infiltration bubble defect series numerical simulation, and combining with capillary number infiltration temperature are studied. The results show that the temperature of 50 °C, the capillary number Ca between 0.007 and 0.008. Infiltrating effect is best. Ca is less than 0.007 is easy to produce large bubbles. Thus, theoretically, the correctness of the numerical simulation and temperature parameter optimization are verified; The siphon experiment was carried out based on the process parameters of numerical simulation to process the strut insulator test pieces. The experimental results were consistent with the numerical simulation results, which further verified the effectiveness of the method and provided theoretical guidance for the design of actual infiltration process.

Acknowledgment
This work was financially supported by the Science and Technology Program of the State Grid Corporation of China (No. 5455DW170027).

References

[1] Verleye B, Lomov S V and Long A 2010 *Composites Part A: Applied Science and Manufacturing* Vol 41, p.29-35.
[2] Krishna M, Pillai and Hua Tan 2007 *Composites Part A: Applied Science and Manufacturing* Vol 38, p.1872-1892.
[3] Vanegas J D, Patiño I D and Vargas C A 2014 *Journal of Composite Materials* Vol 48, p.2603-2629.
[4] De Valve C and Pitchumani R 2013 *Composites Part A: Applied Science and Manufacturing* Vol 51, p.22-32.
[5] Darzi M and Park C 2019 *European Journal of Mechanics - B/Fluids* Vol 74, p.342-350.
[6] Leclerc J S and Ruiz E 2008 *Composites Part A:Applied Science and Manufacturing* Vol 39, p.1859-1868.
[7] Gangloff J J , Daniel C and Advani S G 2014 *International Journal of Multiphase Flow* Vol 65, p:51-60.

Tensile strength of multi-angle composite laminate scarf joints with FEM

R Roy [1], J-H Kweon [2*] and Y Nam [2*]

[1]Research Center for Aircraft Core Technology and ReCAPT, Gyeongsang National University, Jinju, Gyeongsangnam-do, South Korea
[2]School of Aerospace and Software Engineering, Gyeongsang National University, Jinju, Gyeongsangnam-do, South Korea

* Corresponding authors e-mail: jhkweon@gnu.ac.kr, ywnam@gnu.ac.kr

Abstract. This study focus on the tensile strength of angle-ply composite laminate scarf joints and corresponding finite element modelling (FEM) methods. Tensile tests are performed at 25°C and 75°C on scarf joint specimens with unidirectional laminate stacking sequences. The joint tensile strength result values at 75°C represent around 86% of those obtained at 25°C. The tensile test is simulated with finite element modelling including cohesive elements to model failure. The joint adhesive film is represented either with only full thickness cohesive elements, or with thin or zero-thickness cohesive elements combined with solid elements having a plastic adhesive material model. Simulations results show that in the case of a unidirectional laminate, the adhesive modelling method had less than 1% effect on the simulated tensile failure strength. For quasi-isotropic laminates with a given ply angle count but of varying ply order, the simulated tensile failure strength varies up to 15%, which confirms tendencies reported in the literature.

1. Introduction

Advanced fiber reinforced polymer composite materials are now being used for large integrated structural parts. In addition to providing high specific strength and stiffness, good environmental and fatigue durability, these materials can reduce parts count and the number of joints. If these large parts get damaged, it can be advantageous economically and schedule-wise to apply a local repair. If the damage is such that reinforcing fibers are broken, the removal of the damaged material and the application of a repair patch is a common repair solution [1]. There is indeed no applicable method to repair broken fibers in-situ, for in the case of carbon fibers each individual fiber has a diameter in the order of 10 μm. The repair patch can be constructed with new material of the same type as the original structure, however there will be an interface between the patch and the remaining sound original structure (hereafter named parent structure). It is common to use an adhesive between the patch and parent structure, and this adhesive can be tailored to achieve desired properties. Structural epoxy adhesives are frequently used, and their toughness can be several times higher than the epoxy matrix of the parent composite. The composite patch can be either pre-cured separately then bonded in place with a film or paste adhesive (secondary bonding), or if prepreg composite and adhesive film is used, the whole patch can be cured in-place in one step (co-bonding) [2]. Great attention is devoted to the configuration of the repair area on the parent structure following the removal of damaged material and in preparation to patch installation. Two key aspects are the geometry of the repair area and its surface preparation. An angled scarfed joint configuration has proven to be advantageous in smoothly transferring loads better than a lap joint which generally develop stress concentrations in the bond line. The scarf angle is a key parameter, usually defined with the ratio of scarf length to parent laminate thickness (scarf ratio). A scarf ratio of 30:1 has been considered favorable in the literature [3], but a ratio of 50:1 and even up to 100:1 has also been recommended [4]. The repair area surface preparation for bonding is usually done either by sanding, grit blasting, plasma or chemical treatment, and it must ensure that bonding surfaces are free of contaminants [5,6,7]. One goal of surface preparation is also to increase the surface free energy or wettability of the bonding surfaces, which is favorable to strong adhesion, and can be evaluated by measuring the contact angle of liquids. The surface energy is understood to come from the contribution of two main sources: Van der Waals molecular interactions and polar molecular group interactions [8].

Considerations for the geometry of the repair area and its surface preparation can be applicable, with varying parameters, to polymer composite laminates or a more isotropic material like aluminum. A notable difference with angle-ply composite laminates, which can be composed of several unidirectional laminas with different orientations, is the discontinuous nature of their stress state at the ply level. For example, under a given tensile loading, an angle-ply laminate essentially has a constant strain level for every ply at a given point, but the stress level in each ply varies greatly. Locally stiff plies which fibers are aligned with the loading direction are the site of higher stress level than plies with fibers at an angle with the loading direction. In the context of a patch repair or a joint, this stress discontinuity will be transferred at the bond line interface. It has been shown that stress concentrations occur in the bond line of scarf joints in front of plies with fiber orientation aligned with the loading direction (i.e., zero degree plies) [9]. The objective of this study is to evaluate the effect of multi-angle composite laminate adherends on the tensile strength of scarf joints. Laboratory tensile tests were performed on composite laminate scarf joint specimens and material properties were deducted. Finite element modeling (FEM) simulation of laminate scarf joints was carried out with solid 3D and cohesive elements. Cohesive elements were used to model the adhesive layer failure in order to capture the effect of ply angles on the joint tensile strength.

2. Experiments

2.1. Specimen fabrication

One set of laminate scarf joint specimens were prepared for tensile testing. The base composite laminate used is a carbon/epoxy unidirectional lamina prepreg (USN-125B, SK Chemicals, South Korea), with 33% resin weight fraction (approximately 57.5% resin volume fraction) and a nominal ply thickness of 0.12 mm. Laminates with a scarf angle ratio of 30:1 were produced by first laying-up plies at a staggered distance and then sanding this stepped surface flat after curing with #100 grit sandpaper backed by a flat metal bar. The stacking sequence used was $[0°]_{24}$ in order to produce a quasi-uniform shear stress distribution in the bond line and thus better characterize the adhesive strength. A structural epoxy adhesive film with nominal areal weight of 147 g/m^2 (FM-300-2M, Solvay, United States), giving a thickness of 0.12 mm, was used to secondary bond two identical scarfed laminates and form a single scarf joint. A section of this film was cut to a length exceeding the scarf surface by about 2 mm at each ends to ensure a complete coverage. Prior to its installation, the adhesive film section was placed in an environmental chamber at 40°C and below 10% relative humidity for 30 minutes to favor elimination of humidity content in the adhesive. The carbon/epoxy prepreg stack was first cured in a vacuum bag and autoclave oven with a two-step cycle at 80°C/120°C and a pressure of 5 atmospheres. Subsequently, the two cured laminates were secondary bonding also in a vacuum bag and autoclave with a single step cure cycle at 120°C and 5 atmosphere pressure. Glass fiber composite tab panels of 55 mm length with one end chamfered at 45° were then installed on each ends and sides of the secondary bonded scarf joint panel (four tab panels in total). These tab panels were adhered with a room temperature cure epoxy paste and with the assembly enclosed in a vacuum bag. Specimens were cut to dimensions following ASTM standard D3039 [10], with a width of 25 mm and a total length of 310 mm. The distance between tab ends was 200 mm, with the scarf joint being centered at mid distance. The average laminate thickness was 2.89 mm, giving a nominal scarf length of 86.4 mm for a scarf ratio of 30:1 (or a scarf angle (θ) of 1.91°). The average thickness at the scarf center location was 3.01 mm, which reflects the added thickness of the adhesive film and reveals a slight eccentricity in the joint tensile axis.

2.2. Tensile testing

Scarf joint tensile testing was performed with an Instron 5582 universal testing machine equipped with an environmental chamber (Illinois Tool Works Inc., Norwood, USA). The tensile load was applied at a displacement controlled rate of 2 mm/min. The load and machine extension were recorded at a rate of 10 Hz. Twelve specimens were tested until failure: six at room temperature (25°C), and six at 75°C (figure 1). All specimen failed abruptly through the scarf joint bond line.

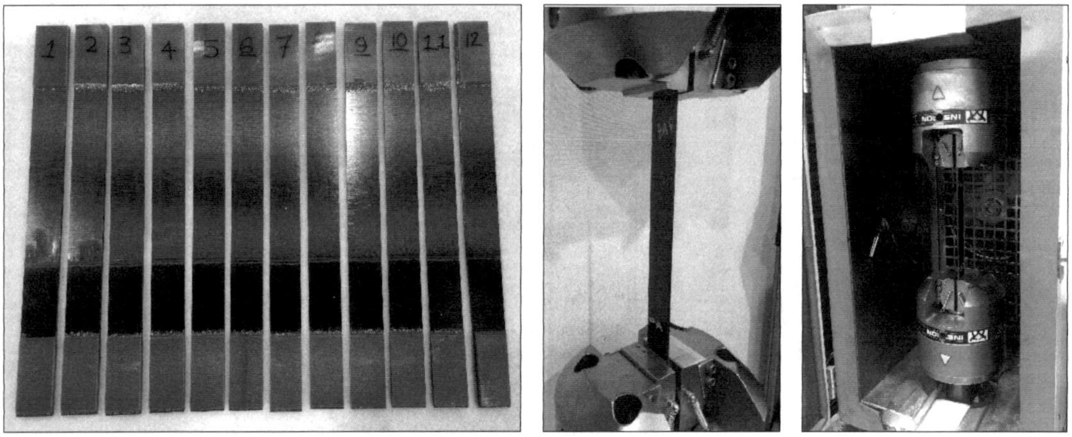

Figure 1. Tensile test scarf joint specimens and test set-up.

3. Test results

The ultimate tensile strength (F_t) of each specimen was calculated by dividing the failure load by the specimen's laminate cross section area (outside of the scarf area). The adhesive bond line average shear (τ_{av}) and peel (σ_{av}) stresses at failure were also calculated with equations (1-2), where θ is the scarf angle. One specimen in each test group was rejected from results compilation due to their apparent premature failure. The section tensile stress versus displacement curves are presented in figure 2. The average test results and corresponding coefficient of variation (CV) are listed in table 1. The failure mode of all specimens was predominantly cohesive failure with some less important areas having the appearance of adhesive failure (see [11] for terminology). The areal percentage of each failure mode was determined by visual inspection and is reported in table 2 for the room temperature tests. The areas of adhesive failure were associated with zones where a bare composite surface is visible; these zones were more predominant at the scarf ends (figure 3). Among the room temperature specimens, there was no clear influence of the failure modes on tensile strength (e.g., specimens 2,3,4 versus specimens 1,5). The elevated temperature specimen 6 had a larger adhesive failure area (29%) which coincided with its relatively lower strength result. The presence of adhesive failure areas can indicate that the tested joints could still benefit from improved surface treatment prior to bonding.

$$\tau_{av} = 0.5 \cdot F_t \cdot \sin(2\theta) \tag{1}$$

$$\sigma_{av} = F_t \cdot \sin^2(\theta) \tag{2}$$

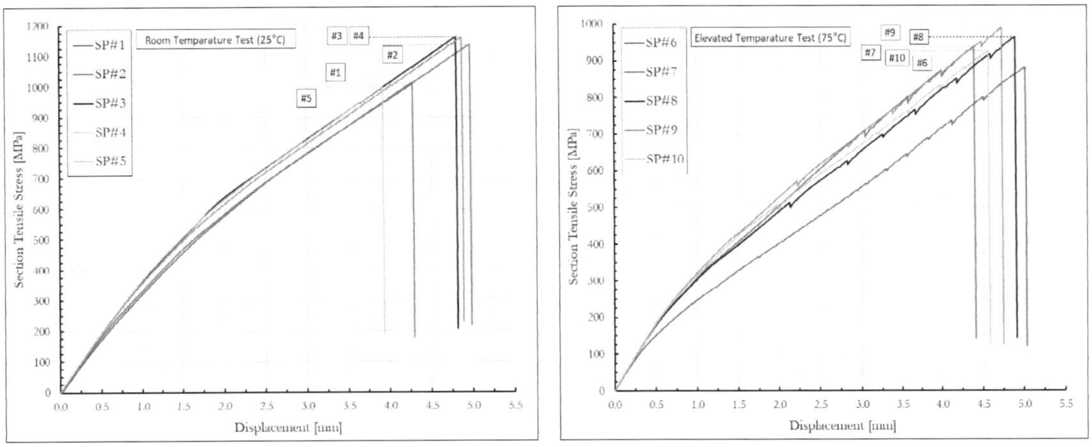

Figure 2. Scarf joint specimen section tensile stress versus displacement curves.

Table 1. Scarf joint specimen tensile test results.

Test condition	Section tensile strength [MPa]	Average bond line shear stress at failure [MPa]	Average bond line peel stress at failure [MPa]
Room temperature (25°C)	1092.7 (CV=7.53%)	36.38 (CV=7.98%)	1.21 (CV=8.47%)
Elevated temperature (75°C)	939.2 (CV=4.32%)	31.57 (CV=4.08%)	1.06 (CV=3.90%)

Table 2. Scarf joint specimen failure mode areal percentages.

Failure mode	Room temperature (25°C)					Elevated temperature (75°C)				
	Sp#1	Sp#2	Sp#3	Sp#4	Sp#5	Sp#6	Sp#7	Sp#8	Sp#9	Sp#10
Cohesive	80%	85%	80%	82%	81%	71%	78%	77%	81%	77%
Adhesive	20%	15%	20%	18%	19%	29%	22%	23%	19%	23%

Figure 3. Failure surface of room temperature specimens #3 and #5.

4. Simulation

4.1. FEM model

A 3D solid half-width symmetric finite element model of the tensile test was built with the software Abaqus/CAE 6.14-2 (Dassault Systèmes SE, Vélizy-Villacoublay, France). Cohesive elements were used for the adhesive material in order to simulate failure. The inputted material mechanical properties were obtained from manufacturer published data sheets and the literature [12,13,14,15,16] (table 3). The adhesive plastic shear properties [16] were entered in the software as converted tensile properties by Von Mises yield criterion theory. The general mesh size was 1 mm, except for one element thick per ply and cohesive elements of 0.12 mm thickness and length [17].

Table 3. Material properties used in FEM.

USN-125B Elastic solid	E_1 [MPa]	E_2 [MPa]	E_3 [MPa]	G_{12} [MPa]	G_{13} [MPa]	G_{23} [MPa]	v_{12}	v_{13}	v_{23}
	131000	7600	7600	5340	5340	3060	0.31	0.31	0.47
FM-300-2M Cohesive (thk = 0.12 mm)	K_I [N/mm^3]	K_{II} [N/mm^3]	K_{III} [N/mm^3]	T_I [MPa]	T_{II} [MPa]	T_{III} [MPa]	G_{IC} [N/mm^3]	G_{IIC} [N/mm^3]	G_{IIIC} [N/mm^3]
	22083	996	996	50	38.5	38.5	1.0	2.45	2.45
FM-300-2M Elastic solid	E [MPa]	G [MPa]	v						
	2650	907.5	0.4						
FM-300-2M Plastic solid	σ_1 [MPa]	ε_{p1}	σ_2 [MPa]	ε_{p2}	σ_3 [MPa]	ε_{p3}	σ_4 [MPa]	ε_{p4}	
	19.01	0.0000	56.37	0.0460	66.68	0.3138	67.72	0.3405	

4.2. Simulation results

We estimated that in this experiment the specimen tensile strength is predominantly governed by the shear strength of the adhesive material. Therefore, in order to calibrate the FEM model, we adjusted by trial and error the adhesive shear damage initiation criteria (T_{II}, T_{III}). We obtained FEM match to room temperature test results (F_t = 1093 MPa) with T_{II} and T_{III} values of 38.5 MPa and a numerical viscous regularization parameter (η) of 1.7×10^{-4} applied to the cohesive elements. The damage initiation value of 38.5 MPa is coherent with the average bond line shear stress at failure test result of 36.38 MPa: in this case of homogeneous material scarf joint, the bond line shear stress distribution is expected to be smooth and parabolic, with its peak value slightly higher than its average. In addition to modeling the adhesive layer entirely with cohesive elements, two alternative configurations were evaluated: thinner cohesive elements (t = 0.02 mm) sandwiched between plastic material solid elements (total of three elements thick), or zero-thickness cohesive elements sandwiched between pairs of plastic material solid elements (total of five elements thick) (figure 4). The simulated scarf joint tensile strength varied little between these three configurations (table 4). Further simulations were performed with the calibrated half-width model, but by implementing various quasi-isotropic laminate stacking sequence angles. The simulated tensile strength varied for the different stacking sequences used (figure 5). The 676 MPa tensile strength result of the $[0/\text{-}45/90/45]_{3S}$ stacking sequence model confirm that zero-degree plies placed on the laminate surfaces cause stress peaks in the adhesive at the joint edges which reduce tensile strength [9].

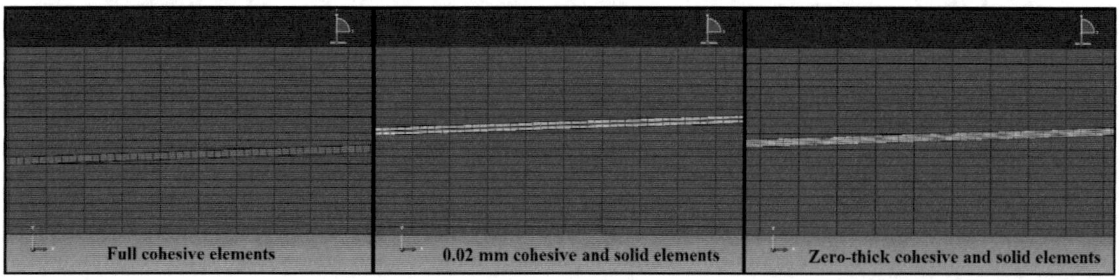

Figure 4. Adhesive layer FEM configurations.

Table 4. FEM simulation tensile strength results.

Adhesive layer model type	F_t [MPa]
Cohesive elements only, half-width model (12.5 mm)	1093
One cohesive element (t = 0.02 mm) between plastic material solid elements, 1 mm width model	1100
One zero-thickness element between pairs of plastic material solid elements, 1 mm width model	1104

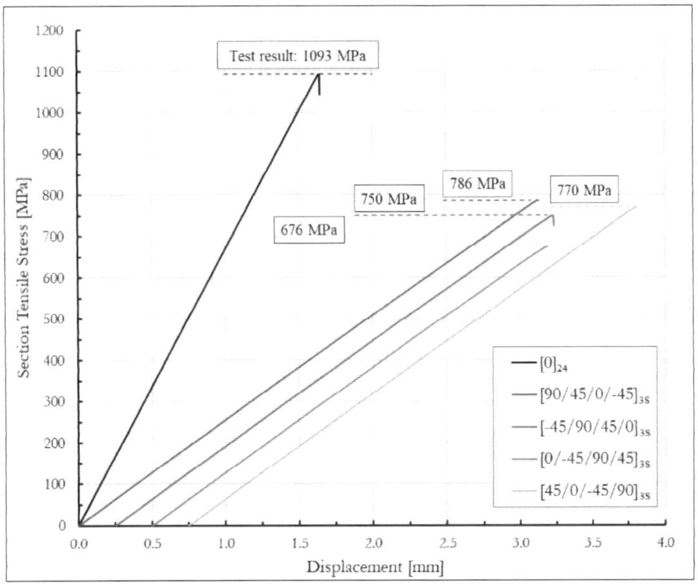

Figure 5. FEM simulation results for the tensile strength of angle-ply scarf joints.

5. Conclusion

In this study, the tensile strength of angle-ply composite laminate scarf joints was evaluated by experiment and FEM simulation. The test results showed predominant failure in the adhesive material. A FEM model with cohesive elements was calibrated to replicate the joint tensile strength test result. This model was then used to evaluate the effect of laminate stacking sequence. It was shown that through the stacking sequence effect on the bond line shear stress distribution, it was possible to capture the stacking sequence effect on scarf joint tensile strength. The different stacking sequences simulated captured tendencies observed in published literature.

Acknowledgements

This work was supported by the National Research Foundation of Korea (NRF) Grant funded by the Ministry of Science and ICT (NRF-2017R1A5A1015311).

References

[1] Katnam K B, DaSilva L F M and Young T M 2013 Bonded repair of composite aircraft structures : A review of scientific challenges and opportunities *Prog. Aerosp. Sci.* **61** 26–42
[2] Budhe S, Banea M D, DeBarros S and DaSilva L F M 2017 An updated review of adhesively bonded joints in composite materials *Int. J. Adhes. Adhes.* **72** 30–42
[3] Yoo J S, Truong V H, Park M Y, Choi J H and Kweon J H 2016 Parametric study on static and fatigue strength recovery of scarf-patch-repaired composite laminates *Compos. Struct.* **140** 417–32
[4] Jefferson A, Arumugam V and Dhakal H 2018 *Repair of polymer composites: methodology, techniques, and challenges* (Woodhead Publishing)

[5] Yang G, Yang T, Yuan W and Du Y 2019 The influence of surface treatment on the tensile properties of carbon fiber-reinforced epoxy composites-bonded joints *Compos. Part B* **160** 446–56
[6] Sorrentino L, Polini W, Bellini C and Parodo G 2018 Surface treatment of CFRP: influence on single lap joint performances *Int. J. Adhes. Adhes.* **85** 225–33
[7] Martínez-Landeros V H, Vargas-Islas S Y, Cruz-González C E, Barrera S, Mourtazov K and Ramírez-Bon R 2019 Studies on the influence of surface treatment type, in the effectiveness of structural adhesive bonding, for carbon fiber reinforced composites *J. Manuf. Process.* **39** 160–66
[8] Encinas N, Oakley B R, Belcher M A, Blohowiak K Y, Dillingham R G, Abenojar J and Martínez M A 2014 Surface modification of aircraft used composites for adhesive bonding *Int. J. Adhes. Adhes.* **50** 157–63
[9] Harman A B and Wang C H 2006 Improved design methods for scarf repairs to highly strained composite aircraft structure *Compos. Struct.* **75** 132–44
[10] ASTM D3039/D3039M-17 2017 Standard test method for tensile properties of polymer matrix composite materials (ASTM International, West Conshohocken, USA)
[11] ASTM D5573-99 1999 Standard practice for classifying failure modes in fiber-reinforced-plastic (FRP) joints (ASTM International, West Conshohocken, USA)
[12] Park S S, Choe H S, Kwak B S, Choi J H and Kweon J H 2018 Micro-bolt repair for delaminated composite plate under compression *Compos. Struct.* **192** 245–54
[13] Truong V H, Kwak B S, Roy R and Kweon J H 2019 Cohesive zone method for failure analysis of scarf patch-repaired composite laminates under bending load *Compos. Struct.* **222** 110895
[14] Mohan J, Ivankovic A and Murphy N 2014 Mode I fracture toughness of co-cured and secondary bonded composite joints *Int. J. Adhes. Adhes.* **51** 13–22
[15] Saunders I, Forte M and Donaldson S 2000 Fracture mechanics characterization of joints with unidirectional and cross-ply adherends (*Durability and intelligent nondestructive evaluation of adhesive composite joints*) ed Y Dzenis (University of Nebraska-Lincoln)
[16] FM® 300-2 2017 Film adhesive technical data sheet (Solvay, Alpharetta, USA)
[17] Buddi S C, Prasanthi P P and Srikanth P 2015 Mechanical properties of fiber reinforced composites using finite element method *Int. J. Mech. Eng. & Rob. Res.* **4** 80–90

IOP Conf. Series: Materials Science and Engineering **784** (2020) 012008 doi:10.1088/1757-899X/784/1/012008

Effect of Sputtering Conditions on Transmittance and Wear Resistance of Diamond-Like Carbon Film

L F Lai, Z Y Xu and J Xu

Ningbo University of Technology, Ningbo 315016, PR China
Email: lailf-325@163.com

Abstract. Diamond-like carbon (DLC) film was prepared by magnetron sputtering method on glass substrate. The microstructure, transmittance, and wear resistance of the diamond-like carbon film were investigated and analyzed. This study clearly showed the significance of sputtering power, Ar pressure and substrate temperature in controlling the transmittance and wear resistance of DLC film. The sp3 bond in the film had a greater impact on the transmittance and wear resistance of DLC film. The optimized deposition parameters obtained through experiments were: Ar pressure of 0.7 Pa, sputtering power of 100 W, and substrate temperature of 150 ℃.

1. Introduction

Carbon-based films are formed by depositing carbon atoms on the substrate. Carbon-based films possess many merits including low coefficient of friction, high wear resistance and chemical stability, as well as good corrosion resistance and optical properties [1-2]. As one of the members of carbon-based films, the Diamond-like carbon (DLC) film has prominent advantages in optical properties, such as good infrared and visible light transmittance, high anti-ultraviolet transmittance. It can be used as anti-reflective protective film for optical instrument and infrared window, practical examples as anti-reflective film for silicon solar cell, anti-wear protective films for infrared optical lenses made by Ge, ZnS and ZnSe, etc [3]. DLC film is a metastable form of amorphous carbon containing fraction of sp3 and sp2 bonds, and the sp2 and sp3 are hybrid bonds between carbon atoms. DLC film has many of the useful properties of diamond and some properties of graphite, and has been one of the hotspots in the field of coating research.

However, there are still many shortcomings in the production of DLC film at present. For example, the higher the sp3 bond in the DLC film, the higher the hardness of the DLC film would be and the closer the film to diamond, but the optical properties of the DLC film are not necessarily good. It is still a problem to make the DLC film have good optical and mechanical properties at the same time. This paper starts with the transmittance and wear resistance of DLC film, and studies the difficult problems. The physical properties of DLC film are mainly controlled by the sp3 bond and sp2 bond, which are closely related to the method of film deposition [4], so the magnetron sputtering technology owning low sputtering temperature, high deposition rate and strong adhesion between film and substrate [5] was chosen to prepare the DLC film. The transmittance and wear resistance of DLC film were investigated by changing Ar pressure, sputtering power and substrate temperature. Optimal experimental parameters had been explored to ensure that the DLC thin film had high transmittance and good wear resistance.

2. Experimental

The DLC film was prepared using GP500 Multi-Function Ultrahigh Vacuum Magnetron Sputtering System. C (99.99% purity) was used as target for sputtering. High purity Ar (99.999% purity) was introduced as the sputtering gas. Float glass was selected as substrate. The specific experimental scheme was shown in Table 1. The Ar pressure as 0.5Pa, 0.7Pa and 0.9Pa, respectively. The sputtering power was 80W, 100W and 120W, respectively. The substrate temperature was room temperature, 150°C and 300°C, respectively. The sputtered time was set at 3600s and the gas flow of Ar was 25sccm.

Table 1. Depositing parameters of carbon-based films

Sample	Pressure (Pa)	Power (W)	Temperature (℃)	Argon flow rate (sccm)	Time (s)	Thickness (nm)
C1	0.5	100	150	25	3600	541
C2	0.7	100	150	25	3600	670
C3	0.9	100	150	25	3600	729
C4	0.7	80	150	25	3600	501
C5	0.7	100	150	25	3600	670
C6	0.7	120	150	25	3600	1086
C7	0.7	100	R.T	25	3600	730
C8	0.7	100	150	25	3600	670
C9	0.7	100	300	25	3600	644

3. Results and discussion

3.1. Argon pressure

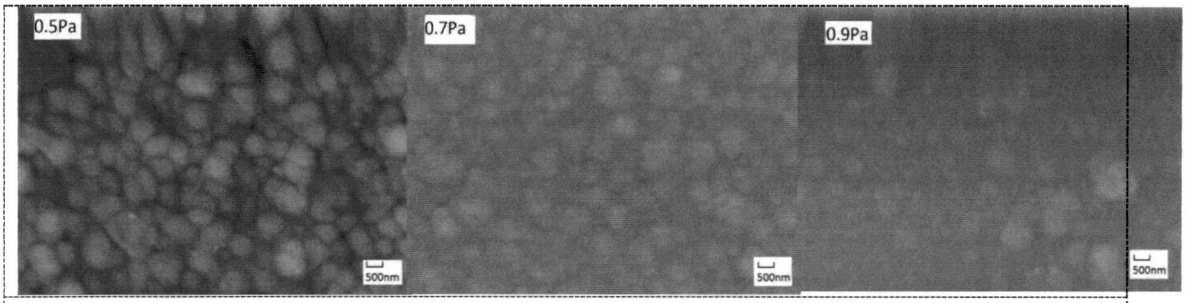

Figure 1. SEM surface images of the carbon-based films under different Ar pressure.

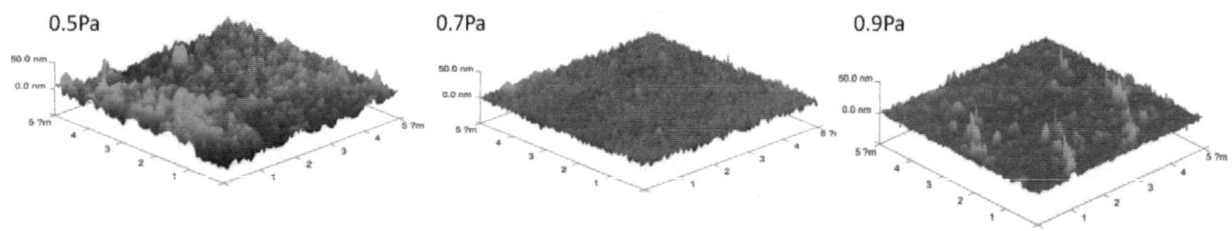

Figure 2. AMF 3D morphologies of the carbon-based films under different Ar pressure.

Fig. 1 and Fig. 2 show the SEM surface images and AMF 3D morphologies of carbon-based films under different Ar pressure, respectively. As can be seen from Fig. 1, the surface structure of the film deposited at 0.7 Pa is more dense, and the particles on the surface of the film are more uniform. In combination with the film 3D morphology of Figure 2, it can be seen that the 0.7Pa film surface is less fluctuating and more uniform than the 0.5Pa and 0.9Pa deposited films. When the Ar pressure was 0.5Pa, the Ar plasma density was low. The low Ar plasma density would result in less carbon particles and a lower deposition rate of carbon. The non-uniform distribution of carbon particles and more defects on the film surface would be caused. When the Ar pressure was increased to 0.7Pa, the density of Ar plasma, the number of carbon particles sputtered and the deposition rate of carbon particles were all increased, so there had enough carbon particles on substrate to construct a smooth and compact film. However, when the Ar pressure was increased to 0.9Pa, the Ar plasma density was too high, thus the collision frequencys between carbon particles and gas molecules were increased, which would weaken the energy of carbon particles. When carbon particles did not have enough energy to migrate on substrate, the defects of thin film were generated.

Figure 3 shows the XRD diagram of carbon-based films under different Ar pressure. There was only a broad peak near 30 degrees. The change of diffraction peaks in carbon-based films were not apparent with the increasing of Ar pressure, which indicated that the film mainly existed in the form of amorphous.

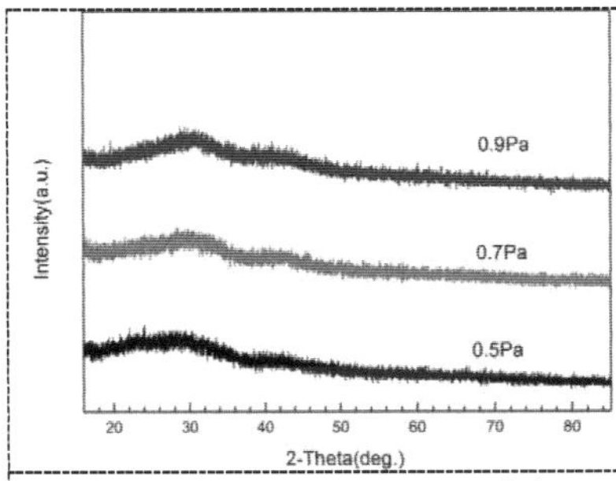

Figure 3. XRD patterns of the films under different Ar pressure.

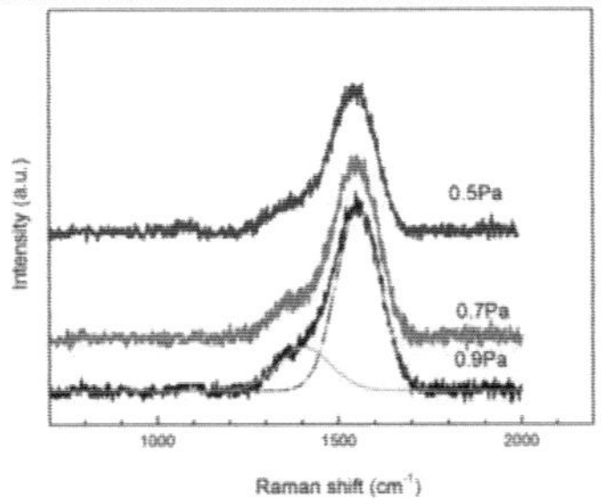

Figure 4. Raman spectra of the films under different Ar pressure.

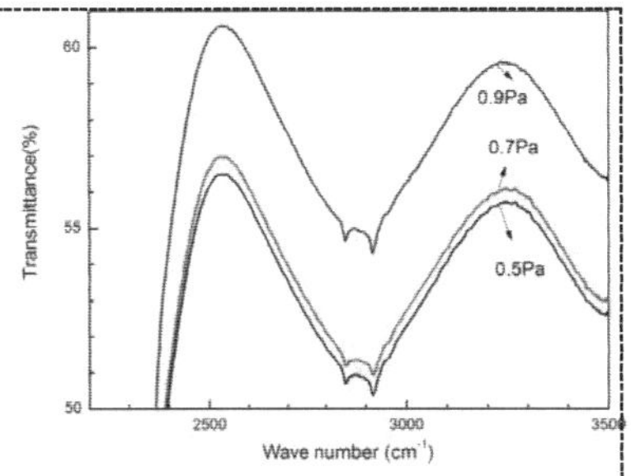

Figure 5. Transmittance spectra of the films under different Ar pressure.

Table 2. The parameters of the C Raman spectra in DLC films

Sputtering conditions	FWHM (G) (cm^{-1})	FWHM (D) (cm^{-1})	R (I_G/I_D)
0.5Pa	130.14	191.32	2.89
0.7Pa	135.18	182.83	3.36
0.9Pa	142.94	172.70	3.40
80W	132.15	201.78	4.10
100W	135.18	182.83	3.36
120W	121.88	349.30	2.67
R.T	148.99	142.67	5.80
150°C	135.18	182.83	3.36
300°C	99.11	126.56	1.19

Fig. 4 displays theRaman spectra of carbon-based films under different Ar pressure. The main feature in the Raman spectra was the G peaks appearing at approximately 1554 cm^{-1}, and a weak D peak at near 1366 cm^{-1}, which accounted for the large amounts of amorphous domain in DLC film. The G peak is due to the bond stretching of all pairs of sp2 atoms in both rings and chains. The G band corresponds to the symmetric E_{2g} vibrational mode in graphite-like materials. The D peak is due to the breathing modes of sp2 atoms in rings caused by A_{1g} non-active vibrations mode [6]. The G and D peaks were fitted with Lorentzian by fitting the data in Table 2. The $R = I_G/I_D$ can reflect the degree of carbon graphitization. The higher R in thin film will lead to more the diamond-like properties, the lower R in thin film will lead to more the graphite-like properties.

As the pressure of Ar increased, the G peak shifted slightly to the high wave number, the G peak was more asymmetrical, and the D peak was less obvious. As can be seen from Table 2, R increased as the Ar pressure increased. Therefore, the sp3 bond fraction in the carbon-based film increased as the Ar pressure increased, and the film structure tended to be diamond-like.

Fig. 5 shows the infrared transmittance of carbon-based films at different Ar pressure. It was clear that the transmittance of the films increased as the pressure of Ar increases. When the Ar pressure was raised from 0.5Pa to 0.7Pa, the maximum transmittance in the film was increased from 56.5% to 57.0%. The increased transmittance was related to the increasing of sp3 bond in the film. In addition, combined with the surface morphology of the film, it can be inferred that the increasing transmittance in the film was related to the surface morphology of the film. Deposited at 0.7Pa, the less defects and roughness of the films reflected that the scattering on the film surface were reduced and the transmittance of the film was increased. When the Ar pressure was risen to 0.9Pa, the maximum transmittance reached 60.5%. The increase of the transmittance of the film was mainly related to the increasing of sp3 bond in the film [7-8]. Since the surface morphology of the film deposited at 0.9Pa was rougher than the surface morphology of the film deposited at 0.7Pa, this impaired the transmittance of the film. In Table 1, as the pressure of Ar increased, the increase in thickness of the film was inversely proportional to the transmittance of the film. In the face of adversity, the transmittance of the film still increased with the increase of Ar pressure, which indicated that the sp3 bond in the film had a greater influence on the transmittance.

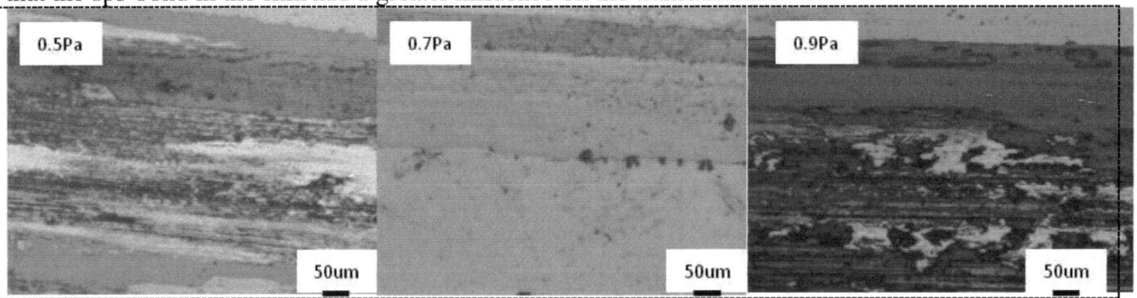

Figure 6. Metallographic microscopic images of the worn carbon-based films under different Ar pressure.

Fig. 6 shows that there was a significant difference between the metallographic microscopic images of the worn carbon-based film at different Ar pressures. When the Ar pressure was 0.5 Pa and 0.9 Pa, respectively, the edge of the worn film peeled off significantly, indicating that the adhesion between the film and the substrate was weak, and the wear resistance was weak. When the pressure was 0.7 Pa, the edges of the worn film were smoother and less debris. In order to maintain good wear resistance and optical properties of the DLC film, an optimum Ar pressure of 0.7 Pa was selected.

3.2. Sputtering power

Fig. 7 and Fig. 8 show the SEM surface topography and AFM 3D topography of carbon-based films at sputtering power of 80W, 100W and 120W, respectively. It can be found that the film surface was tightly bonded and fluctuated uniformly, and the size of the particles on film surface was uniform at 100W. While the surfaces of films deposited at 80W and 120W had some protrusions, and the size of the particles on films surface were not uniform, especially when the sputtering power was 120W.

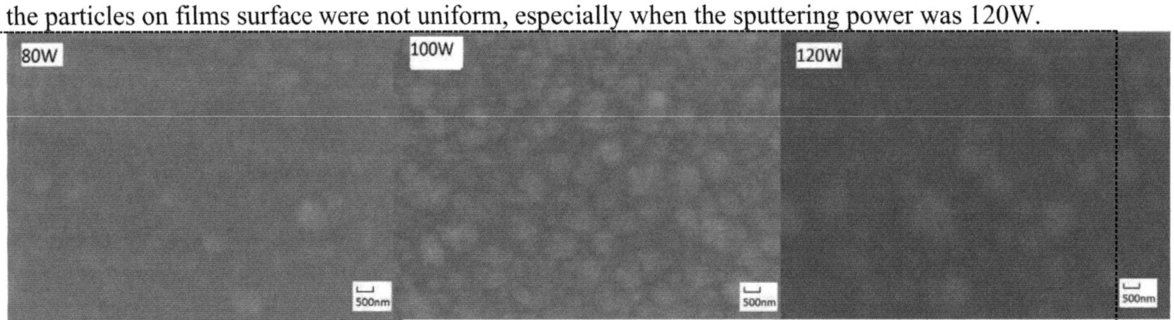

Figure 7. SEM surface images of the carbon-based films under different sputtering power.

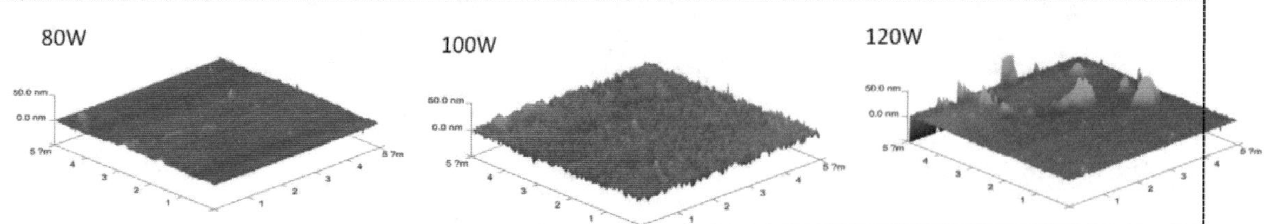

Figure 8. AMF 3D morphologies of the carbon-based films under different sputtering power.

The change of the film structure can be explained as follows: at a sputtering power of 80 W, the sputtering cavity having a low Ar plasma density caused a small amount of sputtering C particles and low energy, which made that the C particles deposited on the substrate did not have enough quantity and energy to migrate, thus creating holes and undulations on the surface of the film. When the sputtering power was risen to 100W, the density and energy of Ar plasma reached a suitable value, which not only improved the deposition rate of C particles, but also gave the C particles enough energy to migrate along the substrate surface. Therefore, the distribution of C particles on the film surface was more uniform, and the film surface was smoother and more compact. When the sputtering power was increased to 120 W, the density and energy of the Ar plasma were so high that some of the C particles on the surface of the substrate had excessive energy, which would collide with the surface of the substrate multiple times and break the film structure before stopping. After the appearance of holes and projections on the surface of the film, it was revealed that the structure of the film was destroyed.

Figure 9 shows the XRD patterns of carbon-based films at different sputtering power. It can be seen from the graphs that the diffraction peaks of the films did not change significantly with the increasing sputtering power, and the diffraction peaks were very broad. The broad diffraction peaks indicated that the carbon-based films existed mainly in amorphous forms under the selected sputtering powers.

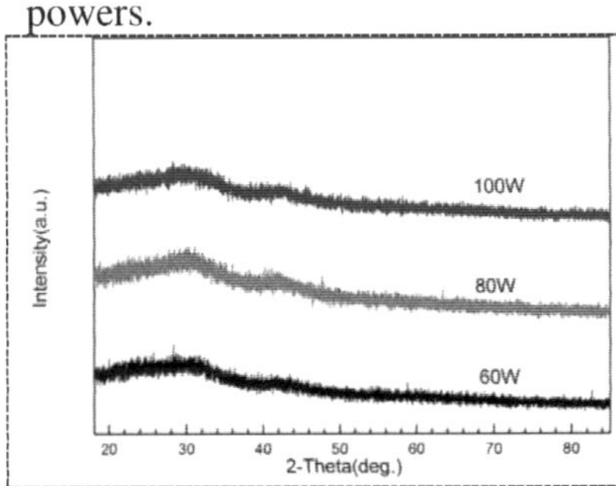

Figure 9. XRD patterns of the films under different sputtering power.

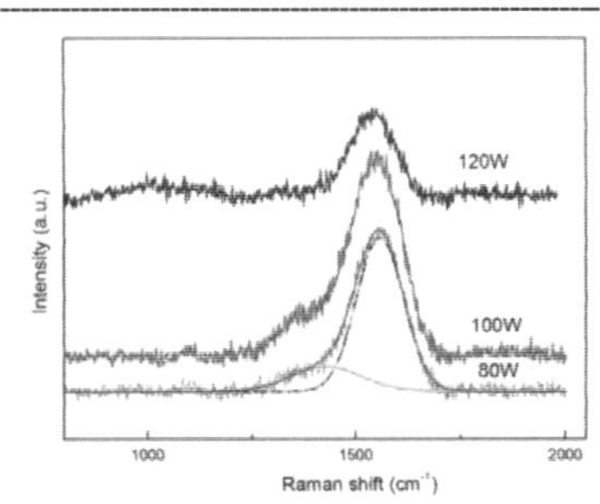

Figure 10. Raman spectra of the films under different sputtering power.

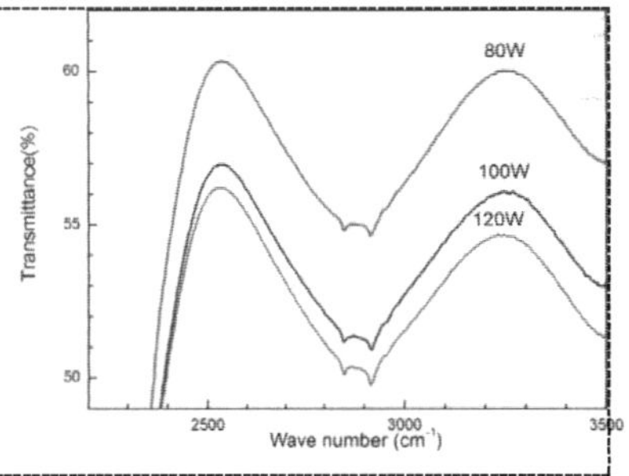

Figure 11. Transmittance spectra of the films under different sputtering power.

Fig. 10 shows the Raman spectra of C-based films at different sputtering power. It was worth noting that the G peak was near 1558 cm^{-1} and the weak D peak was near 1384 cm^{-1}. As the sputtering power increased, the G peak shifted slightly to a lower wave number. Table 2 shows that R decreased with increasing sputtering power, indicating that the degree of graphitization of the film increased as the sp3 bond decreased.

Fig. 11 shows the wave number dependence of the infrared transmittance of C-based films under different sputtering power. It can be seen that as the sputtering power increased, the maximum infrared transmittance of the film decreased from 60.0% to 57.0% and then dropped to 54.7%. The film showed that the red shift for the absorption edge of the films was moved to the low wave number. The result intimated that the sp3 bond in the film was decreased, which was consistent with the previous Raman spectroscopic analysis. The content of sp3 bond was lower, and the transmittance was decreased more [9]. In addition, increasing power would increase the deposition rate of C particles and the thickness of the film (Table 1). Excessive thickness would weaken the transmittance and lead to the decreasing of the film infrared transmittance [10]. When the sputtering power was reached at 120W, the surface roughness of the film was larger, which can enhance the scattering effect of the film and would affect the film transmittance.

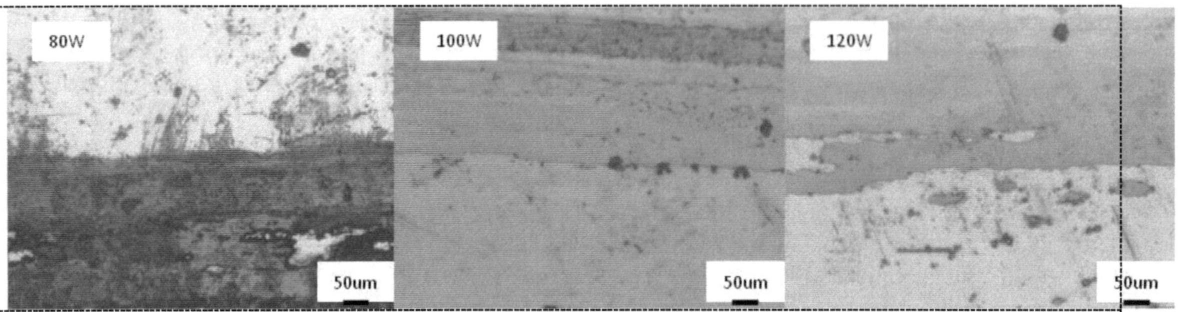

Figure 12. Metallographic microscopic images of the worn carbon-based films under different sputtering power.

Fig. 12 shows metallographic microscopic images of the worn films at different sputtering power. When the sputtering power was 80W and 120W, the carbon-based film at the edge of the wear trace had obvious peeling phenomenon, indicating that the wear resistance of the carbon-based film was poor under the two sputtering powers. While the sputtering power was 100W, the film on the edge of the wear mark was smooth, and the film was tightly bonded to the underlying substrate. The optimum sputtering power was selected to be 100 W in consideration of the abrasion resistance and optical properties of the C-based film.

3.3. Substrate temperature

Fig. 13 and 14 show the SEM and AMF 3D morphologies of carbon-based films at room temperature, 150°C and 300°C, respectively. When the substrate temperature was raised from room temperature to 150°C, the film structure was improved, the fluctuation of film surface became smaller and the particles in film became more uniform. When the substrate temperature reached 300°C, the particles on film surface became smaller, but there were some large clusters and bulges on the film surface. When the energy of C particles arriving at the substrate was constant, substrate temperature was an important factor determining the film morphology [11].

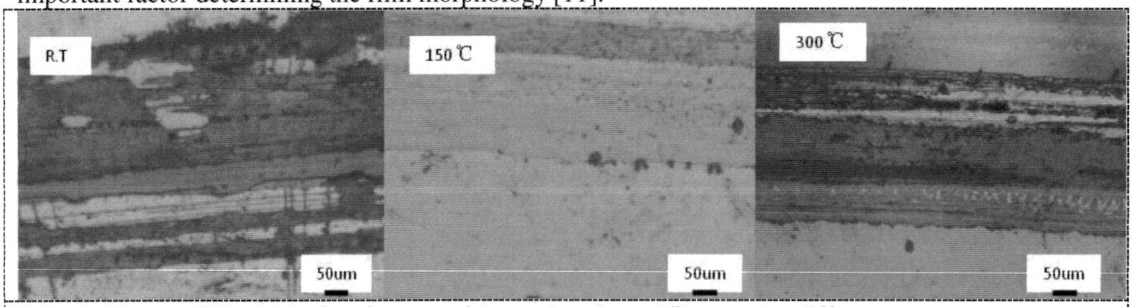

Figure 13. SEM surface images of the carbon-based films under different temperature.

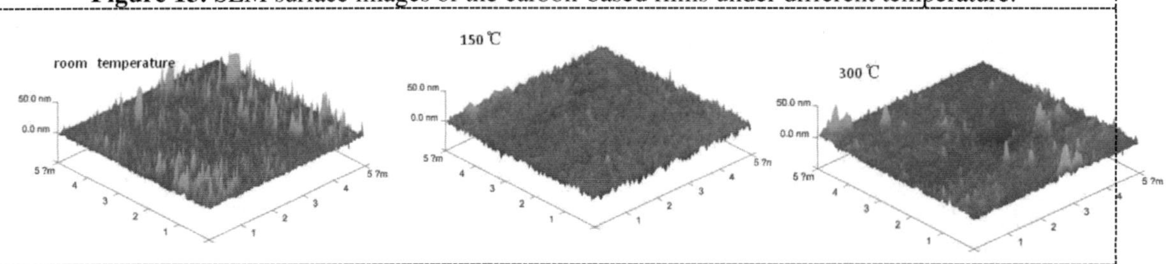

Figure 14. AMF 3D morphologies of the carbon-based films under different temperature.

With the temperature rising from room temperature to 150°C, the kinetic energy of C particles deposited on substrate was increased, and the migration ability of C particles along the substrate surface was improved, thus the film defects were reduced and the film surface became smooth and dense. When the substrate temperature rose to 300°C, the kinetic energy of C particles on substrate was large, which would cause the C particles on the substrate to move in a non-directional direction,

IOP Conf. Series: Materials Science and Engineering **784** (2020) 012008 doi:10.1088/1757-899X/784/1/012008

and did not be easily adsorbed and condensed by the substrate. Some C particles would escape and return to the sputtering chamber, which would destroy the film structure, increase the defects and roughness of the film.

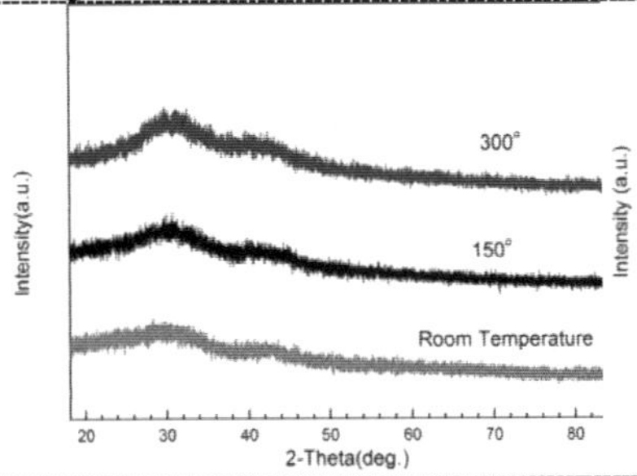

Figure 15. XRD patterns of the films under different temperature.

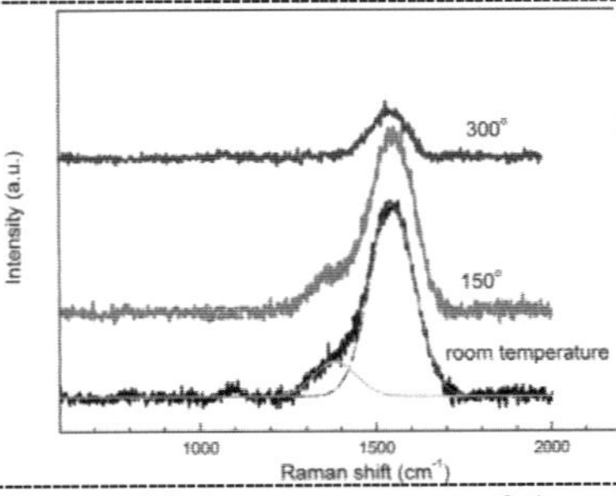

Figure 16. Raman spectra of the films under different temperature.

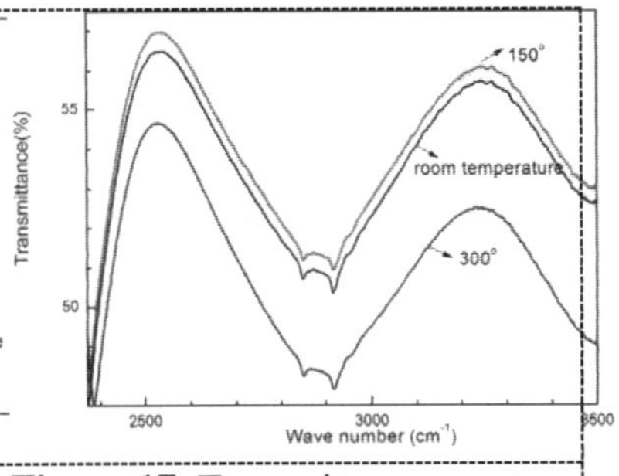

Figure 17. Transmittance spectra of the films under different temperature.

Fig. 15 shows the XRD patterns of C-based films at different substrate temperatures. It can be found that the diffraction peaks of C-based films were increased slightly with the increasing temperature, indicating that the high substrate temperature could promote the crystallization of the films. However, the broad peaks suggested that the films still existed mainly in amorphous form.

Fig. 16 shows the Raman spectra of C-based films at different substrate temperatures. It can be seen that there were obvious G peaks near 1552 cm^{-1} and weak D peaks near 1353 cm^{-1} from the spectra. With the increasing of temperature, the half-width and wave-peak of G-peak were decreased, and the G-peak shifted slightly to the direction of the low wave number. It indicated that the sp3 bond fraction in the films decreased and the sp2 bond fraction increased, which was proved in Table 2 too by the decrease of R with the increasing of substrate temperatures. It can be speculated that the film structure was being transformed to graphite-like structure, but that R was always greater than 1 in table 2 proved that the film was still the DLC thin film.

Fig. 17 shows that the wave number dependence of the infrared transmittance of film under the different substrate temperatures. It can be seen from the graph that the maximum transmittance of the film was increased from 56.5% to 57.0% when the substrate temperature was raised from room temperature to 150°C, which was in contradiction with the Raman test results. The Raman test result showed that the sp3 bond fraction in the film was decreased with the increasing of the substrate temperature, so the transmittance of the film should be decreased. As the substrate temperature increased, the vibration of C particles on the substrate became more intense, the energy dissipation would slow down, and the relaxation time would increase, so the degree of crystallization of films would be improved. The film transitioning from disorder-to-order would lead that the metastable sp3 bond would be turned into the lower binding energy sp2 bond [10,12]. If the inverse relationship between the transmittance and thickness of film was considered, this abnormal behaviour could be explained. Table 1 showed that as the temperature increased, the film thickness decreased, so that the transmittance of the film deposited at 150 ° C can be improved. When the temperature was raised to 300 ° C, the transmittance of the film was lowered to 54.7%, which indicates that the reduced transmittance was caused not only by the decrease of the sp3 bond in the film but also by the irregularity of the surface morphology. The uniform distribution of film particles became worse with the substrate temperature of 300°C, which would enhance scattering effect and weaken the transmittance of the films. The two factors affecting the transmittance of the film in relation to the sp3 bond and the surface morphology of the film were higher than factors that resulted in an increase in film transmittance due to a decrease in film thickness.

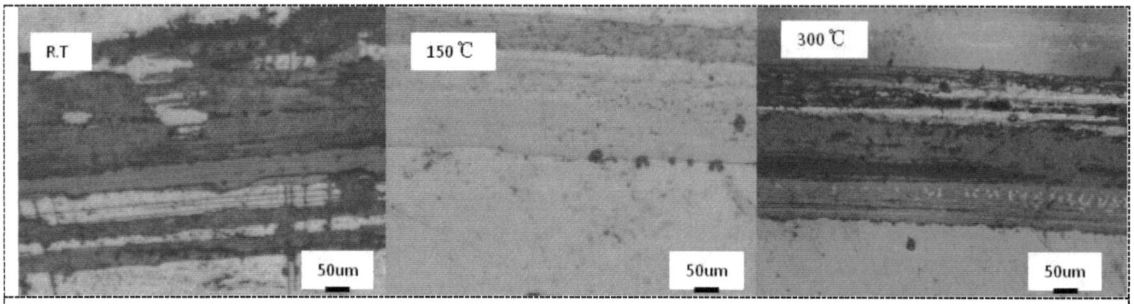

Figure 18 Metallographic microscopic images of the worn carbon-based films under different temperature.

Fig. 18 shows the metallographic microscopic images of the worn films at different substrate temperatures. It can be found that the worn films had obvious differences at different substrate temperatures. When the substrate temperatures were room temperature and 300℃, the carbon-based films at the edge of the wear marks had obvious peeling phenomenon, indicating that wear resistance of the carbon-based film was poor at both temperatures. When the sputtering temperature was selected to be 150℃, the edges of the wear marks were smooth, and were not easily peeled off after rubbing. Considering the wear resistance and optical properties of the DLC thin film, in this experimental scheme, the optimum substrate temperature was easily selected to be 150℃.

4. Summy

In this paper, the effects of Ar pressure, sputtering power and substrate temperature on the transmission and wear resistance of C-based films were studied by magnetron sputtering technique. The main conclusions are as follows:

(1) The carbon-based film prepared under different sputtering conditions belonged to the DLC film. An increase in the sp3 bond in the film is more preferably a diamond-like structure.

(2) The sp3 bond fraction increased as the Ar pressure increased, and decreased as the sputtering power increased or the substrate temperature increased.

(3) The content of sp3 bond had an important influence on the transmittance of DLC films. The higher the content of sp3 bond was, the higher the transmittance of DLC films would. However, when the content of sp3 bond was too large, the higher the internal stress and the worse the wear resistance of DLC films would be.

(4) The optimized deposition parameters were: Ar pressure of 0.7 Pa, sputtering power of 100 W, and substrate temperature of 150 °C.

Acknowledgments

Authors This work was funded by *Ningbo Natural Science Foundation (*No.2018A).

References

[1] SINGH S B, PANDEY M, CHAND N, BISWAS A, BHATTACHARYA D, DASH S, TYAGI A K, DEY R M, KULKARNI S K and PATIL D S 2008 Bull. Mater. Sci. 31 813-818
[2] Liu L, Lu F F, Tian J, Xia S H and Diao Y 2019 Applied Physics A 125 366
[3] Samadi M, Eshaghi A, Bakhshi S R and Aghae A A 2018 i. Opt Quant Electron 50 193
[4] Grill A, Meyerson B S 1994 *Synthetic Diamond: Emerging CVD Science and Technology* (New York: Wiley)
[5] Zhang Y C 2009 *The Technology of Vacuum Coating* (Beijing: Metallurgical Industry Press)
[6] Lin X Y, Zhang K L and Hu K 2011 ECS Trans 34 793-797
[7] FALLON P, VEERASAMY V and DAVIS C 1993 Physical Review B 48 4777-4782
[8] Marcinauskas L, Dovydaitis V, Iljinas A and Mindaugas_A_2019 Thin Solid Films 681 15-22

Temperature Stable and Low Loss Microwave Dielectric Ceramics of $Li_2Mg_{3-x}Sr_xTiO_6$

Xiaolei Shi, Huaiwu Zhang, Dainan Zhang

State Key Laboratory of Electronic Thin Films and Integrated Devices, University of Electronic Science and Technology of China, Chengdu, China

xiaolei_shi316@163.com; hwzhang@uestc.edu.cn; dainanzhang@126.com

Abstract: In this paper, a series of $Li_2Mg_{3-x}Sr_xTiO_6$ (x = 0, 0.02, 0.04, 0.06, 0.08) ceramics were prepared by the conventional solid-state reaction method. The effects of Sr^{2+} substitution on the structure, morphology and microwave dielectric properties of $Li_2Mg_3TiO_6$ ceramics were investigated. The X-ray diffraction (XRD) results indicated that $Li_2Mg_3TiO_6$ and $SrTiO_3$ phases stably coexisted with each other and formed composite materilas. The appearance of porous structure mainly originated from the volatilization of lithium at high sintering temperatures. As the Sr^{2+} content increased from 0 to 0.08, the optimum dielectric constant (ε_r) increased from 13.61 to 17.32, the quality factor ($Q\times f$) decreased monotonously and the temperature coefficient of resonant frequency (τ_f) changed from -34.4 to +16.8 *ppm/°C*. The variations of ε_r, $Q\times f$ and τ_f could be explained by the Lichtenecker logarithmic model and mixing rule, respectively. In particular, optimal microwave dielectric properties of ε_r = 16.08, $Q\times f$ = 82100 GHz, τ_f = +1.7 *ppm/°C* were obtained for $Li_2Mg_{2.94}Sr_{0.06}TiO_6$ ceramic sintered at 1310 °C for 5h.

1. Introduction

Over the past decades, with the rapid development of mobile communications and satellites broadcasting, microwave dielectric materials have attracted more and more attentions due to their widespread applications such as filters, resonators, antennas and oscillator [1-4]. In order to meet the requirements of future microwave communication systems, microwave dielectric materials should possess dielectric characteristics: adjustable dielectric constant (ε_r) for different applications, high quality factor ($Q\times f$) for excellent selectivity and near-zero temperature coefficient of the resonant frequency (τ_f) for thermostability [5, 6].

In recent years, several excellent Li-containing compounds such as $Li_3Mg_2NbO_6$, $Li_2Mg_3Ti_4O_{12}$, $Li_2Zn_3Ti_4O_{12}$, Li_2MgTiO_4, Li_2TiO_3 and $Li_2Mg_3BO_6$ (B = Zr, Sn or Ti) have received considerable attentions as a potential candidate for microwave dielectric materials [7-12]. For example, Wang et al. reported that the strategy of Li non-stoichiometric could be adopted to promote densification and reduce losses of $Li_3Mg_2NbO_6$ ceramics. Zhou et al. reported that the novel low loss $Li_2M_3Ti_4O_{12}$ (M=Mg or Zn) ceramic systems exhibited dense microstructure, pure cubic phase at optimum temperature. Both of them possessed outstanding dielectric characteristics: ε_r=20.2, $Q\times f$=62300 GHz, τ_f=-27.1 ppm/°C ($Li_2Mg_3Ti_4O_{12}$) and ε_r=20.6, $Q\times f$=106700 GHz, τ_f=-48 ppm/°C ($Li_2Zn_3Ti_4O_{12}$). Liu et al. reported the novel cubic structure ceramic systems of $Li_2Mg_3BO_6$. Among these various compounds, $Li_2Mg_3TiO_6$ with rock salt structure has been extensively investigated because of their remarkable microwave dielectric characteristics: ε_r=15.2, $Q\times f$=152000 GHz, τ_f=-39 *ppm/°C*. Up to now, many researchers have focused their attention on the $Li_2Mg_3TiO_6$ material and a number of efforts and attempts have been made to control dielectric characteristics. For instance, Pan et al. established the relationship between intrinsic structure characteristics and dielectric properties by prepared $Li_2Mg_{3-x}Co_xTiO_6$ solid solutions [13]. Yang et al. studied the influences of Zn_{2+} doping on the crystal structure, morphology and microwave dielectric properties of $Li_2(Mg_{1-x}Zn_x)_3TiO_6$ ceramics and found the solid solubility limit was located about 0.15 [14]. Zhang et al. fabricated $Li_2(Ni_{1-x}Mg_x)_3TiO_6$ $(0 \leqslant x \leqslant 1)$ materials through the conventional solid sintering method, the relationship between crystalline structures and microwave dielectric properties were studied [15]. However, the application of $Li_2Mg_3TiO_6$ were hampered due to the bigger negative τ_f. In general, the near-zeroτ_f could be achieved by following methods: (1) directly added the opposite τ_f materials, (2) introduced the suitable ions to obtained corresponding materials with opposite τ_f value [16].

In this article, given the larger positive τ_f of $SrTiO_3$ and remarkable thermal stability of $Li_2Mg_{3-x}Ca_xTiO_6$ composite ceramics, the Sr^{2+} doped $Li_2Mg_3TiO_6$ materials were synthesized by the solid-state reaction and the influences of Sr^{2+} ions doping on the structure, morphology and microwave dielectric properties of $Li_2Mg_3TiO_6$ ceramics were investigated. The excellent microwave dielectric properties such as low loss and near-zero τ_f make $Li_2Mg_{2.94}Sr_{0.06}TiO_6$ ceramic a possible applications in microwave domains.

2. Experimental Procedure

The $Li_2Mg_{3-x}Sr_xTiO_6$ samples were synthesized by the conventional solid-state reaction progress. The high purity starting raw materials of $SrCO_3$, Li_2CO_3, MgO and TiO_2 were weighed accurately according to chemical formula of $Li_2Mg_{3-x}Sr_xTiO_6$ (x = 0, 0.02, 0.04, 0.06, 0.08) and milled with zirconia ball in anhydrous alcohol medium for 6 h. The mixtures were dried and calcined at 1000 °C for 4 h. After re-milling and drying, the binder of polyvinyl alcohol (13 wt.% PVA) was mixed with the finely milled calcined powders. Subsequently, the mixtures were granulated and vertically pressed into cylinders sheets (12mm*6mm). Finally, the samples were preprocessed at 550 °C for 2 h and sintered at 1220–1340 °C for 4 h.

The crystalline structure of $Li_2Mg_{3-x}Sr_xTiO_6$ (x = 0, 0.02, 0.04, 0.06, 0.08) samples were analyzed by X-ray diffraction (DX-2700, XRD). The particle morphology were visualized by scanning electron microscope (JOEL JSM6490LV, SEM). The bulk densities of ceramics were measured by the Archimedes method with medium of distilled water. The dielectric properties of all samples were measured by the network analyzer (N5230A, Agilent, TE011 mode). The temperature coefficient resonant frequency (τ_f) of specimens were calculated with the following formula:

$$\tau_f = \frac{f_2 - f_1}{f_1(85-25)} \times 10^6 \left(ppm/°C \right) \qquad (1)$$

where f_1 and f_2 represent the resonant frequency obtained at 85 °C and 25 °C, respectively.

3. Results and Discussion.

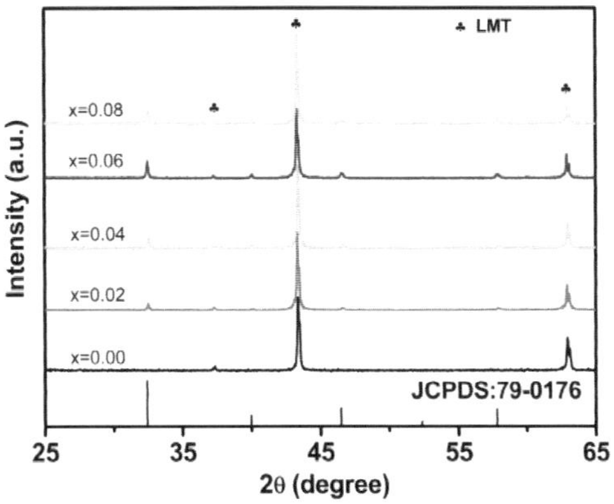

Fig. 1 XRD patterns of pure $Li_2Mg_3TiO_6$ and $Li_2Mg_{3-x}Sr_xTiO_6$ (x = 0, 0.02, 0.04, 0.06, 0.08) ceramics sintered at 1310 °C.

The XRD patterns of the pure $Li_2Mg_3TiO_6$ and $Li_2Mg_{3-x}Sr_xTiO_6$ (x = 0.02, 0.04, 0.06, 0.08) ceramics sintered at 1310 °C for 4 h were shown in Fig. 1. The peaks of pure $Li_2Mg_3TiO_6$ sample were indexed as the rock salt structure with space group of Fm3m (225). On the basis of previous research [17], the structure of $Li_2Mg_3TiO_6$ and $Li_2Mg_3SnO_6$ (JCPDS:39-0932) seemed very similar, which could be ascribed to the similar Shannon's effective ionic radius between Ti^{4+} ions and Sn^{4+} ions (Coordination number = 6, $R_{Ti}{}^{4+}$ = 0.605 Å, $R_{Sn}{}^{4+}$ = 0.69 Å). However, the additional phase compositions of $SrTiO_3$ (JCPDS:79-0176, space group: Pm-3m) could be detected, which may be due to the big difference of radius between Sr^{2+} ions and Mg^{2+} ions (Coordination number = 6, $R_{Sr}{}^{2+}$ = 1.18Å, $R_{Mg}{}^{2+}$ = 0.72 Å). Moreover, with the increase of Sr^{2+} content, the phase content of $SrTiO_3$ gradually increases. The similar results were detected in Ca^{2+} doped $Li_2Mg_3TiO_6$ ceramics. All these results show that Sr^{2+} can react preferentially with Ti^{4+}, which result in the coexist of $Li_2Mg_3TiO_6$ and $SrTiO_3$ structure.

The surface SEM images of pure $Li_2Mg_3TiO_6$ and $Li_2Mg_{3-x}Sr_xTiO_6$ ceramics sintered at 1310 °C for 4h were displayed in Fig.2. Obviously, the pure $Li_2Mg_3TiO_6$ ceramic presents the porous microstructure, which is due to the lithium volatile and evaporate at elevated sintering temperatures [12]. With the increase of Sr^{2+} content, the microstructure of $Li_2Mg_3TiO_6/SrTiO_3$ composite materials turn from porous structure into almost dense structure, which is similar to previous reports about $Li_2Mg_{3-x}Ca_xTiO_6$ [16]. In addition, the average grain size gradually increase with the Sr^{2+} doping content increase as shown in Fig. 2 (b-e). The results indicate that the Sr^{2+} doped could improve densification of samples, which may pave an alternative way to cure the porous structure.

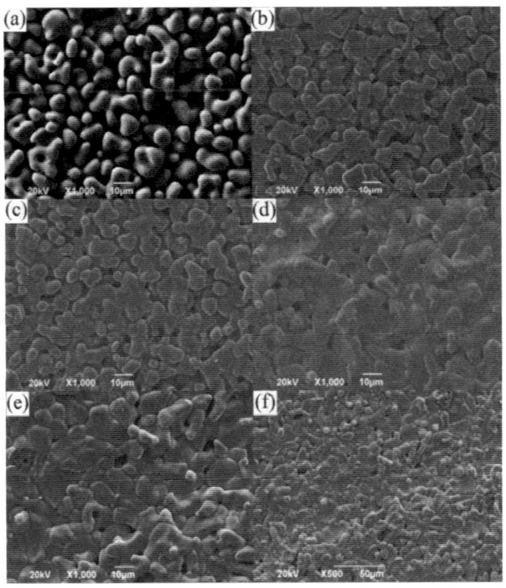

Fig. 2 SEM images of pure $Li_2Mg_3TiO_6$ and $Li_2Mg_{3-x}Sr_xTiO_6$ ceramics sintered at 1310 °C: (a) x=0, (b) x=0.02, (c) x=0.04, (d) x=0.06, (e) x=0.08, (f) x=0.08.

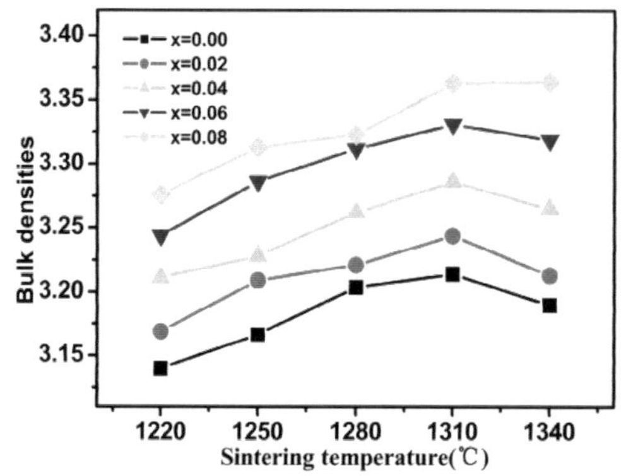

Fig. 3 The bulk densities of the pure $Li_2Mg_3TiO_6$ and $Li_2Mg_{3-x}Sr_xTiO_6$ (x = 0.02, 0.04, 0.06, 0.08) ceramics sintered at different temperatures.

The bulk densities of the pure $Li_2Mg_3TiO_6$ and $Li_2Mg_{3-x}Sr_xTiO_6$ (x = 0.02, 0.04, 0.06, 0.08) ceramics sintered at different temperatures were shown in Fig. 3. It can be seen that the bulk densities increase first and then decrease with the increase of the sintering temperature. In the range of 1220-1310 °C, the gradually increase of bulk densities should be due to the densification process. However, the decreases inversely when sintering temperature is beyond 1310 °C could be ascribed to the lithium evaporation. In addition, the bulk densities increase and relative densities of 92%, 92.4%, 93.1%, 93.9%, 94.3% can be obtained with the increase of x value. The results prove once again the Sr^{2+} doped could eliminate pores of $Li_2Mg_3TiO_6$ ceramics , which was in accordance with the SEM images.

Fig. 4 shows the change trend of ε_r as functions of Sr^{2+} doped content and sintering temperatures. It is noticeable that the trend of ε_r sintered at different temperatures is similar to that of densities with each certain Sr^{2+} concentration, which means the densification play an important role in dielectric constant. In addition, the ε_r values of $Li_2Mg_3TiO_6/SrTiO_3$ composite materials increase with increasing Sr^{2+} concentration. The increasing trend is reasonable related to the incremental relative densities, increasing $SrTiO_3$ (ε_r=290) according to the mixture rule: $ln\varepsilon_r = v_1 ln\varepsilon_{r1} + v_2 ln\varepsilon_{r2}$ [6].

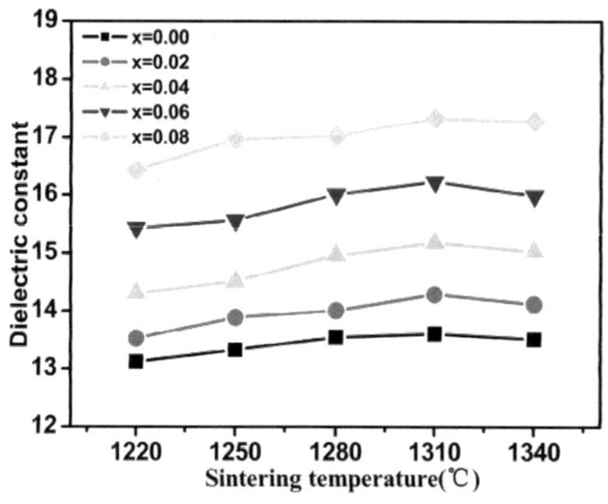

Fig. 4 The change trend of dielectric constant as functions of Sr^{2+} doped content and sintering temperatures.

Fig. 5 displays the change trend of $Q \times f$ as functions of Sr^{2+} doped content and sintering temperatures. It is clear that the $Q \times f$ of $Li_2Mg_{3-x}Sr_xTiO_6$ (x =0.00, 0.02, 0.04, 0.06, 0.08) materials sintered at different sintering temperature displayed a similar trend to bulk densities. As we all know, the $Q \times f$ usually associated with the intrinsic losses and extrinsic losses. The intrinsic losses depend on lattice vibration modes, and the extrinsic losses are depend on porosity, secondary phase and grain sizes. In this paper, given the results of SEM, the gradually reduce of $Q \times f$ mainly because of the increasing secondary phase

$SrTiO_3$ ($Q\times f$ = 4800 GHz). The results demonstrates that, compared with densification, the secondary phase play a more important role on $Q\times f$ value.

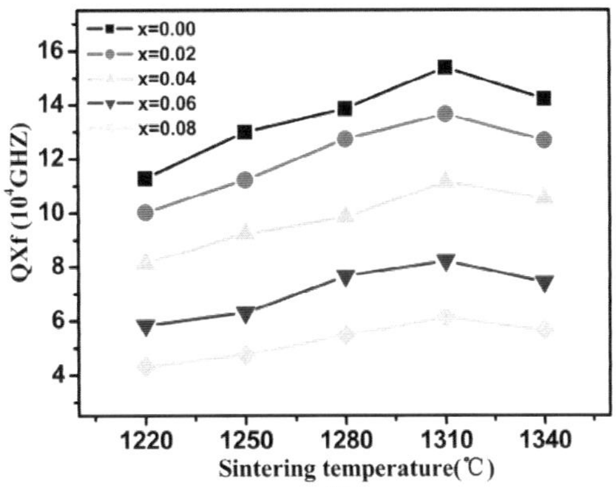

Fig. 5 The change trend of $Q\times f$ as functions of Sr^{2+} doped content and sintering temperatures.

Fig. 6 shows the change trend of τ_f as functions of Sr^{2+} doped content. It can be seen that the τ_f vary from -34.4 to +16.8 ppm/°C with increasing Sr^{2+} concentration. As we all know, the τ_f usually associated with the sample components and secondary phase. According to the XRD patterns, the phase proportion of $SrTiO_3$ in composite materials increase with increasing Sr^{2+} concentration. The near-zero τ_f of $Li_2Mg_3TiO_6/SrTiO_3$ composite materials could be explained by the increasing large positive τ_f $SrTiO_3$ (τ_f =+1650 ppm/°C) based on Lichtenecker empirical logarithmic rule: $\tau_{f}= v_1\tau_{f}+ v_2\tau_f$[17].

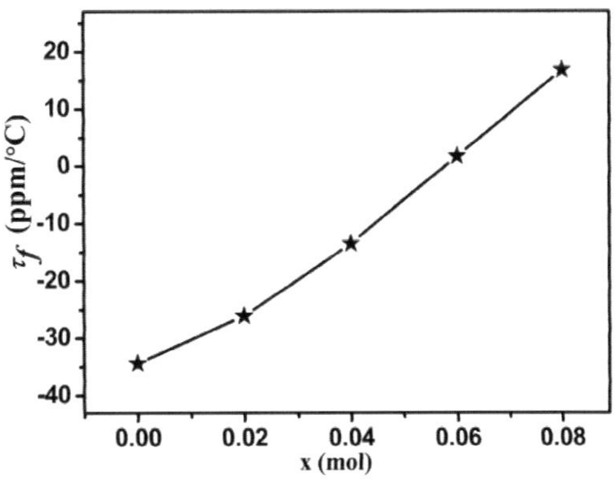

Fig. 6 The change trend of τ_f as functions of Sr^{2+} doped content.

4. Conclusion.

In this paper, the $Li_2Mg_{3-x}Sr_xTiO_6$ (x = 0, 0.02, 0.04, 0.06, 0.08) ceramics were prepared by the conventional solid-state reaction method. The effects of Sr^{2+} substitution on the structure, morphology and microwave dielectric properties of $Li_2Mg_3TiO_6$ ceramics were investigated. The X-ray diffraction (XRD) results indicated that Sr^{2+} ions did not replace Mg^{2+} ions in $Li_2Mg_3TiO_6$ due to the bigger radius of Sr^{2+} ions, which lead to the formation of $Li_2Mg_3TiO_6/SrTiO_3$ composite materials. The appearance of porous structure mainly originated from the volatilization of lithium at higher sintering temperatures. As the Sr^{2+} content increased from 0 to 0.08, the optimum dielectric constant (ε_r) increased from 13.61 to 17.32, the quality factor ($Q\times f$) decreased monotonously and the temperature coefficient of resonant frequency (τ_f) changed from -34.4 to +16.8 ppm/°C. The variations of ε_r, $Q\times f$ and τ_f could be explained by the Lichtenecker logarithmic model and mixing rule, respectively. In particular, excellent microwave dielectric properties of ε_r = 16.08, $Q\times f$ = 82100 GHz, τ_f = +1.7 ppm/°C were obtained for $Li_2Mg_{2.94}Sr_{0.06}TiO_6$ ceramic sintered at 1310 °C for 5h. The excellent comprehensive properties make the $Li_2Mg_{2.94}Sr_{0.06}TiO_6$ ceramic a suitable candidate for microwave applications.

Acknowledgment

The work was supported by National Key Scientific Instrument and Equipment Development Project (51827802).

References

[1] Luo H, Fang L, Xiang H, Tang Y and Li C, Two novel low-firing germanates $Li_2MGe_3O_8$ (M=Ni, Co) microwave dielectric ceramics with spinel structure 2017 *Ceram. Int.* **43** 1622-27

[2] Shi X, Zhang H, Zhang D, Xu F, Zheng Y, Wang G, Liu C, Jin L, Li J and Jia L, Correlation between structure characteristics and dielectric properties of $Li_2Mg_{3-x}Cu_xTiO_6$ ceramics based on complex chemical bond theory, 2019 *Ceram. Int.* **45** 23509-14

[3] Shi X, Zhang H, Zhang D, Xu F, Wang G, Zheng Y, Liu C, Li J, Jin L and Jia L, Temperature stability and chemical compatibility of novel $Li_{1.6}Zn_{1.6}Sn_{2.8}O_8$ ceramics 2019 *Mater. Chem. Phys.* **238** 121960

[4] Tseng C, Tseng P, Chang C and Kao Y, Novel temperature stable Li_2MnO_3 dielectric ceramics with high Q for LTCC applications 2014 *J. Am. Ceram. Soc.* **97** 1918-22

[5] Zhou H, Tan X, Wang K, Sun W, Ruan H and Chen X, Microstructure and sintering behavior of low temperature cofired $Li_{4/5}Mg_{4/5}Ti_{7/5}O_4$ ceramics containing $BaCu(B_2O_5)$ and TiO_2 and their compatibility with a silver electrode 2017 *RSC Adv.* **7** 44706-11

[6] Zhang J, Zuo R, Song J, Xu Y and Shi M, Low-loss and low-temperature firable $Li_2Mg_3SnO_6$-$Ba_3(VO_4)_2$ microwave dielectric ceramics for LTCC applications 2018 *Ceram. Int.* **44** 2606-10

[7] Wang G, Zhang H, Xu F, Huang X,. Lai Y, Gan G, Yang Y, Li J, Liu C and Jin L, Investigation and characterization on crystal structure and enhanced microwave dielectric properties of non-stoichiometric $Li_{3+x}Mg_2NbO_6$ ceramics 2018 *Ceram. Int.* **44** 20539-44

[8] Zhou H, Liu X, Chen X and Fang L, Preparation, phase structure and microwave dielectric properties of a new low cost $MgLi_{2/3}Ti_{4/3}O_4$ compound, 2012 *Mater. Chem. Phys.* **137** 22-25

[9] Zhou H, Liu X, Chen X, Fang L and Wang Y, $ZnLi_{2/3}Ti_{4/3}O_4$: A new low loss spinel microwave dielectric ceramic 2012 *J. Eur. Ceram. Soc.* **32** 261-65

[10] Yang C, Liu Q, Bi J, Tao W and Wu H, Effects of $(Mg_{1/3}Ta_{2/3})$-substitution for Ti-site on the microwave dielectric properties of Li_2MgTiO_4 ceramics 2018 *Ceram. Int.* **44** 5982-87

[11] Du M, Li L, Yu S, Sun Z and Qiao J, High-Q microwave ceramics of Li_2TiO_3 co-doped with magnesium and niobium 2018 *J. Am. Ceram. Soc.* **101** 4066-75

[12] Fu Z, Liu P, Ma J, Zhao X and Zhang H, Novel series of ultra-low loss microwave dielectric ceramics: $Li_2Mg_3BO_6$ (B=Ti, Sn, Zr) 2016 *J. Eur. Ceram. Soc.* **36** 625-29

[13] Pan H, Cheng L and Wu H, Relationships between crystal structure and microwave dielectric properties of $Li_2(Mg_{1-x}Co_x)_3TiO_6$ ($0 \leq x \leq 0.4$) ceramics 2017 *Ceram. Int.* **43** 15018-26

[14] Yang C, Pan H, Yang Y and Wu H, Influence of Zn substitution on the crystal structures and microwave dielectric properties of $Li_2(Mg_{1-x}Zn_x)_3TiO_6$ ($0 \leq x \leq 0.2$) ceramics 2018 *J. Alloys Compd.* **764** 424-30

[15] Zhang P, Yang M and Xiao M, Sintering behavior, crystalline structure and microwave dielectric properties of $Li_2(Ni_{1-x}Mg_x)_3TiO_6$ ($0 \leq x \leq 1$) ceramics 2018 *Ceram. Int.* **44** 21607-12

[16] Fang Z, Tang B, Si F and S.R. Zhang, Temperature stable and high-Q microwave dielectric ceramics in the $Li_2Mg_{3-x}Ca_xTiO_6$ system (x=0.00–0.18) 2017 *Ceram. Int.* **43** 1682-87

[17] Xiang H, Li C, Yin C, Tang Y and Fang L, A reduced sintering temperature and improvement in the microwave dielectric properties of $Li_2Mg_3TiO_6$ through Ge substitution 2018 *Ceram. Int.* **44** 5817-21

An Application of Failure Mode, Effects and Criticality Analysis (FMECA) for Composite Structures of Airplanes' Wings

Qiang Wang, Ruicong Xia, Qiuhan Liu and Qiuping Hu

Air Force Engineering University, Xi'an, Shaanxi, China

rayc0-xia@foxmail.com

Abstract. Composite structures have played a significant role in enhancing airplane's performance However, defects inside can be a potential threat to flight safety. To resolve the hazard. FMECA was applied to the defect analysis of composite structures on airplane's wing. And the failure mode of the highest criticality was found to be the impact damage. Then impact test on typical composite laminates was carried out. The caused damage was qualitatively analysed and proved impact of low energy brings latent risk. Practical solution was put forward to improve the safety or reliability of the applied composites in the airplane.

1. Introduction

At present, the utilization of composite structures in the airplanes' wings has been developed from the secondary bearing parts to the main bearing parts. However, defects inside threatened the flight safety. To reduce the risk, defect analysis of the composite structures.

FMECA has been a well-established method to evaluate defects, or in another way saying, failure modes of products. Duan Zhaobin and Zhang Peng applied FMECA and ontology to the fault diagnosis of flight control system. The share of the fault diagnosis information was realized among aviation manufacturers, users and researchers [1]. Hu Yifan carried out FMECA for an aero-engine spindle subsystem which contributing to the reliability optimization design [2]. So far, few researches on the defect analysis of composites structures by FMECA was studied. Therefore, an approach of FMECA on general and typical composite structures of the wings was advanced. Several failure modes and the causes were concluded. Then the criticality of each mode was evaluated and the highest one came to be impact damage. Thus, impact of different energy level on typical composite laminates was tested and the damage caused was qualitatively studied. Finally, available solution to risk reduction and safety restoration and was proposed.

2. Composite structures of wings

2.1. Distribution and Description of Composite Structures on the Wing

A quite number of composites structures applied in aircraft wing and some of them were listed in Table 1.

Table 1. Composites structures of the wing

Structure	Function
Low and high-speed ailerons	Control aircraft rolling attitude.
Flap track fairing	Wrap flaps and block airflow ahead.
Leading edge flap	Increase the wing area and bending, improve lift coefficient.
Inside and outside flaps	Control aircraft rolling attitude during cruise and link with flaps at low speed.
Spoiler	Disrupt the airflow and create a controlled stall zone on the back wing to dramatically reduce lift.
Winglet	Weaken the effect of airflow around the top surface under wingtip and reduce lift loss

2.2. Common Defects of Composite Structures

Compared with traditional metal material, composites have great advantages in weight reduction but assuring the structural strength simultaneously. However, under the complicated influences of extreme flight environment accompanying material aging and human error, defects or structure failures could emerge with a high possibility. And the most common are delamination and impact damage which could seriously destroy composites' mechanical performance.

Figure 1 is the sketch of delamination. In most case, the delamination was caused by unexpected applied force which leading to the failure of adhesive layer. Then the bonding interface would separate followed by gap generating. That is catastrophic to the composites wing skin in high altitude flying. The most dangerous situation could be a large area of stripped wing skin resulting in instantaneous lift loss during flight. In Figure 2, impact damage refers to surface damage along with internal defect of composites caused by instantaneous exterior force. Due to the characteristics of anisotropy and energy absorption, surface damage could be mild from visual observation while the interior could be rather serious [3, 4]. This feature made impact damage a latent trigger to unexpected structure failure. Therefore, to lessen the harm, adverse effects of the common defects should be evaluated by safety analysis method.

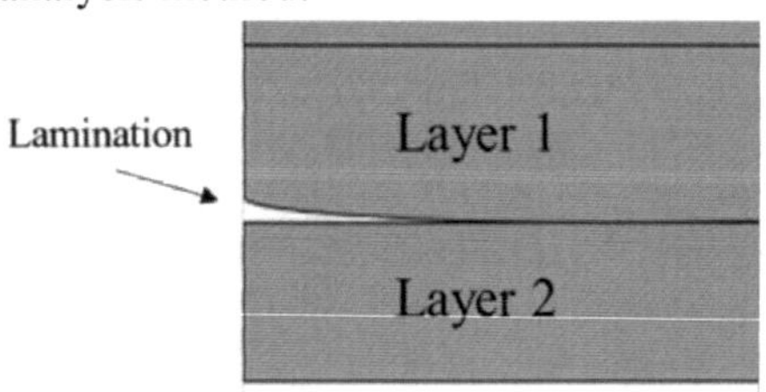

Figure 1. Delamination

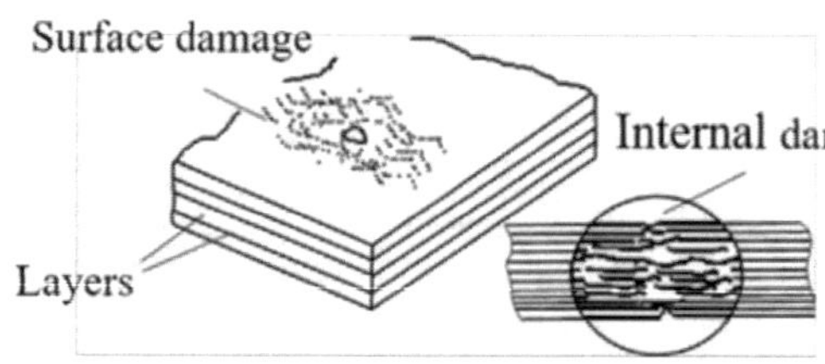

Figure 2. Impact damage

3. An Application Example of FMECA for Composite Structures of Airplanes' Wings

3.1. The Procedures of FMECA

The procedures of FMECA followed below.[5]

1.System definition: analysis of product functions and hierarchy, that is, to determine the general function of the composite structure on the wing is to maintain the overall shape in flight and secure structural integrity. The hierarchy contains structural parts, wing and the aircraft.

2. Failure modes analysis: analyzing possible failure modes of product, that is, to find out every possible failure mode of aircraft wing composite structure.

3. Failure cause analysis: that is to determine the cause of each failure mode.

4. Failure effects analysis: that refers to find out the adverse influence of each mode and determine its severity level.

5. Criticality analysis: comprehensive evaluation of the adverse influence of each mode according to its severity and occurrence level, and proposing improvement or compensation measures.

3.2. Severity and Occurrence Level Definition

The definitions were listed in Table 2 and Table 3.

Table 2. Severity level definition

Severity level	Description	Explanation
I	Catastrophic	Lead to casualty, products destruction, great property damage and environmental damage.
II	Fatal	Lead to human injury, products damage, mission failure, serious property and environmental damage.
III	Medium	Cause moderate injuries to personnel, delayed or degraded missions and damage to products, property and environment
IV	Low	Cause unplanned maintenance or repair

Table 3. Occurrence level definition

Occurrence level	Description	Occurrence probability (P) range
A	Frequently	$P > 10^{-1}$
B	Often	$10^{-3} < P \leq 10^{-1}$
C	Occasionally	$10^{-5} < P \leq 10^{-3}$
D	Rarely	$P \leq 10^{-5}$

3.3. FMEA Worksheet and CA

According to procedures, final FMEA results were shown in Table 4 listed in the next page.
After the FMEA work, CA was carried out according to the severity and occurrence level. The criticality matrix was shown in Table 5.

Table 5. Criticality matrix for composite structures of aircraft wings

Occurrence level \ Severity level	I	II	III	IV
A				
B	5		6	3
C		1		4
D	7		2	

In Table 5, the red area is unacceptable which means failure modes within is of great harm. The yellow represents the selective acceptance region. It refers to that the criticality of failure modes within can be controlled considering actual situation. While the green area is acceptable, that is, the impact of the failure mode inside can be well-controlled. It shows that the criticality of the failure mode No.5 is unacceptable because it's in the red area. The criticality of No.1, No.6, and No.7 failure mode has to be considered according to the actual situation. The rest No.2, No.3, and No.4 are acceptable.

Table 4. Application of FMECA for composite structures of aircraft wings worksheet
Initial indenture level: Aircraft. Indenture level: wing. Mission: A single flight.

No.	Name	Function	Failure mode	Cause	Failure effects			Severity level	Occurrence level
					Local effects	Next higher effects	End effect		
1	Wing flap	Increase wing area and lift	Jam	Flap screw wears	Flaps cannot retract properly	Wing area Reduces	Lift falls	II	C
2	Aileron	Control differential deflection of left and right ailerons to generate rolling torque	Stuck	Actuator keeps stuck	Ailerons cannot deflect	The wing cannot provide rolling torque	Maneuverability reduces	III	D
3	Spoiler	Assist maneuvering system provides lift power and improve the aircraft's maneuverability.	Upturn	Control pulley shaft fouling	The spoiler cannot be retracted to the correct position	The wing cannot take the lift off	Controllability reduces	IV	B
4	Winglet	Reduces drag and fuel consumption during flight	Structural fracture	Material fatigue crack	Structural fracture	The induced drag cannot be reduced	Aircraft induced drag increases	IV	C
5	Flap track fairing	Wrap the exposed flap support structure and drive mechanism	Impact damage	External impact damage	Internal damage of material	Internal structure exposure	Aircraft drag increases	I	B
6	Nacelle fairing	Adjust airflow and protect internal engine components	Crack	Composite delamination	Broken fairing	Engine components exposure	Form threats to the engine	III	B
7	Wing skin	Ensure the aerodynamic shape	Fall off	Composite delamination	Skin falls off	Internal structure exposure and aerodynamic shape damage	Lift reduces dramatically	I	D

So, during the actual use of composite structures in aviation, impact damage with low energy is very common yet a potential threat. Therefore, the impact test of carbon fiber laminates already used in the wings were carried out to probe into the characteristics of the damage in different energy level. And measures were proposed to prevent the impact damage threatening safety in service [6].

4. Impact Damage Test

4.1. Impact test design

In order to make impact energy level the controlled variable, three carbon fiber laminate samples of the same size and thickness were chosen in this test. All of the three were made on aircraft production line. The size of a sample is 150mm×100mm and the thickness is 3mm. A sketch and real photo of these samples were shown in Figure 3.

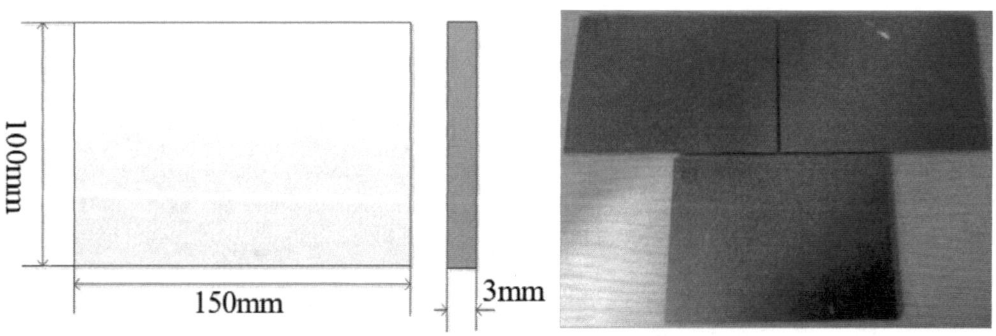

Figure 3. Carbon fiber laminate samples

The chosen platform was CEAST9350 drop hammer impact test machine from Impact Mechanics Laboratory of Xi'an Jiaotong University. To the three samples, three energy level of impact was set and each sample corresponded to each level. By adjusting the velocity of the hammer, the energy of impact could be altered as needed. The weight of the hammer is 5.41kg and other parameters were listed in Table 6.

Table 6. Impact test parameters settings

The impact order of samples	#1	#2	#3
Impact energy (J)	15	20	25
The velocity of the hammer (m/s)	2.35	2.72	3.04

4.2. Analysis of test results

Test results were shown in Figure 4. (a) and (e) are results of sample #1 with 15J energy impact. On the top surface (a) of sample #1, dent of a small size could be hardly seen yet ambient fiber breaks were observable. Meanwhile, from the undersurface (e), the observation of a slight crack along the fiber direction and surface swelling was even harder.

(b) and (f) are the results of the 20J energy impact. Comparing the result (b) of 15J, the dent and damage crack on the top surface could be easily seen. And the crack amount was apparently greater than that of sample #1. Cracks around the dent propagated along with the fiber direction. From the bottom surface (f), a huge fiber crack emerged conspicuously which made great contrast to (e) of sample #1. Moreover, the length of crack also increases dramatically and the bulge was obvious.

(c) and (g) are the results of the 25J energy impact. Because of the highest impact energy applied in this test, damage caused was further aggravated reflecting in the deepest and largest dent as well as the greatest number of cracks emerged in the top surface (c). While, on the back side (g), the bulging area was also of the largest size and the length of the cracks was the longest. What's more, the delamination damage becomes visible to the naked eye.

Above all, the larger impact energy is, the greater damage is. By comparing the test results of different impact energy, it can be found that the low-velocity impact (15J) caused a potential threat because the damage presented differentiate greatly in the top and bottom surface. The defects inside caused by the impact were complicated and invisible to the naked eye. Therefore, corresponding measures must be taken to reduce or even eliminate the threat of the low-velocity impact.

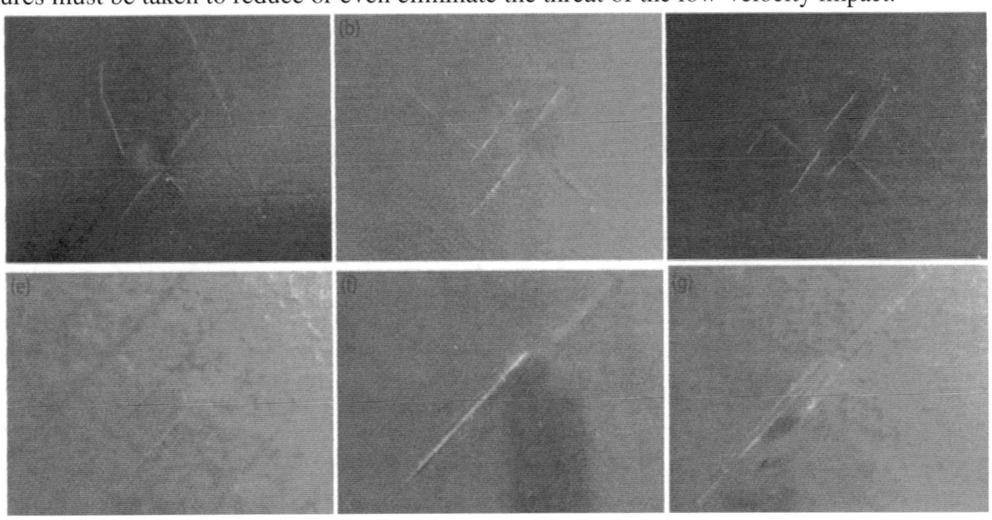

Figure 4. The top and bottom surfaces of samples after impact:
(a) Top surface of #1; (b) Top surface of #2; (c) Top surface of #3;
(e) Bottom surface of #1; (f) Bottom surface of #2; (g) Bottom surface of #3.

5. Corresponding Measures: NDT (non-destructive testing) Technology

At present, a number of the NDT technologies have been actually leveraged in the aviation. S. Gholizadeh reviewed advantages and disadvantages of various NDT methods for the evaluation of composite materials [7]. Among these methods, air-coupled ultrasonic testing technology has realized the defect detection of composite structures in many aircrafts, such as the trailing edge and tail of the Boeing 737s' wing, the Airbus A320 aileron and the Black Hawk helicopter rotor [8]. The aforesaid impact damage of composite structures can be successfully detected by NDT technology to determine the basic information of its location, area, depth, etc. And suitable maintenance measures can be taken by considering the obtained information.

6. Conclusion

To eliminate the potential safety threat of composite structures failure on aircraft, FMECA was used to conduct the defect analysis and failure modes of typical structures were found out. The cause and adverse impact of each failure mode were concluded. According to the severity level and occurrence level, the criticality matrix was drawn and the most critical one was found out to be the impact damage. The FMECA results indicated that unexpected impact on composite structures could be a potential threat to flight safety. Therefore, the impact damage test was carried out on carbon fiber laminates to probe into the damage features in different impact energy. It was found that both the internal and external damage were serious and obvious under the impact energy of 20J and 25J. While when the energy is 15J, damage on the top surface and bottom surface differentiated greatly and the damage inside was invisible which commit to be a safety hazard. Thus, NDT technology were proposed to determine the damage firstly and based on detection results, appropriate maintenance measures could be taken to restore or improve the structure reliability which finally ensures flight safety.

References

[1] Duan Zhaobin, Zhang Peng. 2016 Fault diagnosis method of flight control system based on ontology and FMECA [J]. Journal of Civil Aviation University of China, 34(04):21-26.

[2] Hu Yifan. 2019 Reliability analysis of aviation engine spindle bearing based on vibration signal [D]. Harbin Engineering University.

[3] Liu Songping and Liu Feifei. 2017 Advanced non-destructive testing technology for composite materials (Beijing: Aviation industry press), pp 3-16.

[4] Wang Qiang, Li Xinyi, Chang Tianying, Hu Qiuping and Bai Jinpeng. 2018 Terahertz time-domain Spectroscopic study of aircraft composite and matrix resins, [J]. Spectroscopy and Spectral Analysis, 38(09) 2706-12.

[5] GJB 1391-2006 Guide to failure mode, effects and criticality analysis. 2006

[6] Cai Yushun, Zhu Hao and Qing Xinlin. 2018 Composite material impact damage detection and maintenance requirement analysis to SHM technology, [J] Aeronautical Manufacturing Technology, 61(07) 78-82.

[7] S. Gholizadeh. 2016, A review of non-destructive testing methods of composite materials[J]. Procedia Structural Integrity, 1.

[8] Peters J J, Barnard D J, Hsu D K. 2004 Development of a fieldable air-coupled ultrasonic inspection system[J]. Review of Progress in Quantitative Nondestructive Evaluation, 23(23):1368-1375.

A method of measuring the arc radius of Rockwell diamond indenter

CUI Yuanyuan[1,a], YE Ming[1,b], ZHANG Feng[1,c], LIU Jiping[1,d]

[1]National Institute of Metrology, China, BEIJING 100029

[a]cuiyy@nim.ac.cn, [b]yem@nim.ac.cn, [c]zhangf@nim.ac.cn, [d]liujp@nim.ac.cn

Key words: Rockwell hardness, diamond indenter, radius recognition, simplicity algorithm, uncertainty

Abstract: Rockwell hardness is the most widely used hardness measurement method in industry at present. Taking superficial Rockwell primary standard machine in National Institute of Metrology(China) as an example, the uncertainty caused by the radius of the indenter arc accounts for 50% of the total uncertainty. In view of the important influence of the arc radius at the top of Rockwell diamond indenter on the evaluation of Rockwell hardness uncertainty, a profile projection system is established which is composed of lens, industrial camera and area-source of light. After pictures are obtained with camera, the next task for these pictures is to be calculated to get target parameter. The target parameter is obtained by profile extraction and radius calculation. In this paper, a new method is proposed as an algorithm criterion that sum of absolute value of data points and fitting arcs is used. Simplicity algorithm is introduced to iteratively solve the algorithm criterion. The purpose of using the algorithm is to eliminate the influence of outlier noise points. To verify the validity of the method in this paper, we conduct a series of experiments. The traceability experiment with standard sphere shows that the expanded uncertainty of the system of the instrument can achieve1.0μm,k=2.The uncertainty of measurement result with Rockwell diamond is1.5μm,k=2. The method in this paper bases on a simple-structural system, has high measurement accuracy, and can be used for rapid screening, evaluation, verification and calibration of Rockwell diamond indenters in industry.

1. Introduction

Rockwell hardness is the most widely used hardness measurements in industry.[1] In the various factors that affects its uncertainty, the geometrical parameters of diamond indenter are the most critical impact factors. These factors are conical angle, axis inclination angle, arc radius and et al. Taking the HRN scale of the surface Rockwell hardness primary standard machine as an example, the uncertainty caused by the radius of the indenter arc accounts for 50% of the total uncertainty. The measurement of the arc radius becomes a problem to be studied and solved.[2-5]

Diamond indenter is a transparent object with very small size. Rockwell indenter has a cone angle 120°, a length of generator 0.4mm, a arc radius of 200μm. And the generator is tangent to the arc. Currently, there are several arc measuring method such as curvature sample comparison method, contact probe scanning method[6], confocal microscopy method[7,8], LVDT displacement rotationmeasurement method, contour projection method[9] and so on. For primary reference indenter, it requires a measurement method with high uncertainty level and tolerating high time-consuming. The radius measurement for common indenter should have the characteristics of low cost, fast convenient and high uncertainty level.

2. System Composition and Structure

The optical part of the profile projection measurement system is composed of a surface light source, an imaging lens and an industrial camera, as shown in Fig.1.The surface light source produces uniform projection light to the diamond indenter to be measured, and the parallel light carrying the profile information of the top of the indenterto be measured is incident to the imaging lens and collected by the industrial camera to the computer.In addition, the guide rail is designed on

The imaging lens is composed of two telecentric lenses. Its magnification is 4 x, working distance is 56 mm and depth of field is 0.1 mm.The feature of the lens is that both the image and the object are located at infinite distance. When the measured object moves up and down, the size of the object in the field of view will not change.

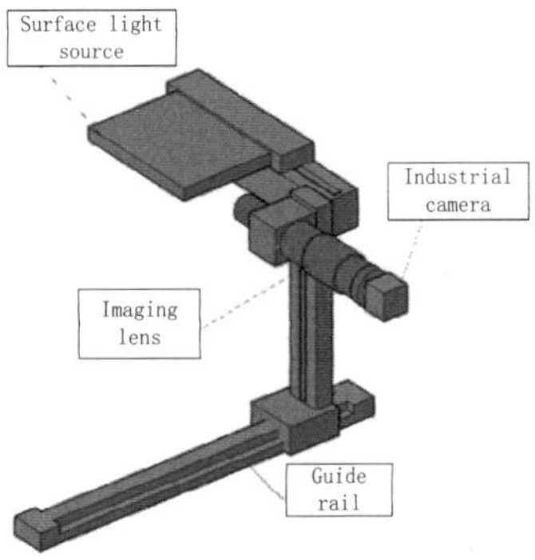

Fig. 1 Structure of profile projection system

3. Arc radiusidentification method

3.1 profile extraction method.

The identification of the arc radius at the top of the indenter includes edge extraction and radius calculation.Here, Gauss function is to construct a image filter, the filter function is as follows:

$$G(x,y)=\frac{1}{\sigma\sqrt{2\pi}}e^{\frac{(x^2+y^2)}{2\sigma^2}}. \quad (1)$$

In the equation， x 、 y 为 is the row number and column number corresponding to the pixel, σ is the filter parameter. The picture $f(x,y)$is filtered smoothing to obtain $f(x,y)*G(x,y)$. Computing the modulus $M$and direction $\theta$of the gradient vector of the picture

$$M(x,y)=\|f(x,y)*\nabla G(x,y)\|, \quad (2)$$

$$\theta(x,y)=\frac{f(x,y)*\nabla G(x,y)}{\|f(x,y)*\nabla G(x,y)\|}. \quad (3)$$

In order to locate the profile accurately, non-maximum suppression is needed to refine the profile band in gradient magnitude image M.Interpolation is carried out along the gradient direction in the 3×3 neighborhood where the pixel point （x，y）is the center. If the gradient magnitude $M(x,y)$at （x，y） is larger than the gradient magnitude of the two adjacent points along the direction, the point will be recorded as candidate edge points, otherwise it will be recorded as non profile image. Then we can get candidate profile image N.

The algorithm detects candidate profile points of candidate profile image N by setting high Th and low Tl thresholds through practical experience. The suspected edge points are further judged according to the connectivity of edges.

3.2 Radius calculation method.

According to the geometrical shape of ideal Rockwell hardness indenter prescribed by CCM-WGH and ISO, the tangent points P1 and P2 of arc and generator are located at ±100μm around the indenter axis on the horizontal direction. By constructing matching template to intercept the target window, a group of data points (x_i, y_i) can be obtained. Using them to fitting the following formula:

$$(x-x_c)^2+(y-y_c)^2=r^2. \tag{4}$$

In the equation, (x_c, y_c) is the center of the circle, R is the radius.The usual least squares algorithm requires the least square of the distance between the data point and the fitting arc. Thus in the following equation

$$SS(x_c, y_c, r)=\sum_{i=1}^{n}\left(r-\sqrt{(x_i-x_c)+(y_i-y_c)}\right)^2, \tag{5}$$

let $SS(x_c, y_c, r)$ be the minimum.First of all, this formula can not be solved analytically. Secondly, the square term of the least square is very sensitive to outlier noise points. The algorithm is very effective when the data points fully conform to the normal distribution, but it is not so effective when in machine vision there are some points that deviate along a certain direction.Therefore, the absolute value of the sum of the distance between the data point and the fitting arc is chosen as the criterion.It is as follow

$$SSA(x_c, y_c, r)=\sum_{i=1}^{n}\left|r-\sqrt{(x_i-x_c)+(y_i-y_c)}\right|. \tag{6}$$

Let $SSA(x_c, y_c, r)$ be the minimum, we can get

$$\nabla SSA(x_c, y_c, r)=\nabla\left[\sum_{i=1}^{n}\left|r-\sqrt{(x_i-x_c)+(y_i-y_c)}\right|\right]=0. \tag{7}$$

Formula (7) also has no analytic solution.Thus, we introduce Nelder-Mead algorithm, i.e. simplicity method, to solve the problem iteratively. A tetrahedron composed of four vertices is used to approximate the optimal point$O(X_c, Y_c, R)$ step by step. Here, let O_1, O_2, O_3, O_4 be the vertex of a polyhedron, and satisfy $SSA(O_1)\le SSA(O_2)\le SSA(O_3)\le SSA(O_4)$ 。Fig.2 shows the flow chart of iteratively updating the tetrahedron to make it move and contract to approximate the optimal solution.

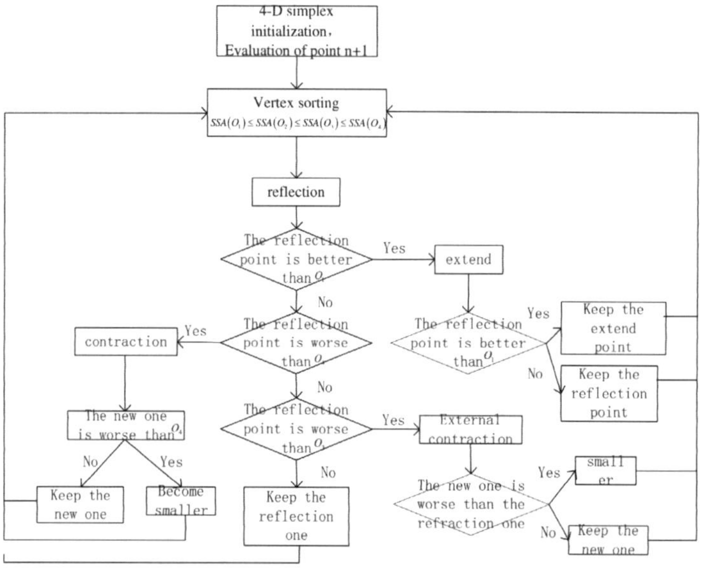

Fig.2 Flow chart of iterative algorithm for radius solution

4. Experiment

4.1 Traceability method and experiment.

The uncertainty of arc radius measurement is evaluated by standard sphere. Take a standardspherewith the radius of 200.6 as an example, let's analyze the uncertainty of the instrument.

According to the calibration result of the standard sphere ball, the average deviationΔR and the standard uncertainty$u(R_M)$can be calculated, as show as in Table1.

Table 1 Results of radius measurement

Standard value[μm]	Test values[μm]			Averagedvalue[μm]	ΔR[μm]	$u(R_M)$[μm]
	1	2	3			
200.6	200.8	200.2	200.5	200.5	-0.1	0.2

The combined uncertainty of standard spheremeasurement is calculated using the following equation:

$$u(R) = \sqrt{u^2(R_{RS}) + u^2(R') + u^2(R_M)}. \tag{8}$$

In the equation, $u(R_{RS})$is measurement uncertainty from standard sphere calibration certificate, $k = 1$,$u(R')$is the relative measurement uncertainty caused by instrument resolution,$u(R_M)$is the standard uncertainty of radius measurement. The measurement uncertainty of sphere is $U(R_{RS}) = 0.8\mu\text{m}(k = 2)$, the effect of instrument resolution can be brought into repeatability, so it is neglected. From the information above, we can calculated the uncertainty of radius measurement, as shown as in Table 2.

Table2 Uncertainty calculation of instrument radius measuremt

quantit yX_i	Distribut-ion type	Standard uncertain-ty $u(x_i)$[μm]	Sensitivityc oefficientc_i	Uncertainty $u_i(R)$[μm]
R_{RS}	normal	0.4	1	0.4
R'	rectangle	0	1	0
R_M	normal	0.2	1	0.2
Combined standard uncertainty $u(R)$				0.5
Expanded uncertainty $U(R)(k=2)$				1.0

4.2 Physical experiment.
Fig.3 and Table 3 show the contour projection and measurement result obtained from the experiment with standard Rockwell diamond indenters. The calculation of the results is:

$$\begin{aligned} U_l(R_l) &= 2\times\sqrt{u^2(R)+u^2(R_l)} \\ &= 2\times\sqrt{0.5^2+\left[(202.54-197.77)/2.97\right]^2/9} \\ &= 1.5(\mu m). \end{aligned} \quad (9)$$

Fig.3 The contour projection of Rockwell diamond indenter

Table 3 Radius measurement results of Rockwell diamond indenter

Location angle[°]	Test value of radius [μm]			
	1	2	3	The average
0	197.77	197.77	198.64	198.06
5	202.54	202.52	202.41	202.49
90	199.31	198.68	198.65	198.88
135	201.63	200.91	200.50	201.01
The average	200.1, $U_l(R_l)$=1.5μm, $k=2$			

5. Concludes
This paper presents a method for identifying the arc radius of Rockwell diamond indenter, in which the absolute value of the sum of the distance and Nelder-Mead method are introduced. The traceability experiment with standard sphere shows that the expanded uncertainty of the system of the instrument can achieve1.0μm,k=2.The uncertainty of measurement result with Rockwell diamond is1.5μm,k=2.

IOP Conf. Series: Materials Science and Engineering **784** (2020) 012011 doi:10.1088/1757-899X/784/1/012011

The system has simple structure, fast identification process and high measurement uncertainty. It can be used for rapid screening, evaluation, verification and calibration of Rockwell diamond indenter in industry. In industry application, there are some interferences sources that affect the algorithm accuracy, which need to be investigated later. In summary, the method in this paper has strong robustness, high reliability and a worth of popularization.

6. References

[1] B. Bhushan, B.K.Gupta: Handbook of Tribology, New York, McGraw-Hill, 1991.
[2] M.R.Vanlandingham, T. Juliano, M.Hagon: Measuring tip shape for instrumented indentation using atomic force microscopy, Measurement Science and Technology. 2005, vol.16(11),p.2173-2185.
[3] D.J. Munoz=Paniagua, M.T.Mcdermott, P.R.Norton, et al: Direct tip shape determination of a Berkovich indenter: effect on nanomechanical property measurement and description of a worn indenter, Nanotechnology, IEEE Transactions on, 2010, vol.9(4), p.487-493.
[4] J.Krier, J.Breuils, L.Jacomine, et al: Introduction of the real tip defect of Berkovich indenter to reproduce with FEM nanoindentation test at shallow penetration depth, Journal of Materials Research, 2012, vol.27(1): p.28-38.
[5] M.A.Monclus, S.Lotfian, J.M.Molina-aldareguai: Tip shape effect on hot nanoindentation hardness and modulus measurements, International Journal of Precision Engineering and Manufacturing, 2014, vol.15(8), p.1513-1519.
[6] J.Song, S.Low, A.Zhang, et al: Geometrical measurements of NIST SRM Rockwell hardness diamond indenters. IMEKO 2010 TC3, TC5 and TC22 Conferences Metrogogy in Mondern Context, Chonburi, Thailand, 2010, p.137-140.
[7] G.Dai, K.Hermann, F.Menelao: Progress in the characterization of the geometry of Rockwell diamond indenters. XIX IMKO World Congress Fundamental and Applied Metrology, Lisbon, Portugal, 2009, p. 985-990.
[8] S.Talagi. Determination of the geometrical parameters of Rockwell diamond indenters by iterative regression method. XX IMEKO World Congress Metrology for Green Growth, Republic of Korea,2012,p.9-14.
[9] A.Germak: Investigation of new possibilities in the calibration of diamond hardness indenters geometry. Measurement, 2011,44, p.351-358.

Molecular dynamics simulations the effect of temperature on the plastic deformation mechanism in aluminum single crystal

Zhao Wang[1,a], Yalin Li[2,b], Ting Hu[3,c], Jun Cai[*]

[1,2,3] School of Nuclear Science and Engineering, North China Electric Power University, Beijing 102206, People's Republic of China

*School of Nuclear Science and Engineering, North China Electric Power University, Beijing 102206, People's Republic of China

2735339069@qq.com; caijun@ncepu.edu.cn

Abstract: Molecular dynamics simulations are performed to investigate the effect of temperature on the plastic deformation mechanism in aluminum single crystal. It is found that as temperature increases the Yield strength and Young's modulus of the aluminum under compressive and tensile strain will reduce. Moreover, it is found that the higher temperature is, the easier the dislocation emission is. Under compressive strain, the proportion of 1/6<112> Shockley type of dislocations to total dislocations is found to increase with the temperature increasing. It is also found that only a large amount of dislocation occurring incipiently the strength of the material can be yield.

1. Introduction

The generation and quantitative analysis of dislocation are particularly important for the study on mechanical properties of materials when an external force acts on it [1-3]. The emergence of dislocations, as well as the dislocation types, is influenced by many factors, such as temperature, grain boundaries, secondary phases, vacancies and the mode of external force. Many researchers have kept paying attention to the effect of temperature on generation and evolution of dislocation [4]. For example, a tensile test was performed at different temperatures ranging from 1 to 600 K to assess the yield strength and Young's modulus of copper-silver nanocomposites. In this study dislocation extraction analysis was implemented to reveal the temperature-based underlying mechanism of plastic deformation [4]. On the other hand, limited researches are reported to study how temperature, at the atomic scale, affect the mechanical properties of materials by the generation and evolution of dislocation in aluminum in detail. In present work, molecular dynamics simulations were employed to investigate the temperature-based plastic deformation mechanism under uniaxial compression/tension in aluminum single crystal.

2. Theory and Method

In this work, the Embedded Atom Model (EAM) potential was utilized to describe the interaction between aluminum atoms. In this method the total energy of a monoatomic system is represented as

$$E_{tot}=1/2\sum V(r_{ij})+\sum F(\rho_i) \quad (1)$$

Here $V(r_{ij})$ is a pair potential as a function of the distance r_{ij} between atoms i and j, and F is the "embedding energy" as a function of the host "density" ρ_i induced at site i by all other atoms in the system. The unstable stacking fault energy is a crucial parameter of the barrier for partial dislocation nucleation [5]. Thus, the EAM potential employed for atomistic simulations must accurately reproduce pertinent features of the resulted energy curves, especially the unstable stacking fault energy [6]. The accurate characterization of the dislocation nucleation and interaction relies on a qualified EAM potential [7]. The potential for aluminum reported by Mishin et al. [8] describes the characterization of the dislocation nucleation and interaction well and has been widely used for simulation research [7]. Hence, the EAM potential [8,9] for aluminum is chosen in this work to carry out the simulations.

All simulations were performed using the Large-scale Atomic/Molecular Massively parallel simulator (LAMMPS) code [9] developed by Sandia National Laboratory. The calculated supercell containing 256000 atoms with size of $80a_0 \times 80a_0 \times 10a_0$ was built and periodic boundary conditions were applied along <100>(X), <010>(Y), <001>(Z) directions. In order to study the temperature effect on plastic deformation mechanism, simulations were carried out at different temperature of 300, 400, 500, 600, and700K. The temperature controlling was realized via Nose-Hoover thermostat [10]. According to LAMMPS user manual, the Nose-Hoover thermostat could bring the undesirable oscillation of pressure and temperature to the simulated system [7]. Thus, the uniaxial compressive or tensile simulation can be performed at expected temperature and pressure at a certain rate. In all simulations, energy minimization is realized by the conjugate gradient algorithm to minimize the energy of the simulation system, followed by relaxation to achieve an equilibrium state. Based on above, the compressive or tensile strain were loaded as the increase of time steps along <100> (X) direction, the strain rate (2e9/s), within the scope of general adoption and applied in previous work [11-13], was employed in this work. The analysis works were implemented by OVITO software package. Using the OVITO software package, the dislocation extraction algorithm (DXA) developed by Stukowski and Albe [14] was employed to identify the nature of the dislocations, such as the type of dislocations and the dislocation density.

3. Results and Discussion

In order to investigate the effect of temperature on the mechanical properties under uniaxial compression and tension, the curves of stress-strain at different temperatures of 300K, 400K, 500K, 600K, and700K are shown in Fig.1. According to the presented results in Fig.1 (a), all the stresses at various temperatures have a peak (Yield Point) with the compressive strain loading. At T=300K, the compressive strain at Yield Point is about 0.1, and the compressive stress is 6.27GPa. At all considered temperature, the compressive stress increases linearly from the start of the compressive strain loading to the Yield Point and then has a rapidly decline. For uniaxial tension, the similar variation trend is also found. The difference is that the tensile stress and the tensile strain corresponding to the Yield Point are larger than the compressive stress and the compressive strain in magnitude. At the atomistic scale, the Yield strength is defined as the stress corresponding to the Yield Point. Also, the Young's modulus can be obtained via the slope of linear section of the stress-strain curve at low strain stage. As shown in Fig.1, the Young's modulus and Yield strength both decrease with the temperature increasing. The Young's modulus was calculated to be 64.7GPa at 300K, it decreases by 14.5% and drops to 55.3GPa as the temperature rises to 700K. Under compressive strain loading, the Yield strength decreases from 6.27GPa to 4.4GPa as the temperature rises from 300K to 700K. The situation for tensile loading is from 7.7GPa to 4.8GPa. The weakening effect of temperature on the Young's modulus and Yield strength was also found in Cu single crystal and Cu/Ag nanocomposites in previous works.

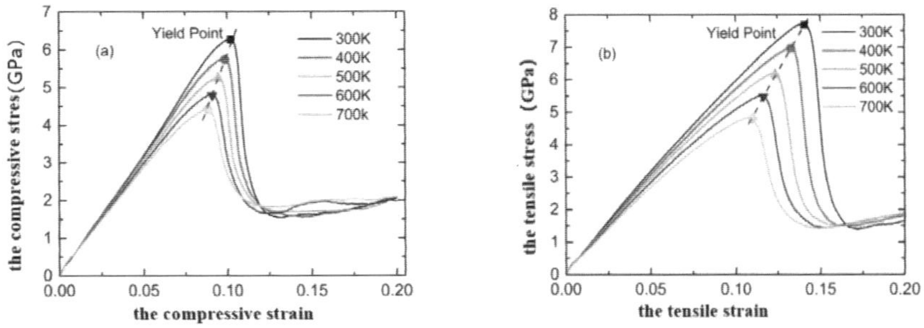

Fig.1 The curves of stress-strain under uniaxial compression and tension at different temperatures of 300K, 400K, 500K, 600K, and 700K.

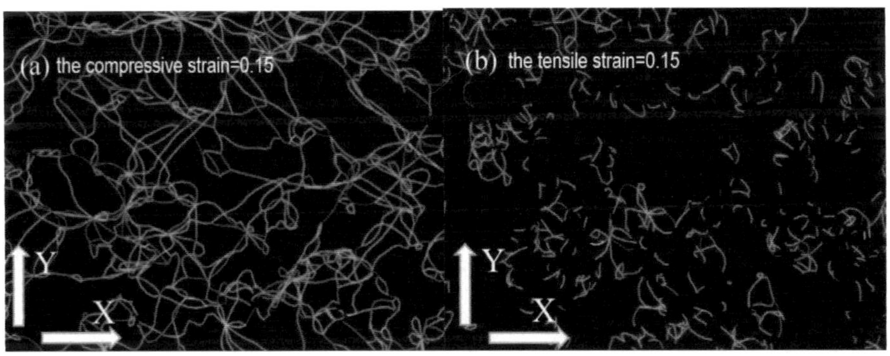

Fig.2 Snapshots obtained from(a) compressive and (b) tensile MD simulations at 300K, where the different types of dislocations lines, including screw and edge, at strain of 0.15 are shown. The Green, blue, purple, yellow, and cyan lines are for 1/6<112> Shockley, 1/2<110> Perfect, 1/6<110> Stair-rod, 1/3<001> Hirth, and 1/3<111> Frank dislocations, respectively. The red lines are for other types of dislocations.

As the strain is loaded to the Al, the emergence of dislocation marks the beginning of the strain hardening region. When the strains are loaded continuously the stress increases linearly and reaches an ultimate strength and then the stress drops down rapidly, as shown in Fig. 1. This leads to the yield of strength in the Al. If the strain is not unloaded at this situation the stress will change to keep a stable level (see Fig.1). In fact, during the deformation there are large of dislocations occurring. For example, at strain of 0.15, the distribution of total dislocations in compressed and stretched aluminum single crystal are shown in Fig.2 (a) and (b), respectively. Under a big strain, the dislocations move, accumulate, and interact into net-like structure, as shown in Fig. 2. The similar case is also reported by Kardani et al. [15]. From Fig. 2 it can be seen intuitively that the 1/6<112> Shockley types of dislocations account for a large proportion, while the 1/2<110> perfect types only have a small part. This may be attribute to a most common dislocation reaction ($1/2[110] \rightarrow 1/6[211] + 1/6[12\bar{1}]$). As well known, Al is a fcc crystalline where a 1/2[110] perfect dislocation often is produced firstly and then the perfect dislocation is easy to dissolved into two partial dislocations of 1/6[211] and 1/6[121] under a deformation. This also explain why 1/6<112> Shockley types of dislocations occupy a large proportion in a deformed Al crystalline. The existing of a large amount of 1/6<112> partial dislocation in Al also indicates that the 1/6<112> Shockley types of dislocations have a most stable structure in fcc aluminum under external force, as compared with the other types of dislocations. In addition, it is worth noting that the aluminum single crystal under tensile strain loading has a smaller total dislocation density than the compressed one at the strain of 0.15. The total dislocation density at different temperatures was analyzed to investigate the temperature effect on plastic deformation. Fig.3

shows the variation of total dislocation density vs compressive/tensile strain at different temperatures. For all the conditions of the temperatures it is found that the total dislocation density increases slowly at a small amount level at its incipient deformation stage, and then suddenly increases at its slightly later deformation stage. We call this point as transition point later, where dislocation density varies from slowly to fast with the strain increasing. Compared Fig. 3 with Fig. 1, we can see that the point of strength yield (shown in Fig. 1) is just corresponding to the transition point of variation of dislocation density vs. Strain. This indicate that a large amount of dislocation occurring incipiently the strength of the material can be yield. When such a strain is not unloaded the dislocation density continuously increases till a maximum value and then it falls down rapidly and converges to a stable value as shown in Fig. 3. From Fig. 3 it can be seen that at the incipient and rapidly taking off stages(strain is smaller than 0.10 and 0.14 for compressive and tensile load, respectively) of dislocation concentration increasing, the dislocation density increases with the temperature increasing, while at the convergence stage (strain is larger than 0.11 and 0.15 for compressive and tensile load, respectively) the dislocation density decreases with the increasing of the temperature. Perhaps, it is due to that a higher temperature can provide a larger thermal activation energy and more vacancies for dislocation emission at the former two stages [15,16]. While at the stable stage of the variation of dislocation density vs. strain, the higher temperature, in turn, perhaps is benefit to weaken the resistance of dislocation slipping and make it easier for the annihilation of dislocations [17]. In this situation the dislocation density reduces with the increasing of temperature, as shown at the final stage of variation of dislocation density vs strain in Fig.3. By comparing Fig 1 with Fig 3, it can be seen that the strains corresponding to Yield Point in Fig.1 is identical to the ones where the variation of dislocation density vs. strain takes off incipiently and rapidly in Fig.3. This indicates that only the dislocations accumulate to an enough amount, the yield of a stress in a material takes place.

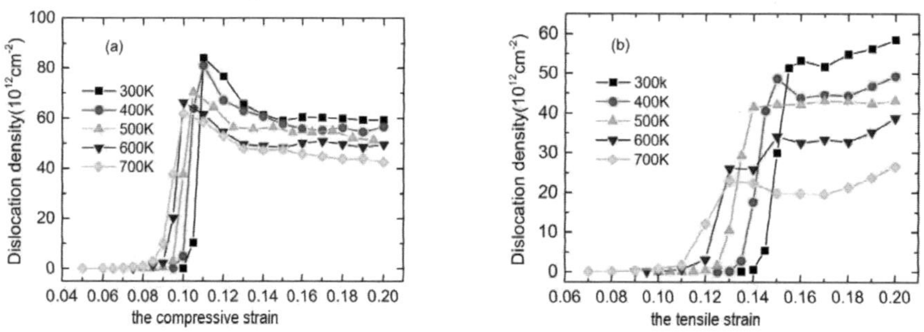

Fig.3 The variation of total dislocation density at different temperatures with (a) the compressive and (b) the tensile strain.

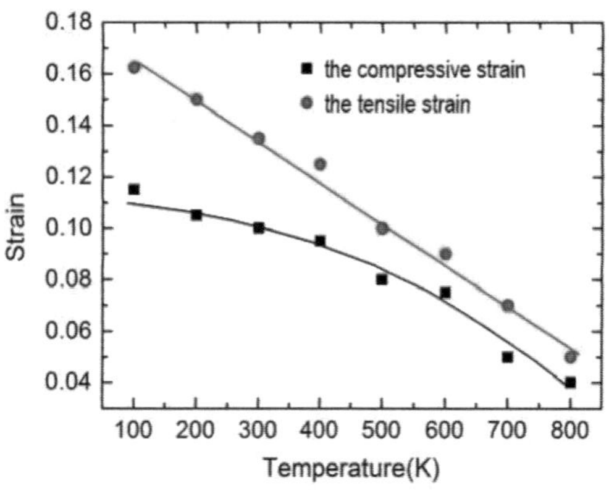

Fig.4 The variation of the strain at transition point with the temperature during the compressive/tensile strain loading.

As we know, the large amount of dislocations to form incipiently is a sign of the beginning of strain strengthening. As presented in Fig.3, the strain for the dislocation forming incipiently at large amount is seriously affected by temperature during the compressive and tensile loading. In order to describe this variation trend more clearly and intuitively, the strains, where the dislocations form incipiently at large amount, at the temperatures from 100K to 800K are calculated and shown in Fig.4. It is seen that at the point (we call it transition point, where an amount of dislocations form incipiently) the compressive strain is smaller than the tensile strain at a certain temperature. Due to the higher temperature provides larger thermal activation energy and more vacancies, at the transition point the compressive strain and tensile strain both decrease with the increasing temperature. Moreover, the tensile strain decreases linearly with the temperature increasing while the compressive strain decreases nonlinearly with the temperature increasing.

Conclusion

Molecular dynamics simulations are performed to study the temperature-based effect on plastic deformation mechanism in aluminum single crystal. The Yield strength and Young's modulus are assessed in the temperature range of 100K-800K. It is also found that they both decrease with the increasing temperature. Based on the analysis of dislocations at different temperatures, it is also found that temperature can promotes the dislocation emission and reduce the strain at the transition point. The compressive strain at transition point has a nonlinear decline, while the tensile strain decreases linearly with the temperature increasing. In turn, at a larger strain loading a high temperature goes against accumulation of dislocations. Moreover, the proportion of 1/6<112> (Shockley) type of dislocations to total dislocations is found to increase as the temperature increasing.

Acknowledgments
This work is supported by Central University Fund of China (No. 2018ZD10).

Reference

[1]L.P. Dávila, P. Erhart, E.M. Bringa, M.A. Meyers, V.A. Lubarda, M.S. Schneider, R. Becker,M. Kumar Appl. Phys. Lett. 86 (2005) 161902.
[2]Hirth, J.P., Lothe, J., 1982. Theory of dislocations, Wiley & Sons, New York.
[3] Krasnikov, V.S., Kuksin, A.Yu., 2010. Solid State 52 (7), 1386–1396.
[4] A. Kardani, A. Montazeri, Computational Materials Science 144 (2018) 223–231.
[5] J.R. Rice, Dislocation nucleation from a crack tip: An analysis based on the Peierls concept, J. Mech. Phys. Solids 40 (1992) 239–271.
[6] J.A. Zimmerman, H. Gao, F.F. Abraham, Sci. Eng. 8(2000)103.
[7] Y. Cui, Z.T Chen, Molecular dynamics simulation of the influence of elliptical void inter action on the tensile behavior of aluminum Comput. Mater.Sci.108 (2015) 103.
[8] E.T. Seppälä, J. Belak, R.E. Rudd, Phys. Rev. B 71 (2005) 064112.
[9] Lu Wang, Jian qiu Zhou, David Hui, Shu Zhang, International Journal of Mechanical Sciences 79 (2014) 168–175.
[10] Y. Mishin, D. Farkas, M.J. Mehl, and D.A. Papaconstantopoulos, Phys. Rev. B 59, 3393 (1999).
[11] J. Marian, J. Knap, M. Ortiz, Nanovoid deformation in aluminum under simple shear, Acta Mater. 53 (2005) 2893–2900.
[12] Y. Cui, Z.T Chen, Mater.Sci.108 (2015) 103.
[13] A. B. Swantek, J. M. Austi, J. Fluid Mech. 649(2010)399.
[14] S. Plimpton, Fast parallel algorithms for short-range molecular dynamics, Computational Physics 117 (1995) 1–19.
[15] S. Melchionna, G. Ciccotti, B. Lee Holian, Hoover NPT dynamics for systems varying in shape and size, Mol. Phys. 78 (1993) 533–544.
[16] A. Stukowski, K. Albe, Simul.Mater. Sci. Eng. 18 (2010) 085001.
[17] A. Kardani, A. Montazeri, TComputational Materials Science 144 (2018) 223–231.

IOP Conf. Series: Materials Science and Engineering **784** (2020) 012013 doi:10.1088/1757-899X/784/1/012013

Stress intensity factor of circumferential periodic cracks on tube with stiffened plates under tension

W J Yuan[1, 2, a], **Y J Xie**[2, b*]

[1] College of Pipeline and Civil Engineering, China University of Petroleum (East China), Qingdao, Shandong 266555, PR China;
[2] School of Mechanical Engineering, Liaoning Shihua University, Fushun, Liaoning 113001, PR China
[b] yjxie@lnpu.edu.cn

Abstract. Fracture is one of the most common failure of engineering structure and the stress intensity factor is the key parameter used to decide whether the structure will crack or not. This essay is based on the concept of conservation law and the tension theory in mechanics of material. It analyzes the stress intensity factor of circumferential periodic cracks on tube with stiffened plates, it obtains the specific expression of stress intensity factor of circumferential periodic cracks on tube with stiffened plates under tension. Finite element method is used for checking the stress intensity factor of cracks and proving the accuracy of expression's results. The method which proposed in this essay is easy to calculate and can get the closed solution.

1. Introduction

For the crack problems of infinite two-dimensional and three-dimensional elastic bodies, a closed analytical solution of the stress intensity factor (SIF) can be given relatively conveniently, when considering the object boundary influence on the crack, which is limited in 2D and 3D cracks, such as unilateral crack, tube with stiffened plates, to give the analytical solution which can satisfying the boundary conditions is very difficult, this must depend on numerical methods. In recent years, the J_2 -integral theory was proposed by Xie et al[1], which is derived from the three-dimensional conservation law and the principle of virtual work. Because of the finite and infinite boundary crack problem, the J_2-integral method greatly facilitates the analysis of stress intensity factors[2-7].

The tube with stiffened plates is the components which commonly used in engineering, while circumferential periodic cracks on tube with stiffened plates and circular cracks on tube are typical belong to complex three-dimensional crack structures. In this paper, A method of solving the SIFs of the circumferential periodic cracks on tube with stiffened plates and circular cracks on tube by the conservation law and the tension theory in material mechanics is established.

2. Configuration of circumferential periodic cracks on tube with stiffened plates

There has $2m$ stiffened plates which full of the periodic cracks in a circular tube. Figure 1 gives the case of m=2. For the thin-walled tube, it has the characteristics of the three-dimensional shell and the slender beam, and the crack on the stiffened plate has the characteristics of the two-dimensional stress field. So the stress intensity factor can be obtained by the conservation law and the tension theory in the material mechanics.

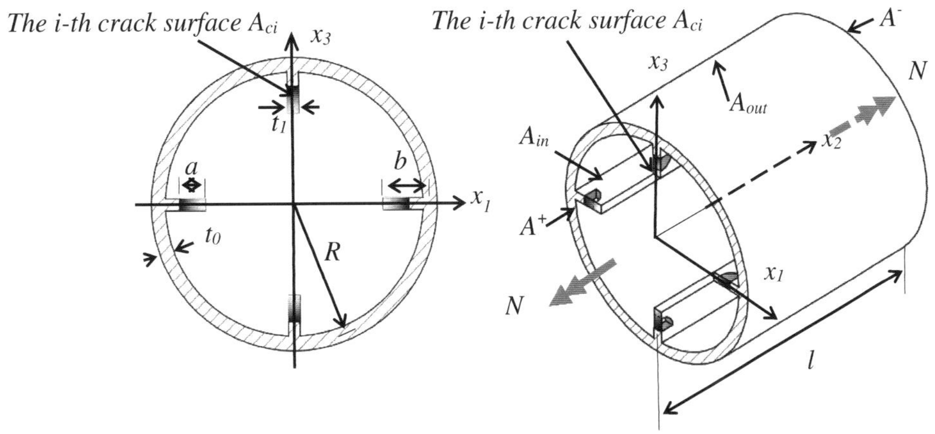

Figure 1. Periodic cracks in eight fins of fin-shaped shells under tension.

3. J_2-Integral and stress intensity factors

Considering a three-dimensional strain field, for which the displacement vector u_i depends on x_1, x_2, x_3, the conservation law J_j-integral can be defined as [7]:

$$J_j = \oiint_A \left(wn_j - T_i u_{i,j}\right) dA = 0,\ j = 1, 2, 3. \tag{1}$$

The A in equation (1) is an arbitrary closed surface, in the solids closed by which there are no defects; w is the strain energy density; T_i is the stress vector acting on the outer side of A; n is the unit outward normal to A. For a two-dimensional deformation field. Figure 2 gives a two-dimensional crack model of unit thickness (the integral path is a curve in the x_1-x_2 plane). Let S represent to a closed contour S_{abca} within the K-dominant region around upper crack tip. Note that S_{ab} is a straight line and S_{bc} is a quarter of circle. As such, the following results can be obtained by calculating the stress and displacement next to crack tip

$$J_2 = \int_{S_{abc}} \left(wn_2 - T_i u_{i,2}\right) ds = \frac{\left(1-\mu^2\right) K_I^2}{2\pi E}. \tag{2}$$

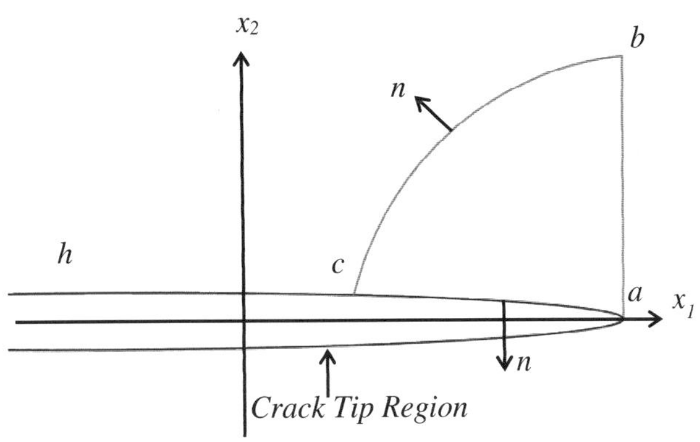

Figure 2. Integration path around crack tip region.

For the closed path $S_{abca} = S_{abc} + S_{ac}$, from the conservation law [6]. It follows that

$$\int_{S_{abca}} \left(wn_2 - T_i u_{i,2}\right) ds = \int_{S_{abc}} \left(wn_2 - T_i u_{i,2}\right) ds + \int_{S_{ac}} \left(wn_2 - T_i u_{i,2}\right) ds = \frac{\left(1-\mu^2\right) K_I^2}{2\pi E} + \int_{S_{ac}} \left(wn_2 - T_i u_{i,2}\right) ds = 0, \tag{3}$$

then

$$J_2 = \int_{S_{ac}} \left(wn_2 - T_i u_{i,2}\right) ds = -\frac{\left(1-\mu^2\right)K_I^2}{2\pi E}, \tag{4}$$

where μ represents the Poisson's ratio and E the elastic modulus.

4. Circumferential periodic cracks on tube with inner stiffened plates under tension

4.1. A tube full of periodic cracks on the stiffened plates

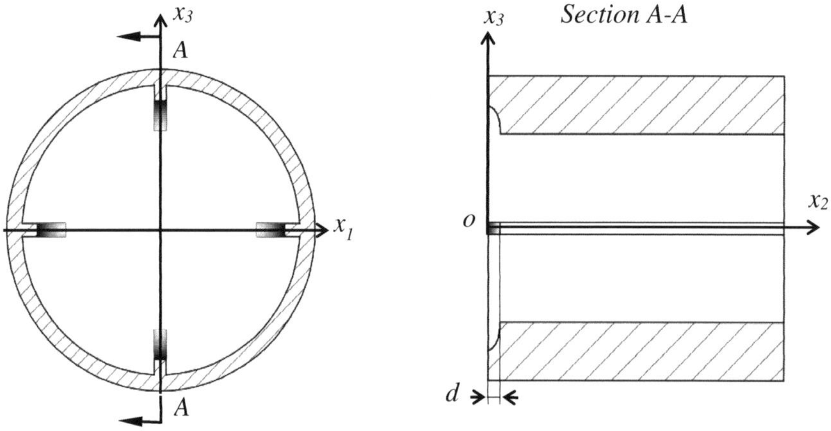

Figure 3. Elliptical modelling for cracks with d→0.

The Figure 1 show a tube full of periodic cracks on the stiffened plates, b is the height of the stiffened plate, t is the thickness of the stiffened, a is the length of the periodic crack, $2R$ is the intermediate diameter of tube, t_0 is the thickness of tube. N is the axial force. The Figure 1 also show a closed surface A_{closed},

$$A_{closed} = A^+ + A^- + A_{in} + A_{out} + A_c \tag{5}$$

A^+ is the cracked cross-section; A^- is the remote uncracked cross-section; A_{in} is the inner surface of shell; A_{out} is the outer surface of shell; A_c is the sum of the eight crack surfaces. On the free surface A_{in} and A_{out}, $T_i = 0$ and $n_2 = 0$. Then the following results can be found

$$\iint_{A_{in}} \left(wn_2 - T_i u_{i,2}\right) dA = 0, \tag{6}$$

$$\iint_{A_{out}} \left(wn_2 - T_i u_{i,2}\right) dA = 0. \tag{7}$$

As the fin-shaped shells possess beam characteristics, stresses and displacements in the J_2-integral can be given by elementary mechanics. Then for the surfaces A^+ and A^-, it is not difficult to get:

$$\iint_{A^-} \left(wn_2 - T_i u_{i,2}\right) dA = -\frac{N}{2}\tilde{\mu}_{2.2}^-, \tag{8}$$

$$\iint_{A^+} \left(wn_2 - T_i u_{i,2}\right) dA = \frac{N}{2}\tilde{\mu}_{2.2}^+, \tag{9}$$

$\tilde{u}_{2.2}^-$ is the axial strain of fin-shaped shells, $\overline{w}$ is the strain energy density per unit length. All the quantities in equations (8) and (9) can be calculated by the beam theory. The axial strain of remote uncracked cross-section is

$$\tilde{\mu}_{2.2}^- = \frac{N}{EA}, \tag{10}$$

where the moment of inertia $A= 2\pi Rt_0+2mbt$. If the crack is formed from an ellipse with $d\rightarrow 0$ as shown in Figure 3, the fin-shaped shell can be regarded as a variable cross-section beam, and the average axial strain at the cross section becomes

$$\tilde{\mu}^{+}_{2.2}=\lim_{d\to 0}\frac{1}{d}\int_0^d\frac{N}{EA}\frac{dx_2}{1-\gamma_1\left(\frac{a}{b}\right)\left(1-\left(x_2/d\right)^2\right)^{1/2}}=\frac{N}{EA}\left[\frac{2}{\gamma_1\left(\frac{a}{b}\right)\left(1-\gamma_1\left(\frac{a}{b}\right)^2\right)^{1/2}}\arctan\left(\frac{1+\gamma_1\left(\frac{a}{b}\right)}{1-\gamma_1\left(\frac{a}{b}\right)}\right)^{1/2}-\frac{\pi}{2\gamma_1\left(\frac{a}{b}\right)}\right],\quad(11)$$

where

$$\gamma_1(a/b)=\frac{2mta}{2mRt_0+2mtb}=\frac{\left(\frac{a}{b}\right)}{1+\left(\frac{\pi}{m}\right)\left(\frac{R}{b}\right)\left(\frac{t_0}{t}\right)}.\quad(12)$$

Then, from equations (3)-(12), the J_2-integral over the closed surface can be expressed as

$$\frac{mt\left(1-\mu^2\right)K_I^2}{\pi E}+\iint_{A_{ch}}wdA=\frac{N}{2}\left(\tilde{\mu}^{+}_{2.2}-\tilde{\mu}^{-}_{2.2}\right).\quad(13)$$

4.2. Stress intensity factor

Substituting the equation (10) and equation (11) into equation (13), the J_2-integral over the closed surface can be expressed as

$$\frac{mt\left(1-\mu^2\right)K_I^2}{\pi E}+2m\iint_{A_{ch}}wdA=\frac{N^2}{2EA}\left[\frac{2}{\gamma_1\left(\frac{a}{b}\right)\left(1-\gamma_1\left(\frac{a}{b}\right)^2\right)^{1/2}}\arctan\left(\frac{1+\gamma_1\left(\frac{a}{b}\right)}{1-\gamma_1\left(\frac{a}{b}\right)}\right)^{1/2}-\frac{\pi}{2\gamma_1\left(\frac{a}{b}\right)}-1\right].\quad(14)$$

Because of the free action of the crack surface with A_{ch} just out of the K-dominant region, the integral in the left hand side of equation (14) is a small quantity and can be neglected. Hence, the normalized stress intensity factors can be found as

$$f_1(a/b)=\frac{K_I}{\sigma_0\sqrt{\pi b}}=\left[\frac{\left(\frac{a}{b}\right)}{\gamma_1\left(\frac{a}{b}\right)\left(1-\mu^2\right)}\left(\frac{2}{\gamma_1\left(\frac{a}{b}\right)\left(1-\gamma_1\left(\frac{a}{b}\right)^2\right)^{1/2}}\arctan\left(\frac{1+\gamma_1\left(\frac{a}{b}\right)}{1-\gamma_1\left(\frac{a}{b}\right)}\right)^{1/2}-\frac{\pi}{2\gamma_1\left(\frac{a}{b}\right)}-1\right)\right]^{1/2},\quad(15)$$

where $\sigma_0=N/A$.

4.3. Comparison between numerical examples and results

This section will give a discussion on the comparison between the normalized stress intensity factor expressed by equation (15) and the results from the finite element method (EFM). Computations were used XFEM in the commercial software Abaqus [8-9], in which the C3D20 type element was employed. The elastic modulus is $E=200GPa$ and Poisson's ratio is $\mu=0.3$. The length of tube with stiffened plates is $500mm$, $m=2$, $2R=200mm$, $t=10mm$, $t_0=3mm$, $b=60mm$, Figure 5 is show the finite element model and meshes and Figure 6 is used to compare the results of equation (15) with the finite element analysis.

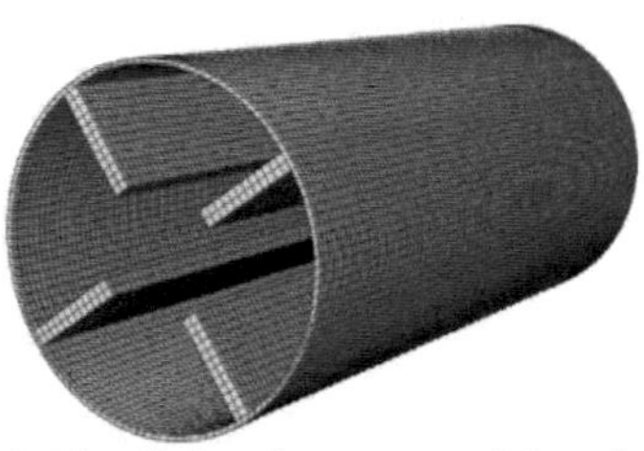

Figure 4. The finite element model and meshes.

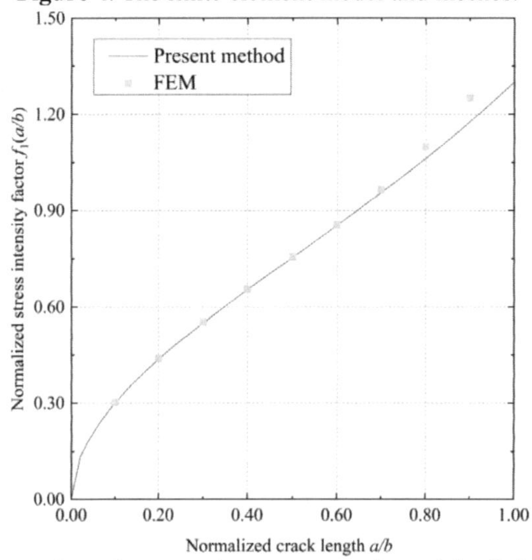

Figure 5. The comparisons between the present results and the finite element results (R/b=1.67, π/m=1.57, t_0/t=0.3).

5. Circular crack on inner surface of tube under tension

5.1. Circular crack on inner surface of tube

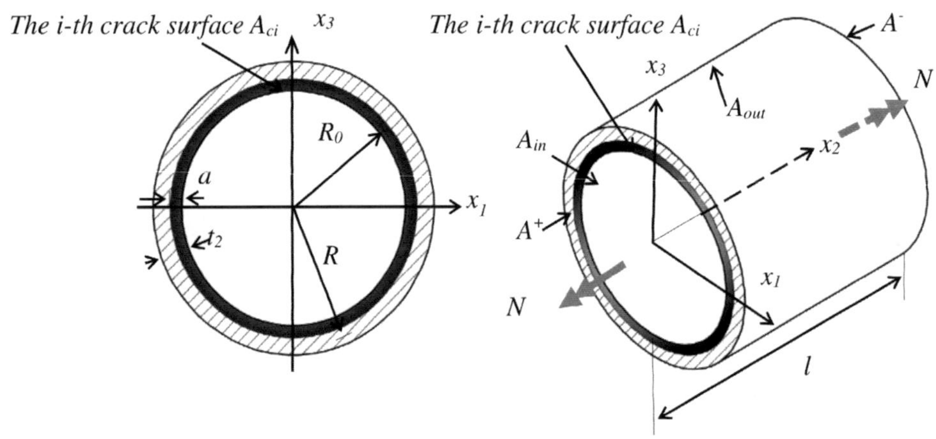

Figure 6. Circular crack on inner surface of tube under tension.

When the number of stiffened plates in the tube tends to be infinite ($m\to\infty$), the stiffened plates cover the inner surface of the tube, and the cracks on each stiffened plate will become circular cracks on inner surface of tube. As shown in Figure 6, $2R_0$ is the intermediate diameter of tube, t_2 is the thickness of tube. $2\pi R_0 a$ is the area of the cracks on the tube and $2\pi R t_2$ is the area of remote uncracked cross-section. Hence,

$$\gamma_1(a/b)=\left(\frac{R_0}{R}\right)\left(\frac{a}{t_2}\right). \qquad (16)$$

The normalized stress intensity factor is the same as equation (15).

5.2. Compare results

Tada's[8] solution to the stress intensity factor of the circular crack on inner surface of tube under tension is:

$$K_I=\sigma_0\sqrt{\pi a}\,f\left(\frac{r_i}{r_0},\frac{a}{b}\right) \qquad (17)$$

To facilitate the comparison between the solution of this paper and the Tada's, equation (17) can changed to

$$K_I=\sigma_0\sqrt{\pi t_2}\,f_1\left(\frac{r_i}{r_0},\frac{a}{t_2}\right) \qquad (18)$$

Where

$$f_1\left(\frac{r_i}{r_0},\frac{a}{t_2}\right)=\sqrt{\frac{a}{t_2}}\,f\left(\frac{r_i}{r_0},\frac{a}{b}\right) \qquad (19)$$

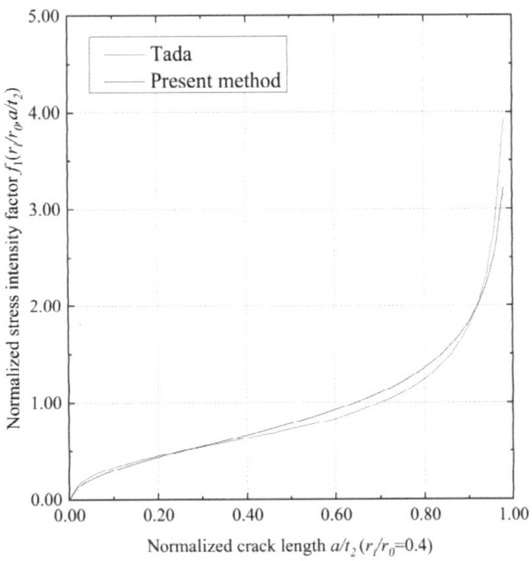

Figure 7. The Comparisons between the present results and Tada's.

When the number of stiffened plates in the tube tends to be infinite (m→∞), the comparisons between the present results and Tada's are shown in Figure 7, which shows that the stress intensity factor solution obtained in this paper agrees well with the existing results.

6. Conclusions

Circumferential periodic cracks on tube with stiffened plates and circular cracks on surface of tube is a typical three-dimensional problem, in this paper, the conservation law and tension theory of material mechanics are used to study the SIFs of Circumferential periodic cracks on tube with stiffened plates, then, used XFEM in Abaqus checking the expression's results. Based on the expression of the stress intensity factor of Circumferential periodic cracks on tube with stiffened plates, it's easy to deriving the expression of stress intensity factor of circular cracks on surface of tube and the results agree well with the existing results.

7. References:

[1] Xie, Y. J. A theory on cracked pipe. *Int J Pres Ves Pip* 75.12(1998):865-869

[2] Xie, Y. J., P. N. Li, and H. Xu. On KI, estimates of cracked pipes using an elliptical hole model and elementary beam strength theory of cracked beams. *Eng Fract Mech* 59.3(1998):399–402.

[3] Xie, Y. J., et al. Micro - indentation fracture from flat - ended cylindrical indenter. FATIGUE FRACT ENG M 35.1(2012):45–55.

[4] Xie, Y. J., et al. A theoretical note on mode-I crack branching and kinking. *Eng Fract Mech* 78.6(2011):919-929.

[5] Xie, Y. J., X. Z. Hu, and X. H. Wang. Frictional contact induced crack initiation in incompressible substrate. *Eng Fract Mech* 78.17(2011):2947-2956.

[6] Xie, Y. J., et al. Applications of conservation integral to indentation with a rigid punch. *Eng Fract Mech* 76.7(2009):949-957.

[7] Xie, Y. J. An analytical method on circumferential periodic cracked pipes and shells. *Int J Solids Struct* 37.37(2000):5189-5201.

[8] Tada H. The stress analysis of cracks handbook, 1973[M]. ASME Press.

[9] Sukumar, N, et al.Extended finite element method for three - dimensional crack modelling. Int J Numer Meth Eng 48.11(2015):1549-1570.

[10] Zeng pan. Finite Element Analysis and ApplicatIon. 2004[M]. TSINGHUA UNIVERSITY PRESS.

Online fatigue damage evaluation method based on real-time cycle counting under multiaxial variable amplitude loading

Long Xue[a], De Guang Shang[b], Luo Jin Li[c]

College of Mechanical Engineering and Applied Electronics Technology, Beijing University
of Technology, Beijing, 100124, China

[a]xuelongbjut@163.com, [b]shangdg@bjut.edu.cn, [c]liluojin788@163.com

Keywords: Multiaxial fatigue; Damage evaluation; Real-time cycle counting; variable amplitude loading;

Abstract. To realize on-line multiaxial fatigue damage evaluation for the servicing mechanical components, a real-time multiaxial cycle counting method is proposed, meanwhile, an online multiaxial fatigue damage evaluation method based strain-critical plane model is given and validated by the fatigue experimental data of En15r steel. Investigation showed that the proposed on-line multiaxial fatigue damage evaluation method could provide satisfactory evaluation results, and the real-time performance of the proposed cycle counting method is path-dependent.

1. Introduction

Most of the engineering components in service are subjected to multiaxial cycle loading, and multiaxial fatigue damage evaluation, especially for online fatigue damage evaluation under multiaxial variable amplitude loading, still is a challenging problem despite numerous of investigations have been conducted in the past several decades[1-4].

In order to realize on-line multiaxial fatigue damage evaluation, a real-time multiaxial cycle counting method is necessary. For multiaxial cycle counting, some methods have been proposed by Bannantine and Socie, Wang and Brown, Meggiolaro, Wei, Anes and so on[5-9]. However, all of this method cannot used in real-time. Chen and Shang[10] provided an online multiaxial fatigue damage evaluation algorithm based on a new real-time multiaxial cycle counting method. Nevertheless, for most of the multiaxial random loading paths, the evaluate result is quite poor.

The object of this paper is to propose a real-time multiaxial cycle counting method and give a satisfactory and acceptable online multiaxial fatigue damage evaluation method. A set of fatigue experimental data for En15r steel under multiaxial variable amplitude loading is employ to verify the proposed method.

2. The proposed method

2.1 The proposed real-time cycle counting method

To realize on-line multiaxial fatigue damage evaluation, a real-time cycle counting method is proposed for multiaxial variable amplitude loading. The proposed method can be descripted as follows:

(1) Receive data in real-time by the data acquisition system with given frequency;

(2) With respect to the first data point, calculate relative equivalent strain of each received data point in time by Eq. (1), then, a relative equivalent strain time-history for all the data points having been received is obtained in real-time;

$$\varepsilon_{eq,i}^{R} = \sqrt{(\varepsilon_i - \varepsilon_i^r)^2 + \frac{1}{3}(\gamma_i - \gamma_i^r)^2} \tag{1}$$

where i $(i = 1, 2, 3...n)$ represents the i-th point of the time history;

(3) Seek out the first peak point $\varepsilon_{eq,k}^{R*}$ $(k = 1)$ of the relative equivalent strain time-history using the following equation and memory its value timely, meanwhile, set corresponding data point of the multiaxial loading history as the reference point T_{ref} to determine the time-history interval will be intercepted;

$$\varepsilon_{eq,k}^{R*} = \varepsilon_{eq,i}^{R} = \begin{cases} \varepsilon_{eq,i}^{R} > \varepsilon_{eq,i+1}^{R} \\ \varepsilon_{eq,i}^{R} > \varepsilon_{eq,i\text{-}1}^{R} \end{cases} \tag{2}$$

where $\varepsilon_{eq,k}^{R*}$ is the k-th $(k = 1, 2, 3...n)$ peak value.

(4) Continue to receive data to repeat step (2), and determine and memory the peak point by Eq. (2) in time;

(5) Once another peak point $\varepsilon_{eq,k+n}^{R*}(n = 1, 2, 3...)$ of the relative equivalent strain time-history is checked, and the condition that $\varepsilon_{eq,k+n}^{R*} \geq \varepsilon_{eq,k}^{R*}$ is met, then, a time-history interval, [T1,T1'] as shown in Fig. 1 can be gained in rain-flow counting method. Otherwise, repeat step (4)-(5);

(6) For the time-history interval intercepted: a) Take the first point of intercepted interval as the reference point $(\varepsilon_i^r, \gamma_i^r)$; b) Calculate the relative equivalent strain of each points by the Eq.(1); c) Seek out every peak value of the relative equivalent strain history, then, a counted reversal can be obtained by combining the sequence of data points where $\varepsilon_{eq,i}^{R}$ is increasing and reaches the maximum peak value. Concurrently, one or more fragmentations of the loading strain history may occur whenever $\varepsilon_{eq,k}^{R*}$ starts to decrease; d) Reset the reference point $(\varepsilon_i^r, \gamma_i^r)$ as the first point of each fragmented data block; e) Repeat step (b) to (d) for each block, then, all the reversals can be counted within the intercepted interval.

(7) Reset the reference point T_{ref} as the corresponding data point of the multiaxial loading history where maximum peak value max($\varepsilon_{eq,k+n}^{R*}$) is located presently;

(8) Repeat step (4)-(7) until the last received data point is processed. Subsequently, the largest reversal can be counted by combining the rest sequence of data points where $\varepsilon_{eq,i}^{R}$ is increasing and reaches the maximum peak value $\varepsilon_{T_{max}}^{r}$ $(\max(\varepsilon_{eq,K+n}^{R*}))$ for the multiaxial loading history;

(9) The fragmented loading time-history interval, which is intercepted from the data point where the maximum peak value $\varepsilon_{T_{max}}^{r}$ begins to decrease to the end of the multiaxial loading history, is also treated as step (6).

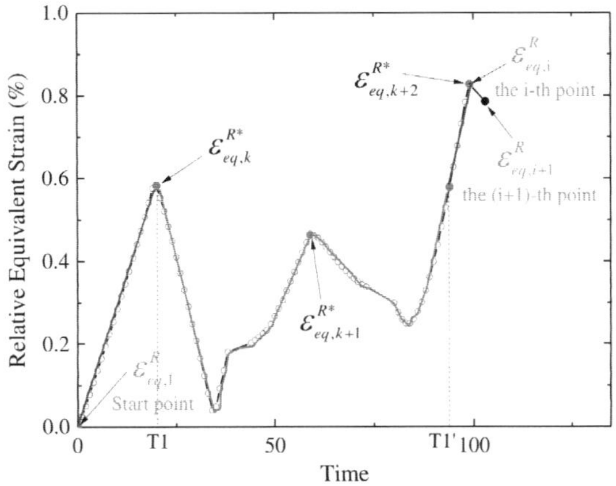

Fig. 1 Illustration of the intercepted time-history interval in real-time.

2.2 Multiaxial fatigue damage calculating method

As illustration in Fig. 2, the stress and strain of thin-walled tube specimen on the plane angled θ with the axial direction under axial-torsion loading can be expressed as

$$\varepsilon_\theta = \frac{1-\nu}{2}\varepsilon_x + \frac{1+\nu}{2}\varepsilon_x \cos(2\theta) + \frac{\gamma_{xy}}{2}\sin(2\theta)$$

(3)

$$\gamma_\theta = (1+\nu)\varepsilon_x \sin(2\theta) - \gamma_{xy}\cos(2\theta) \quad (4)$$

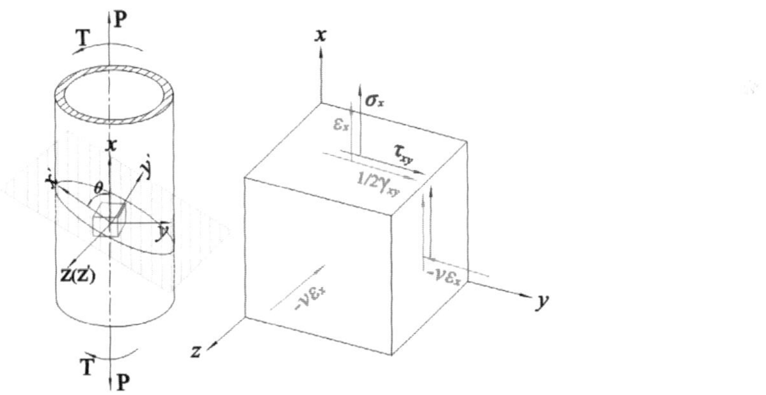

Fig. 2 Stress-strain states analysis for thin-walled tube specimen under axial-torsional loading

The Shang-Wang's[11] fatigue damage model given as Eqs. (6) is employed in the proposed online damage evaluation method, parameters are defined on the critical plane which is regarded as the plane experiencing the maximum shear strain rang. Calculation of fatigue damage for each counted reversal are processed as the following procedure in real-time:

(1) Once a reversal is counted, calculate the shear strain for each data point of an angle from 0 to 90 degrees by Eqs. (4), and the step length is considered as 1 degree;

(2) Obtain the shear strain range of each angle, meanwhile, seek out the maximum shear strain range $\Delta\gamma_{\max}$ and the angle θ and ($90^\circ + \theta$) corresponding to the maximum shear strain range plane, respectively;

(3) The normal strain amplitude ranges, $\Delta\varepsilon_\theta$ and $\Delta\varepsilon_{90^\circ+\theta}$ corresponding to θ and ($90^\circ + \theta$) plane, are computed, respectively. And then, determining the orientation of the critical plane as θ_{cr}

where the larger normal strain range is located;

(4) The two adjacent turning points of the maximum shear strain rang, $\Delta\gamma_{max}$ on the critical plane are found and flagged as t_{min} and t_{max} , respectively. By Eqs. (3), the normal strain excursion ε_n^* between two turning points on the critical plane, θ_{cr} can be given as

$$\varepsilon_n^* = Max(\varepsilon_{\theta_{cr}}(t))\Big|_{t_{min}}^{t_{max}} - Min(\varepsilon_{\theta_{cr}}(t))\Big|_{t_{min}}^{t_{max}} \quad (5)$$

(5) Calculate the fatigue damage of this counted reversal by Eqs. (6) and (7), concurrently, update the value of total fatigue damage in real-time by Eqs. (8)

$$\sqrt{\varepsilon_n^{*2} + \frac{1}{3}\left(\frac{\Delta\gamma_{max}}{2}\right)^2} = \frac{\sigma_f'}{E}(2N_f)^b + \varepsilon_f'(2N_f)^c \quad (6)$$

$$D_j = \frac{1}{2N_{f,j}} \quad (7)$$

$$D_{total} = D_{total} + D_j \quad (8)$$

3. Experimental validation and discussion

Fatigue experimental data for En15R steel under multiaxial variable amplitude loadings were employed to validate the capability of the proposed method. The fatigue tests were conducted by M.W. Brown et al and were reported in Ref.[12], altogether eight multiaxial variable amplitude loading paths for twenty tubular specimens controlled by strain history in blocks were used to validate the Wang-Brown method and the proposed method. As shown in Fig. 3, satisfactory predictions can be obtained for both the two method with an error factor of 2.

Analysization shows that the proposed method can be successfully applied in multiaxial variable amplitude loading in real-time. For the application of the proposed method, the real-time performance of the proposed method is path-dependent. As illustrated in Fig.4(a), the multiaxial variable amplitude loading path may have the highest real-time performance because the location of the maximum relative equivalent strain with respect to the beginning point is same to the last peak value of the relative equivalent strain history. Contrariwise, as shown in Fig.4(b), when the maximum relative equivalent strain happens in the location where the first peak value of the relative equivalent strain history is located, the real-time performance of the proposed method is poorest.

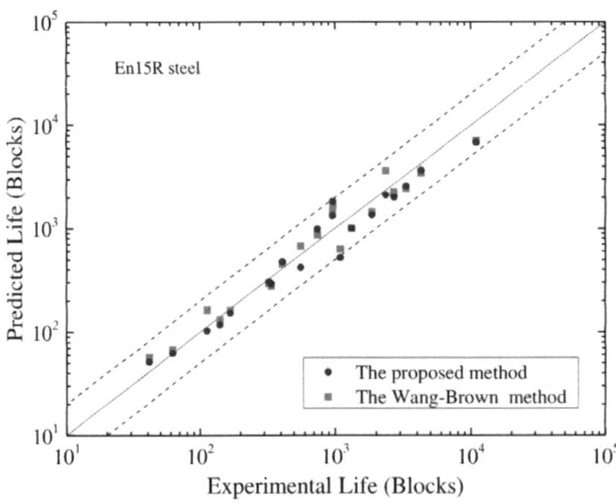

Fig. 3 Comparison between experimental and predicted lives for the proposed method and the WB method.

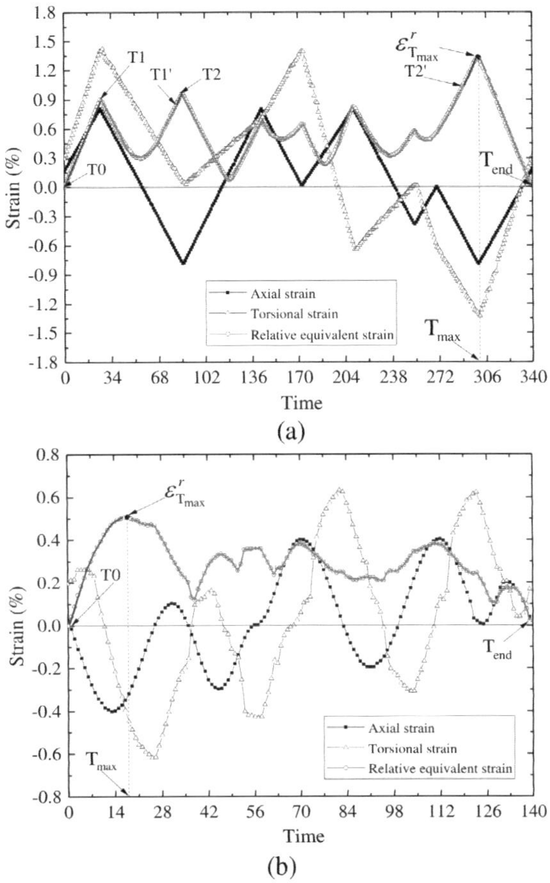

Fig.4 Real-time performance analysis illustrations of the proposed cycle counting method

4. Conclusions

The real-time performance of the proposed cycle counting method is path-dependent, and the proposed on-line multiaxial fatigue damage evaluation method seems to be a more convenient and better choice for the development of the on-line multiaxial fatigue damage monitoring system.

5. Acknowledgements

Grateful acknowledgement for the financial support of the National Natural Science Foundation of China (NO.51535001).

6. References

[1] Carpinteri A, Spagnoli A, Vantadori S. A review of multiaxial fatigue criteria for random variable amplitude loads. *Fatigue Fract. Eng. Mater. Struct.* 2017;40(7):1007-1036.

[2] Karolczuk A, Macha E. A review of critical plane orientations in multiaxial fatigue failure criteria of metallic materials. *Int. J. Fract.* 2005;134(3):267-304.

[3] Fatemi A, Shamsaei N. Multiaxial fatigue: An overview and some approximation models for life estimation. *Int. J. Fatigue.* 2011;33(8)948-958.

[4] Ince A, Glinka G. A generalized fatigue damage parameter for multiaxial fatigue life prediction under proportional and non-proportional loadings. *Int. J. Fatigue.* 2014;62:34-41.

[5] Bannantine JA. A variable amplitude multiaxial fatigue life prediction method. Ph.D. thesis, University of Illinois at Urbana-Champaign, USA, 1989.

[6] Wang CH, Brown MW. Life prediction techniques for variable amplitude multiaxial fatigue-part I: Theories. *J. Eng. Mater. Technol.-Trans. ASME* 1996;118(3):367-370.

[7] Meggiolaro MA, de Castro, JTP. An improved multiaxial rain-flow algorithm for non-proportional stress or strain histories-Part II: The Modified Wang-Brown method. *Int. J. Fatigue*. 2012;42:194-206.

[8] Wei, ZG, Dong PS. A rapid path-length searching procedure for multi-axial fatigue cycle counting. *Fatigue Fract. Eng. Mater. Struct.* 2012;35(6):556-571.

[9] Anes V, Reis L, Li B, et al. New cycle counting method for multiaxial fatigue. *Int. J. Fatigue*. 2014;67:78-94.

[10] Chen H, Shang DG. An on-line algorithm of fatigue damage evaluation under multiaxial random loading. *Int. J. Fatigue*. 2011;33(2):250-254.

[11] Shang DG, Wang DJ. A new multiaxial fatigue damage model based on the critical plane approach. *Int. J. Fatigue*. 1998;20(3):241-245.

[12] Wang CH, Brown MW. Life prediction techniques for variable amplitude multiaxial fatigue-part II: Comparison with experimental results. *J. Eng. Mater. Technol.-Trans. ASME* 1996; 118(3): 371-374.

Life prediction of multi-pit-crack interaction in pressure vessel

Zou Zijie [1,a], **Qin Zhongbao** [1,b], **Liu Qi** [1,c]

[1]Rocket force university of engineering ,China

[a]799449341@qq.com, [b]13572208634@163.com, [c]2095103496@qq.com

Abstract: During the service of pressure vessel, there are many kinds of structural damage. The main defect is pitting caused by corrosion of loaded materials, cracks are formed because of stress concentration in the pit position, and crack propagation leads to crack leakage. As an important parameter to characterize cracks, fatigue life is widely used in engineering to judge the safety and reliability of mechanical structures. In this paper, based on the two-parameter pit-crack model, the fatigue life of pressure vessel with multiple pits-cracks is discussed by using ANSYS finite element analysis software. The results show that the fatigue life of pressure vessels decreases with the existence of multiple pits and cracks, but with the increase of pit-crack spacing, the effect decreases gradually until it returns to a stable state. The effect of pit parameters and the ratio of crack length to depth on fatigue life has obvious interval effect. This paper can be used as a reference for failure assessment of multi-pit-crack pressure vessels.

1. Introduction

Point corrosion pits of different sizes and shapes will be formed on the inner surface of pressure vessel due to the corrosion of gas or solution. These pits will cause local stress concentration in the vessel, and stress concentration is the main cause of crack formation [1-4]. The existence of these cracks will lead to the decrease of local deformation resistance of pressure vessel, even damage and leakage failure, which greatly affects the safety and reliability of the equipment [5-7]. Fatigue life, as an important index of crack evaluation, directly affects the crack propagation mechanism and fracture failure criterion [8-10]. Li Xiubo et al. [11] innovatively put forward a two-parameter pit-corrosion fatigue crack model in the study of crack life prediction of corrosion pits in offshore engineering structures, and found that the location parameters of corrosion pits will determine the location of the initiation cracks. NEWMAN et al. [12] studied the variation of stress intensity factors of cracks in cylindrical pressure vessel. The results show that the pressure vessel is mainly affected by open-mode crack growth under internal pressure. Yu Jianxing et al. [13] Through the analysis of stress intensity factor of pit-crack on pipeline surface, the conclusion that pit can obviously reduce the threshold of crack growth is obtained, and the influence trend of pit parameters on stress intensity factor is different. Yao Anlin et al. [14] used finite element method to explore the influence of crack interaction on stress intensity factor in high pressure gas pipeline, and used the concept of nominal stress intensity factor to demonstrate the relationship between the numerical value and the related parameters of crack.

Because of the existence of pits, the change rule of stress intensity factor and the propagation mechanism of cracks will change in some properties, especially in the case of multiple pits and cracks, how cracks propagate at the same time needs to be studied. Therefore, on the basis of defining the reliability of finite element simulation calculation, this paper establishes a multi-pit-crack model, which is devoted to discussing the influence of pit-crack parameters on nominal stress intensity factor. This provides a theoretical reference for safety assessment and failure identification of pressure vessels with defects in practical engineering applications.

2. Establishment of Finite Element Model

In order to explore the law of interaction between multiple pits and cracks in pressure vessels, a finite element model of pressure vessels with inner diameter of 600 mm and wall thickness of 60 mm was established by SolidWorks according to the actual dimensions of common cylindrical pressure vessels in engineering. The material parameters of the model are shown in Table 1. Because the model is a continuous symmetrical and uniform entity, in order to reduce the computational workload of the finite element method, the pressure vessel is divided into four parts along the central symmetry, and one fourth of them is imported into ANSYS for finite element simulation. In this paper, a two-parameter method is used. (α=d/f , $\beta = b/f$) .A pit model [11] is established on pressure vessel, which treats pits as uniform ellipsoidal pits. The model schematic is shown in Fig. 1.

Tab.1 Material Properties of Pressure Vessels

Material	Modulus of elasticity (MPa)	Poisson's ratio	Yield strength (MPa)	ultimate strength (MPa)
15MnMoVN	200000	0. 3	690	750

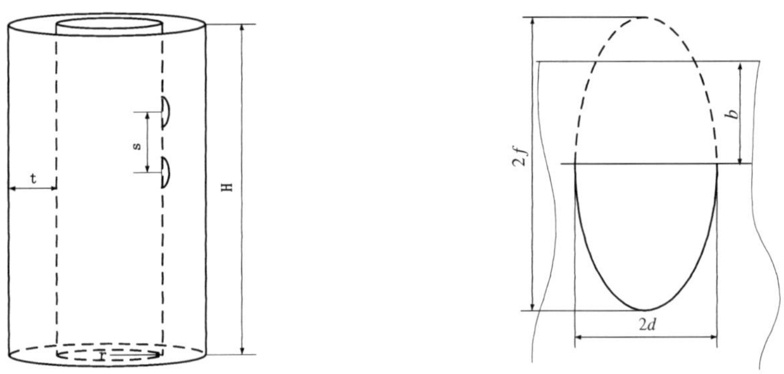

(a) Pressure Vessel Model (b) Erosion pit model

Fig. 1 Pressure vessel and its pit geometry

Among them, the radius of the inner wall of the pressure vessel is r ,the thickness of the wall is t ,the height of the vessel is H ,the depth of the pit is f , the radius of the pit is d ,the distance between the two pit centers is s .The vertical position of ellipsoid center is changed by changing the value of β (b/f) ;and the curvature of pit bottom is changed by changing the value of α (d/ f) ;thus different pit models are obtained.

According to Yu Jianxing's research, when the parameter is less than 0.2, the maximum stress concentration always appears at the bottom of the pit. Because stress concentration is the main cause of local cracking of components, this paper mainly aims at adding different length-depth ratio (a/c) cracks at the bottom of the pit under the condition of $\beta = 0$. The crack schematic is shown in Fig. 2.

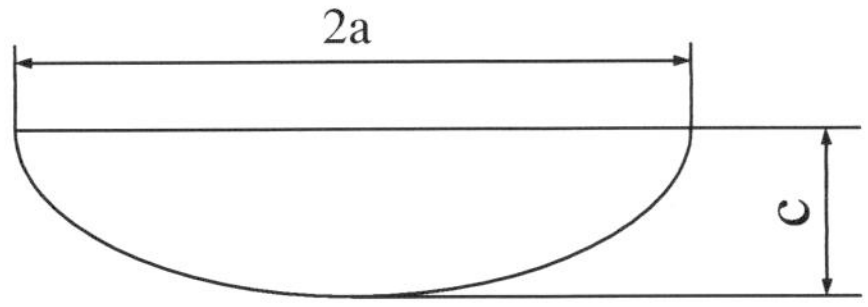

Fig. 2 crack geometry

When meshing the model in ANSYS, because the model is a complete uniform continuum, the model is meshed by the sweeping meshing method and two-node SOLID186 element. At the same time, the CONTACT174 contact element is used to ensure the connection between the crack mesh and the whole mesh. In order to ensure the accuracy of crack stress intensity factor calculation, six integral contours are set up; displacement along the y-axis at the bottom of pressure vessel is restrained, and friction-free constraints are set at the two cutting surfaces of the vessel as boundary conditions for finite element calculation; 10 Mpa pressure is fixed on the inner surface. Its meshing is shown in Figure 3.

(a) Pressure Vessel Mesh (b) Pit-crack Mesh

Fig.3 Pressure Vessel and Pit-Crack Mesh Generation

3. Analysis of the interaction between double pit-cracks

3.1 Parameter Setting of Pit-Crack Model

When studying the influence of different pit parameters and the ratio of crack length to depth on the life of pressure vessel, we can change the parameters by fixing the pit depth to change the radius of the pit; fixing the crack depth to change the length of the crack, so as to change the ratio of the length to depth of the crack; and changing the distance between the two pits and cracks (to ensure that the two pits and cracks are independent intact bodies which do not cross each other, and the value needs to start from 5) to obtain the value of stress intensity factor under different conditions. The related parameters are shown in Table 2.

Tab. 2 Pit-Crack Parameters

pit parameter β	Depth of pit f (mm)	pit parameter α	Crack depth c (mm)	Crack length-depth ratio a/c	Pit-crack spacing s (mm)
0	2	0.9, 1.0, 1.1, 1.2	1	0.6, 0.8, 1.0, 1.2, 1.4	5, 6, 7, 8, 9, 10, 11, 12, 13

3.2 Fatigue Life Analysis of Pit-Crack Interaction

The fatigue life of pressure vessel with single and double pits-cracks is calculated and analyzed in the range of pre-set pit parameters and crack length-depth ratio. In order to embody the principle of data normalization, the nominal fatigue life K_{IN} is defined as:

$$K_{IN} = K_{UN} / K_0$$

Among them, the fatigue life of pressure vessel in the presence of double pits and cracks is K_{UN} (MPa. mm$^{1/2}$) and that of pressure vessel in the presence of single pit and crack is K_0(MPa. mm$^{1/2}$).

By comparing the finite element results of single pit-crack fracture parameters and double pit-crack fracture parameters, the curves of fatigue life varying with pit-crack spacing under different pit parameters are obtained as shown in Fig. 4.

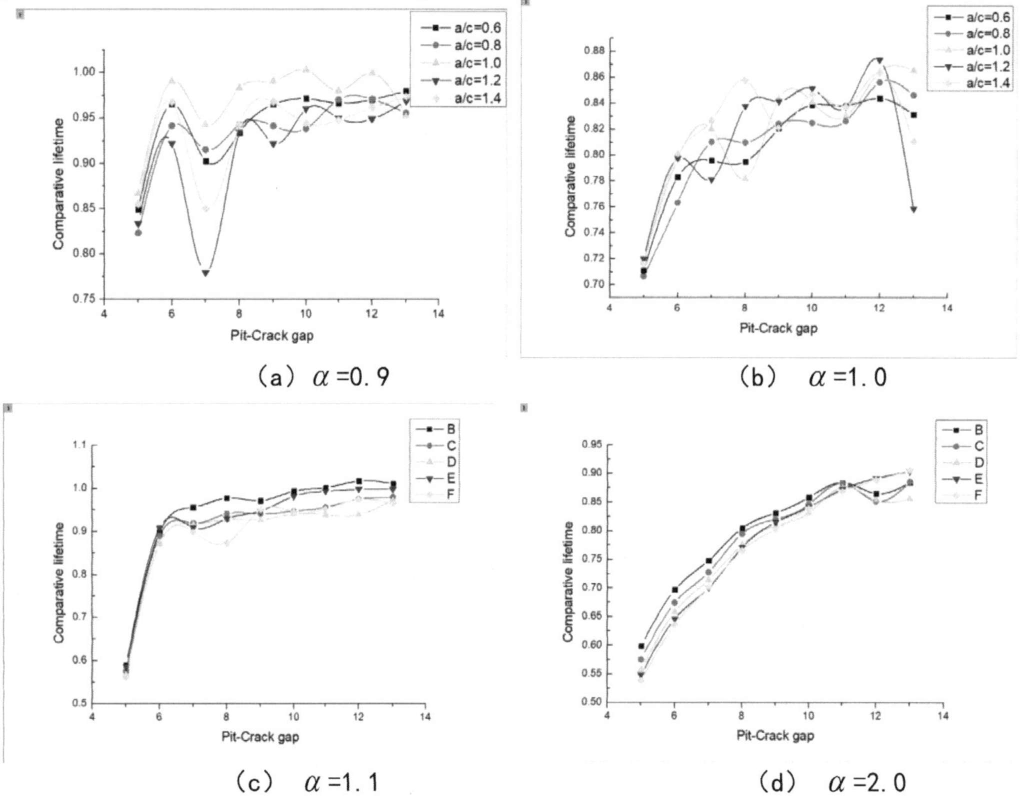

Fig.4 Nominal fatigue life K_{IN} curve of pressure vessel with spacing

It can be seen from the graph that the fatigue life of the pressure vessel with damage under the interaction of pit and crack is similar to that of the pressure vessel with pit parameters, the ratio of crack length to depth is less than 1, equal to 1 or greater than 1, and the trend of change with the spacing S is basically similar, showing a gradual increase and the final fluctuation is stable. The results show that the effect of pits and cracks on the life of pressure vessel is affected by the distance between them, and the larger the distance, the smaller the effect until stability. However, the nominal fatigue life is always less than 1. It can be inferred that due to the existence of double pits-cracks, the effect of each other on the fatigue life of pressure vessels is always affected, resulting in a decrease in the numerical value, reducing the service life of pressure vessels, and making pressure vessels more vulnerable to structural failure. In addition, with the increase of pit parameters, the pit-crack spacing required for the nominal fatigue life of pressure

IOP Conf. Series: Materials Science and Engineering **784** (2020) 012015 doi:10.1088/1757-899X/784/1/012015

vessels to stabilize gradually increases. That is to say, the larger the distance, the longer the distance is needed to achieve a stable state.

In summary, the interaction between pits and cracks promotes the effect of cracks on the fatigue life of pressure vessels, and the effect decreases gradually until stable with the increase of pit-crack spacing. In addition, the pit parameters will determine the spacing required for the transition to a stable state, and the larger the spacing, the larger the spacing required.

In order to clarify the effect of crack parameters on fatigue life of pressure vessels, it is necessary to study the variation of nominal intensity factor with the ratio of crack length to depth under different pit parameters, as shown in Fig. 5.

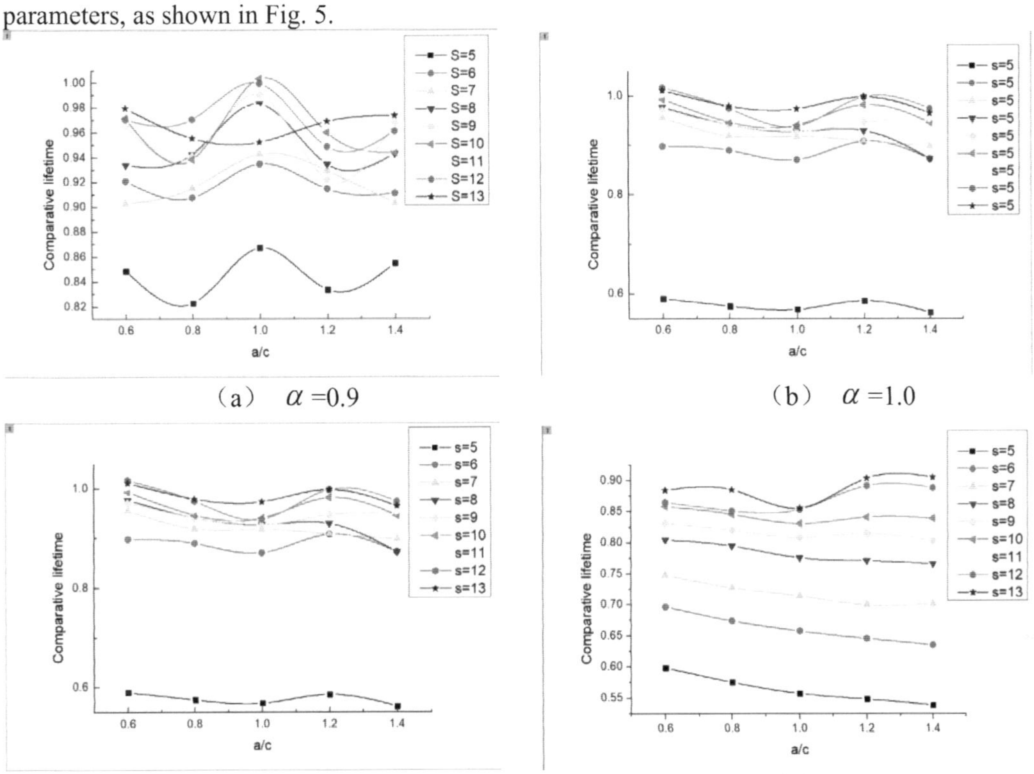

(a) α =0.9 (b) α =1.0

(c) α =1.1 (d) α =2.0

Figure .5nominal fatigue life curve of pressure vessel with crack length-depth ratio

It is found that the nominal stress intensity factor decreases first, then increases and then decreases under the condition of pit parameters, and tends to a straight line under the condition of> 1. It shows that with the increase of pit parameters, the effect of crack length-depth ratio on stress intensity factor becomes smaller and smaller, and when it increases to a certain value, it has almost no effect. In addition, the spacing between adjacent curves becomes smaller and smaller with the increase of the pit-crack spacing, which indicates that when the pit-crack spacing increases to a certain value, the change of the stress intensity factor of the crack will not be affected.

4. Summary

Based on the two-parameter pit-crack model, the distribution of local stress and the change trend of nominal stress intensity factor of pressure vessel in the presence of multi-pit-crack are discussed by using ANSYS finite element analysis software. Through comparative analysis, the following conclusions are drawn:

(1) The fatigue life of pressure vessel varies with the distance s, showing a trend of decreasing gradually and finally fluctuating smoothly. The larger the crack parameters are, the larger the spacing required to reach a stable state is.

(2) The ratio of crack length to crack depth has a significant effect on nominal stress intensity factor, but has little effect on nominal stress intensity factor at > 1.

5. References

[1] GUAN C P. Three - dimensional stress concentration analysis of surface ellipsoidal defect[J]. Journal of Guangdong Polytechnic Institute of Light Industry, 2011, 4(10):13-17.
[2] Nishida Masako. Stress Concentration[M]. Beijing: Machinery Industry Press, 1986.
[3] Liang R, Zhang X Y, Li S X, etc. Influence of semi-ellipsoidal pit on stress concentration of round bar[J]. Journal of Chinese Society for Corrosion and Protection, 2013(6): 532-536.
[4] CERIT M, GENEL K, EKSI S. Numerical investigation on stress concentration of corrosion pit[J]. Engineering Failure Analysis, 2009(16): 2467-2472.
[5] Rokhlin S I, Kim J Y, Nagy H, et al. Effect of pitting corrosion on fatigue crack initiation and fatigue life[J]. Engineering Fracture Mechanics, 1999(62): 425-444.
[6] WANG A Q, QIN T Y. Fracture and reliability assessment on submarine pipelines with axial surface cracks[J]. Journal of Mechanical Strength, 2009, 31 (1) : 144-149.
[7] SHU H M. Stress intensity factors for surface cracks in thin pipes[J]. Oil & Gas Storage and Transportation. 2001, 20(03): 25-28.
[8] LIU Z H, XIE D. Engineering fracture dynamics[M].Wuhan: Huazhong University of Science & Technology Press, 1996: 5-8.
[9] BAI X, XIE L Y. The stress intensity factor variation of two cracks interaction in nuclear piping[J]. Nuclear Science and Engineering. 2015(1):174-180.
[10] LI Z N, GUAN Z D. Applied fracture mechanics[M]. Beijing: Beijing University of Aeronautics and Astronautics Press, 2012: 111-112.
[11] LI X B, YU J X, FAN Z Y, etc. Prediction of Initial Life of Corrosion Fatigue Cracks Based on Two-parameter Corrosion Pits[J]. Oil & Gas Storage and Transportation, http://kns.cnki.net/kcms/detail/13.1093.TE.20190411.1126.002.html.2018-04-09/2019-04-15.
[12] NEWMAN J C, RAJU I S. Stress-intensity factors for internal surface cracks in cylindrical pressure vessels[J]. Journal of Pressure Vessel Technology, 1980, 102 (4) : 342-346.
[13] YU J X, LI X B, TAN Y N, etc. Analysis of the Stress Intensity Factor of a Pipeline Surface with a Pit-Crack[J]. Journal of Tianjin University(Science and Technology), 2019, 52 (5) : 522-528.
[14] YAO A L, HU Y P, ZENG X G, etc. Finite element numerical analysis of multiple interacting cracks in high pressure gas pipeline[J]. Pressure Vessel Technology, 2011, 28 (3) : 28-32.

IOP Conf. Series: Materials Science and Engineering **784** (2020) 012016 doi:10.1088/1757-899X/784/1/012016

Research on tool path generation for integrated off-axis four-mirror group in single point diamond turning

Peixing Ning[1], Ji Zhao[1,2*], Shijun Ji[1*], Jingjin Li[1], Handa Dai[1]

[1]School of Mechanical and Aerospace Engineering, Jilin University, Changchun 130025, China
[2]School of Mechanical Engineering and Automation, Northeastern University, Shenyang 110819, China
[a]ningpx17@mails.jlu.edu.cn, [b]jzhao@jlu.edu.cn, [c]jishijun97@126.com, [d]lijj1416@mail.jlu.edu.cn, [e]dhd@jlu.edu.cn

Abstract. Multi-free-surface mirror integrated system is the development trend of optical system, but its ultra-precision machining is very difficult, this paper mainly studies the tool path generation problem of multi-free-surface reflector integrated system. In this paper, two tool path generation methods for the off-axis four-mirror group machining with SPDT are presented, one tool path generation method is separate machining and the other is integrated machining. Both methods are based on cylindrical spiral and generate tool path by projecting points onto surfaces. For the tool path generation method of integrated machining, the linear transition is used to realize the smooth transition of the mirror surfaces. By comparing the simulation machining of two kinds of tool path generating methods, it is found that the integrated tool path generating method is more suitable for off-axis four-mirror group machining and has higher machining efficiency.

1. Introduction

Ultra-precision machining is a very promising method, which can provide high efficiency, high flexibility, and low cost for creating the high quality surface with achievable sub-micrometric form accuracy and nanometric surface roughness [1], such as off-axis aspheric surfaces. The off-axis four-mirror group is a free-form surface group composed of several off-axis aspheric surfaces, which is very popular in airborne field.

At present, there are many researches from various aspects about ultra-precision manufacturing, and most of them have achieved great improvement [2-4]. Because tool path generation is an important part of ultra-precision machining, a lot of researches on tool path generation methods have been carried out in recent years. Khaghani, A etc. proposed an innovative approach to tool path generation for ultra-precision machining of freeform optic surfaces based on the principle of Automatic Dynamics Analysis of Mechanical Groups, generated the tool path for very complex freeform surfaces [5]. Liu, YZ etc. proposed a partition method for T-spline surface based on its flexible topology and face-by-face sampling, which improved the partition efficiency [6]. Fountas, NA etc. dealt with an original approach of globally optimizing tool paths to CNC-machine sculptured surfaces [7]. Nevertheless, there are few researches on tool path generation for multi-free-surface group manufacturing. At present, more and more attentions are paid to the global optimization of multi-free-surface group. Therefore, it is necessary to propose a tool path generation method for integrated machining of the off-axis four-mirror group.

IOP Conf. Series: Materials Science and Engineering **784** (2020) 012016 doi:10.1088/1757-899X/784/1/012016

In this paper, two tool path generation methods for the off-axis four-mirror group machining with SPDT are presented. This paper gives a detailed introduction of the tool path generation process of these two methods, and simulates the machining of the two tool paths by using MATLAB software, and compares the machining efficiency of the two methods.

2. Design of the off-axis four-mirror group

The off-axis four-mirror group can make the optical group obtain a larger imaging field of view and make the group structure more compact. Based on Zernike free-form polynomial, four aspheric free-form surfaces (M1- M4 surface) are designed and combined into an off-axis four-mirror group, as shown in Fig.1. The expression of Zernike polynomial is:

$$z = \frac{c(x^2+y^2)}{1+\sqrt{1-(1+k)c^2(x^2+y^2)}} + \sum_{m,n} C_{m,n} x^m y^n \tag{1}$$

Where 1/c is the Y radius, k is the conic constant, $C_{m,n}$ is the coefficients for $x^m y^n$.

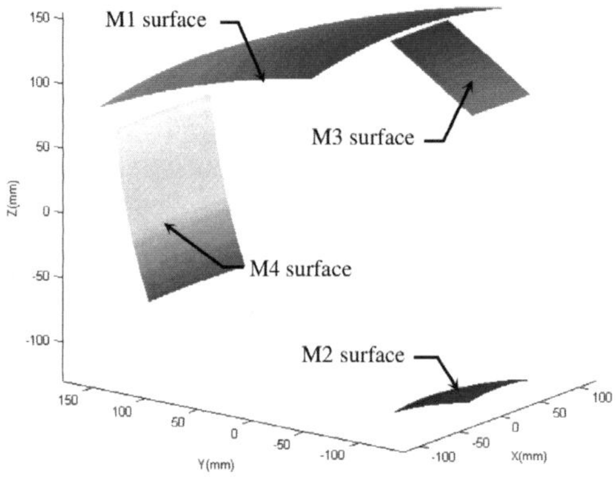

Figure 1. The design of the off-axis four-mirror group

3. Tool path generation for separate machining

The first method is to process each surface in the off-axis four-mirror group separately. The path of tool contact points is generated by projecting the points on a cylindrical spiral tool path onto the surfaces. So first of all, we should to generate a cylindrical spiral tool path which tangent to the point closest to the axis in the off-axis four-mirror group. The expression of the cylindrical spiral is:

$$\begin{cases} y = \rho \cdot \cos\theta \\ z = \rho \cdot \cos\theta \\ x = x \end{cases} \tag{2}$$

where ρ is the radius of the cylindrical spiral tool path, θ is the angle of each point on the cylindrical spiral tool path.

Then the points on the cylinder helix can be projected onto the surface (M1- M4 surface), and the tool path of each mirror processing is shown in Fig.2:

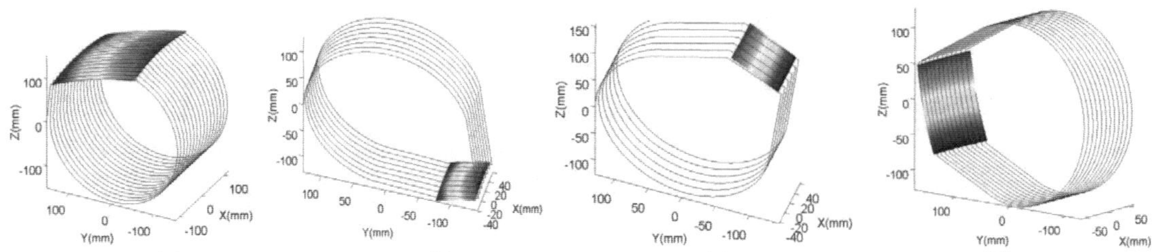

(a) (b) (c) (d)

Figure 2. Tool path generation for separate machining of the off-axis four-mirror group: (a) M1 tool path generation; (b) M2 tool path generation; (c) M3 tool path generation; (d) M4 tool path generation; (e) the off-axis four-mirror group tool path generation.

In this paper, MATLAB software was used to simulate the machining of the off-axis four-mirror group with the separate machining method, and the machining time of tool path generated by different kinds of pitch are obtained, as shown in table 1. Supposed the cutting speed is v=40mm/s and the no-load speed is v'=200mm/s. In this paper, six kinds of pitch are set for tool path generation to simulate the cutting time of the off-axis four-mirror group. The six kinds of pitch are set as 0.05mm, 0.06mm, 0.08mm, 0.10mm, 0.12mm, 0.15mm.

Table 1. The processing time of the off-axis four-mirror group with the separate machining method.

Group number	M1 processing time [s]	M2 processing time [s]	M3 processing time [s]	M4 processing time [s]	Total processing time [s]
1	6.5492×10^4	3.2953×10^4	3.4763×10^4	4.4283×10^4	1.7749×10^5 s
2	5.6181×10^4	2.5245×10^4	2.8969×10^4	3.6907×10^4	1.4730×10^5
3	4.0930×10^4	1.8933×10^4	2.1726×10^4	2.7680×10^4	1.0927×10^5
4	3.2742×10^4	1.5146×10^4	1.7379×10^4	2.2144×10^4	8.7441×10^4
5	2.7283×10^4	1.2621×10^4	1.4482×10^4	1.8450×10^4	7.2836×10^4
6	2.1825×10^4	1.0096×10^4	1.1585×10^4	1.4760×10^4	5.8266×10^4

4. Tool path generation for integrated machining

Same as to the first method, the integrated machining method of the off-axis four-mirror group also need to project the points on a cylindrical spiral tool path onto each surface to obtain the overall and continuous tool path. The cylindrical spiral tool path generated in the off-axis four-mirror group tangent to the nearest point off the axis is shown in Fig.3:

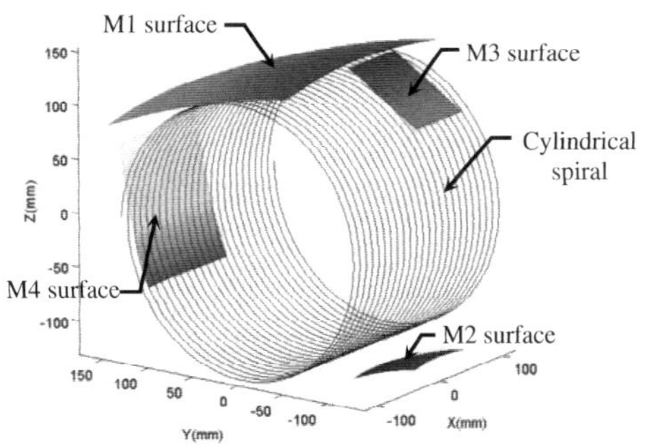

Figure 3. The cylindrical spiral tool path generation

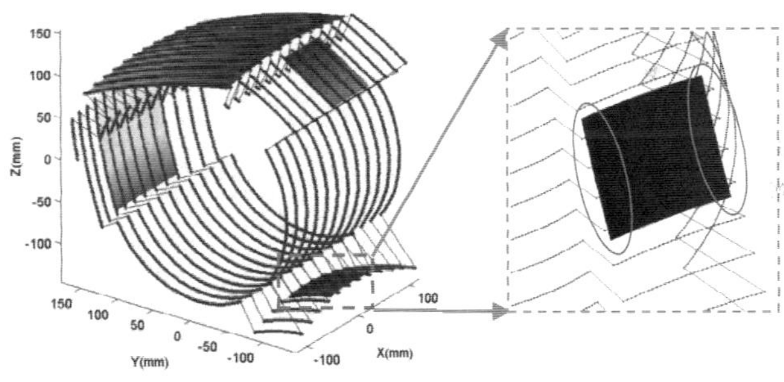

Figure 4. The tool path generation for integrated machining

Then, according to the angle range of each surface in the off-axis four-mirror group, the points on the cylinder spiral in this range are projected onto the surfaces. In order to ensure the processing quality of surface boundary in the group, the tool path on the surface boundary is widened to generate a transition zone [9], as shown in Fig.4:

However the generated tool path is not smooth between the cylinder surface and the off-axis four-mirror group surface, as shown in Fig.4. In this case, during the actual processing, the machine tool will produce a large acceleration change, resulting in the vibration of the machine tool and even cause the machine tool to stop. Therefore, it is necessary to process the tool path of the transition between the surfaces of the off-axis four-mirror group, so as to realize the smooth transition.

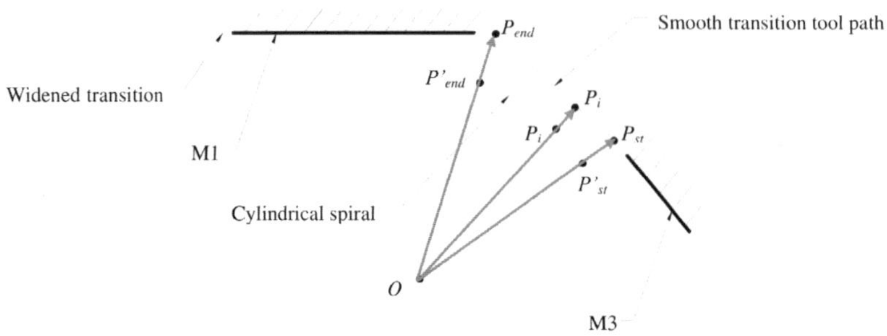

Figure 5. The principle of tool path transition between surfaces of the off-axis four-mirror group

The principle of tool path transition between surfaces of the off-axis four-mirror group is shown in Fig.5. Because the start points P_{st} and end points P_{end} of the transition tool path are on the surface of the Off-axis four-mirror group. So the coordinates of these two points $P_{st}(x_{st}, y_{st}, z_{st})$ and $P_{end}(x_{end}, y_{end}, z_{end})$ can be obtained according the Eq.1. The transition line between surfaces is:

$$\frac{x - x_{st}}{x_{end} - x_{st}} = \frac{y - y_{st}}{y_{end} - y_{st}} = \frac{z - z_{st}}{z_{end} - z_{st}} \tag{3}$$

The points $P_i^{'}$ $(x_i^{'}, y_i^{'}, z_i^{'})$ on the cylindrical spiral can be projected onto the transition lines in the direction of radius to obtain the point $P_i(x_i, y_i, z_i)$:

$$\begin{aligned} x_i &= x_i^{'} \\ y_i &= \rho_i \cdot \cos\theta_i \\ z_i &= \rho_i \cdot \sin\theta_i \end{aligned} \tag{4}$$

where $\theta_i = \theta_i^{'}$. Substitute Eq.4 into Eq.3 to calculate the radius ρ_i of projection points on the transition line. Then according to Eq.4 can get the coordinates of the point $P_i(x_i, y_i, z_i)$ and the smooth transition between the surfaces of the off-axis four-mirror group can be completed.

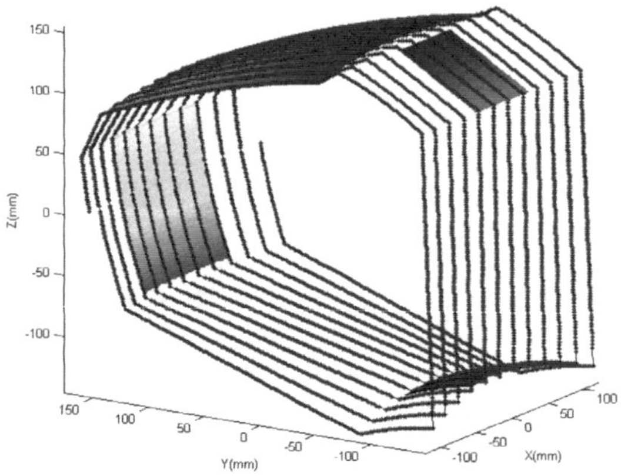

Figure 6. The smooth transition tool path generation for integrated machin-ing of the off-axis four-mirror group

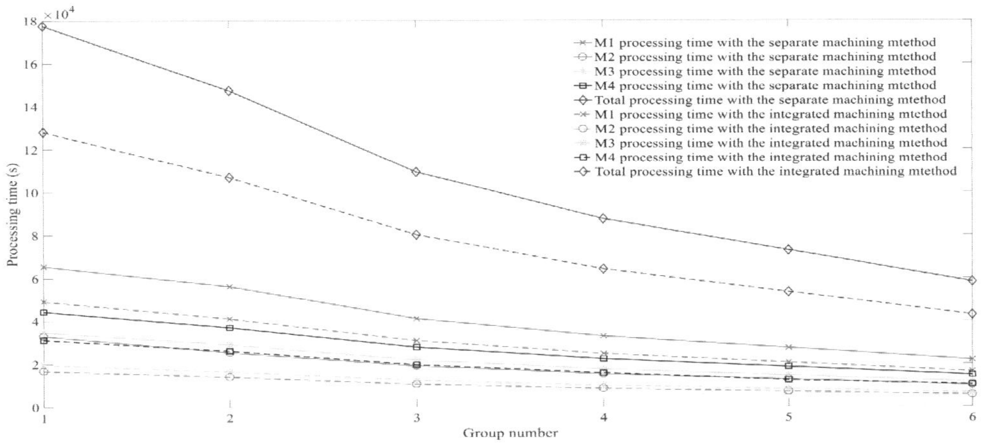

Figure 7. The simulation processing time of the two methods

Similar to the first method, this paper also simulates simulate the machining of the off-axis four-mirror group with the integrated machining method, and sets the same kinds of pitch to generate tool path. The specific simulation machining time is shown in Table 2:

Table 2. The processing time of the off-axis four-mirror group with the separate machining method.

Group number	M1 processing time [s]	M2 processing time [s]	M3 processing time [s]	M4 processing time [s]	Total processing time [s]
1	4.9138×10^4	1.6713×10^4	1.9628×10^4	3.1194×10^4	1.2817×10^5
2	4.0949×10^4	1.3926×10^4	1.6357×10^4	2.6009×10^4	1.0682×10^5
3	3.0712×10^4	1.0442×10^4	1.2268×10^4	1.9509×10^4	8.0118×10^4
4	2.4569×10^4	8.3513×10^3	9.8141×10^3	1.5607×10^4	6.4092×10^4
5	2.0474×10^4	6.9577×10^3	8.1784×10^3	1.2999×10^4	5.3402×10^4
6	1.6379×10^4	5.5642×10^3	6.5427×10^3	1.0402×10^4	4.2721×10^4

5. Analysis of two tool path generation methods

The simulation processing time of the two methods is compared in Fig.8. It can be seen from Fig.8 that the integrated processing method of off-axis four-mirror group takes less time and has higher processing efficiency than the separate processing method. Therefore, the tool path generation method of integrated machining is more suitable for off-axis four-mirror group machining.

6. Conclusion

This paper introduces two kinds of tool path generation methods for off-axis four-mirror group machining. For the tool path generation method of integrated machining, a linear transition method is adopted to realize the smooth transition of tool path between the mirror surfaces, which is suitable for SPDT of slow tool servo. This paper also compares the processing time of the two methods, and concludes that the integrated tool path generation method has short processing time. Therefore, the tool path generation method of integrated machining studied in this paper can be applied to the machining of off-axis four-mirror group and improve the machining efficiency.

7. Acknowledgments

This work is supported by Key R&D Projects of the Ministry of Science and Technology of China (Grant Nos. 2018YFB1107600 and 2017YFA0701200), National Natural Science Foundation of China (Grant No 51775237).

References

[1] Zhang, SJ; Zhou, YP; Zhang, HJ; etc. Advances in ultra-precision machining of micro-structured functional surfaces and their typical applications. International Journal of Machine Tools and Manufacture. 142 (2019) 16-41

[2] Zhang, Quanli; Guo, Nan; Chen, Yan; etc. Simulation and Experimental Study on the Surface Generation Mechanism of Cu Alloys in Ultra-Precision Diamond Turning. Micromachines. 10 (2019)
[3] Wu, DX; Wang, B; Fang, FZ. Effects of tool wear on surface micro-topography in ultra-precision turning. The International Journal of Advanced Manufacturing Technology. 102 (2019) 4397-4407
[4] Li, ZX; Fang, FZ; Chen, JJ; etc. Machining approach of freeform optics on infrared materials via ultra-precision turning. Optics Express. 25 (2017) 2051-2062
[5] Khaghani, A; Cheng, K. Investigation on multi-body dynamics based approach to the toolpath generation for ultraprecision machining of freeform surfaces. Proceedings of the Institution of Mechanical Engineers Part B-Journal of Engineering. 2019
[6] Liu, YZ; Zhao, G; Zavalnyi, O; etc. Toolpath generation for partition machining of T-spline surface based on local refinement. International Journal of Advanced Manufacturing Technology. 102 (2019) 3051-3064
[7] Fountas, NA; Vaxevanidis, NM; Stergiou, CI; etc. Globally optimal tool paths for sculptured surfaces with emphasis to machining error and cutting posture smoothness. International Journal of Production Research. 57 (2019) 5478-5498
[8] Yang, Tong; Cheng, Dewen; Wang, Yongtian. Design method of nonsymmetric imaging systems consisting of multiple flat phase elements. Optics Express. 26 (2018) 25347-25367
[9] Zhang, Lin; Naples, Neil J.; Zhou, Wenchen; Yi, AY. Fabrication of infrared hexagonal microlens array by novel diamond turning method and precision glass molding. Journal of Micromechanics and Micro engineering.

Additional bending moment of thick-walled elliptical tube under internal pressure

Gang Chen, Guan-Nan Chu*, Cai-Yuan Lin

School of Materials Science and Engineering, Harbin Institute of Technology at Weihai,Weihai 264209, P.R.China
E-mail address: chuguannan@hit.edu.cn

Abstract. With the development of industry, the elliptic-sectional tube is widely used in some special situation. In this paper, a theoretical model is developed to calculate the additional bending moment of thick-walled elliptical tube under the internal pressure, based on the structural static force method. The effects of internal pressure and ellipticity on the additional moment were discussed. At last, finite element analyses were conducted and the results were compared with the theoretical results to verify the correctness of this theoretical model. This model can provide a guideline for determining the maximum additional bending moment of elliptic, and help in estimating the failure of the elliptical tube under the internal pressure.

1. Introduction

At present, the pressure pipeline and vessel typically adopt the circular cross-section, but the elliptical section is also widely used in some special industrial fields, such as electricity, petroleum and chemical industry [1]. Compared with the circular section, the elliptical section tube has shorter service life. The plastic failure controlled by limit load is the general failure mode. Because of the large additional bending moments generated by the internal pressure applied on the elliptical section, the circumferential stress caused by the bending moments is the main factor causing the failure of the elliptical section tube [2-3]. Therefore, it is great significant to accurately obtain the stress distribution on the tube section for the judgment of tube failure.

Much of the researchers have put forward the mathematical method to calculate the ultimate bearing capacity of the pressured tube. For example, Chen [4] developed a simplified numerical method for both lower and upper bound limit loads of the 3-D pipeline with one or two part-through slots of various geometrical configurations are calculated by the proposed method, using the finite element method and mathematical programming techniques. Lei [5] reviewed the limit load solutions for axially cracked cylinders, and developed new limit solutions for thick-walled cylinders with axial cracks under internal pressure to overcome problems in the existing solutions. In addition, Kanninen et.al [6] put forward the net section collapse criterion, considering the strain hardening phenomenon of tube material. Then, Kurihara et.al [7] proposed several simple limit-load criteria to predict in a light water reactor pressure boundary piping the failure bending moment of a tough pipe. Kim et.al [8-9] proposed plastic limit load solutions for thin-walled branch junctions under internal pressure and in-plane bending, based on finite element limit loads resulting from three-dimensional FE limit analyses using elastic-perfectly plastic materials.

The finite element method is an effective method to solve the tube stress, but it is not easy to implement under any conditions due to the restrictions of element division, data preparation and computer capacity. In this paper, a theoretical model was developed to estimate the additional bending moment of thick-walled elliptical tube under internal pressure, and the effects of internal pressure and ellipticity on the additional moment were discussed. At last, the result was compared with finite element analysis. This model can be used to calculate the additional moment at any point on the elliptical tube is obtained by a short program, and help in estimating the failure life of the elliptical tube.

2. Theoretical analysis

Figure 1(a) shows the schematic diagram of an elliptical tube under internal pressure. Due to the symmetry of its shape and stress state when the internal pressure p was applied, the model can be simplified to a quarter of the section for stress analysis, as shown in figure 1(b). Section A is the intersection of the major axis with the coordinate x-axis. In order to facilitate the subsequent calculation process, in this paper, the major and minor half axis of the geometric neutral layer of the ellipse is defined as a and b; the major and minor half axis of the outer layer is a_o and b_o, the inner layer is a_i and b_i; the wall thickness is t. Obviously, there is a geometric relationship as:

$$a_i = a - t/2,\ b_i = b - t/2;\ a_o = a + t/2,\ b_o = b + t/2 \tag{1}$$

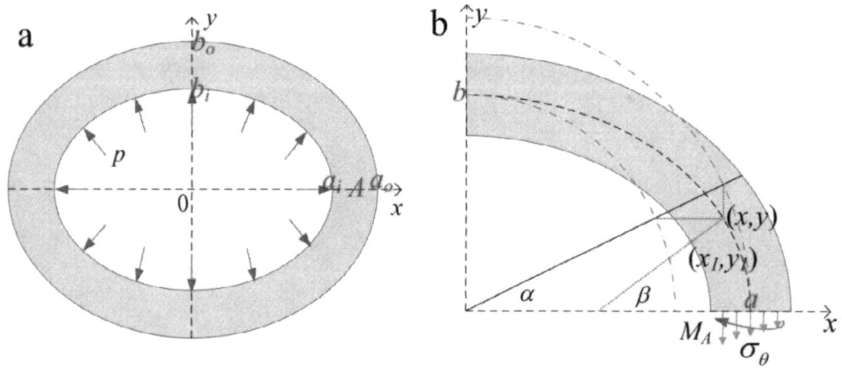

Figure 1. Analyzed model of ellipse (a) the whole model and (b) simplified model.

The curve of the neutral layer can be expressed by the parametric equation of the ellipse:

$$\begin{aligned} x &= a\cos(\alpha) \\ y &= b\sin(\alpha) \end{aligned} \tag{2}$$

The point (x, y) was defined by parameter α. The angle between the normal line of the point (x, y) and the x-axis is β, and the point (x_1, y_1) is the intersection of the normal line and the inner surface, as shown in Fig. 1b.

$$\begin{aligned} x_1 &= x - t/2\cos(\beta) \\ y_1 &= y - t/2\sin(\beta) \end{aligned} \tag{3}$$

The arc length of the neutral layer is:

$$ds = \sqrt{x'^2 + y'^2}d\alpha = \sqrt{(a\sin(\alpha))^2 + (b\cos(\alpha))^2}d\alpha \tag{4}$$

There are circumferential stress σ_θ and the additional bending moment M_A in section A. For the circumferential σ_θ, the elliptical tube can be regarded as a thick-walled cylinder with internal radius a_i and external radius a_o. So the circumferential stress σ_θ can be expressed as:

$$\sigma_\theta = \frac{pa_i^2}{a_o^2 - a_i^2}\left(1 + \frac{a_o^2}{\rho^2}\right) = \frac{pa_i^2}{2at}\left(1 + \frac{a_o^2}{\rho^2}\right) \tag{5}$$

Where ρ is the distance from any point at the section A to the coordinate origin.

The additional bending moment M_A is the maximum because the section A is on the major axis. Consequently, the moment M_A is needed to be solved, and it can be obtained from the canonical equations of force method. Since the structure is statically indeterminate at this time, the canonical equation can be obtained by using the deformation compatibility condition with zero rotation angle of section A.

$$\int\left(M_\sigma+M_p\right)ds+M_A\int ds=0 \tag{6}$$

In the Eq. (6), M_p is the bending moment acting on the point (x, y) by internal pressure, and M_σ is the bending moment caused by circumferential stress. Then M_A can be expressed as:

$$M_A=-\int\left(M_\sigma+M_p\right)ds\times\left(\int ds\right)^{-1} \tag{7}$$

The moment M_σ relative to the point (x, y) can be expressed as Eq. (8), where the clockwise as positive.

$$M_\sigma=\int_{a_i}^{a_o}\sigma_\theta\left(\rho-x\right)d\rho=\frac{pa_i^2}{2at}\left(at-\frac{2atx}{a_i}+a_o^2\ln\left(\frac{a_o}{a_i}\right)\right) \tag{8}$$

The bending moment M_p can be decomposed into the moment M_{px} caused by internal pressure in horizontal direction and the moment M_{py} caused by internal pressure in vertical direction, that is:

$$M_{px}=-\frac{1}{2}p\left(a_i-x\right)^2+\frac{1}{2}p\left(x-x_1\right)^2=\frac{pt^2}{8}\cos^2\left(\beta\right)-\frac{1}{2}p\left(a_i-x\right)^2 \tag{9}$$

$$M_{py}=-\frac{1}{2}py^2+\frac{1}{2}p\left(y-y_1\right)^2=\frac{pt^2}{8}\sin^2\left(\beta\right)-\frac{1}{2}py^2 \tag{10}$$

Thus,

$$M_p=M_{px}+M_{py}=\frac{pt^2}{8}-\frac{p}{2}\left(\left(a_i-x\right)^2+y^2\right) \tag{11}$$

Finally the M_A can be solved by Matlab program based on the Eq. (12), and the direction is counter clockwise.

$$M_A=-\left(\int_0^{\pi/2}M_\sigma ds+\int_0^{\pi/2}M_p ds\right)\left(\int_0^{\pi/2}ds\right)^{-1} \tag{12}$$

For other section, the additional moment $M(\alpha)$ is:

$$M\left(\alpha\right)=M_A+M_p+M_\sigma \tag{13}$$

3. Validation of the analytical model

To validate the analytical model, the finite element analysis (FEA) was used and the results were compared with those of the theoretical solution. To simulate the process, the software ABAQUS 6.13-1 and static, general step was used. Considering the symmetry of the forming parts, the right part along the axisymmetric line is taken as the research object, and the plane strain model is established. The tube pipe is defined as isotropic elastic-plastic model with CPE4R 4-node bilinear plane strain quadrilateral elements. The maximum element size was chosen to be 0.25mm to accurately represent the bending effect. The idealized material was used for the simulation and the Young modulus is assumed as E=1000GPa.

Five simulation models were performed with the different parameter and the detailed dimensions are shown in Table 1. Take the Model 1 as the example, the major half axis a=10mm and the minor b=10mm, thickness t=2mm and the internal pressure p=1MPa. The finite element model is shown in figure 2. In the post processing of Abaqus, the moment and force in the section A can be shown by using view cut, the simulation result as shown in figure 3. It is shown that the moment of FEA result M_A =9.53N/mm, while the theoretical result is M_A =9.32N/mm, and the error is about 2.2%.

Table 1. Detailed dimensions and M_A for five models.

Model	*a/mm*	*b/mm*	*t/mm*	*p/MPa*	M_A(Theory)	M_A(FEA)	error
1	10	8	2	2	9.32	9.53	2.2%
2	10	6	2	5	88.11	88.02	-0.1%
3	10	6	3	5	86.08	89.91	3.5%
4	30	20	2	2	273.96	265.09	-3.3%
5	30	25	2	5	357.66	335.87	6.5%

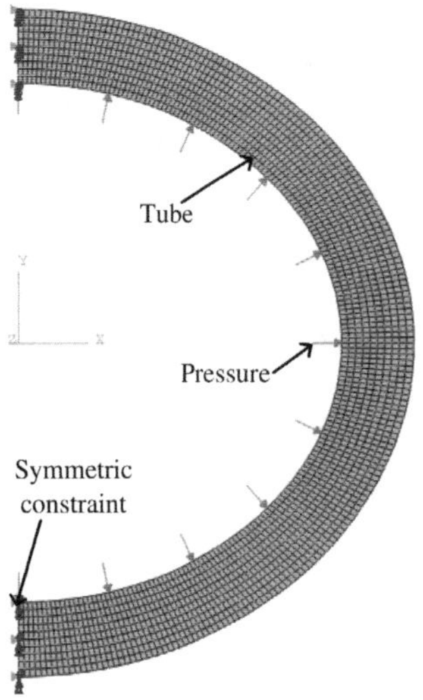

Figure 2. Finite element model.

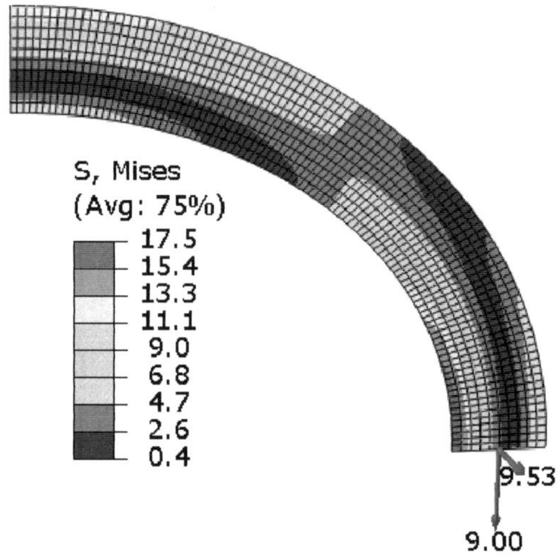

Figure 3. The FEA result of model 1.

The results of models 2, 3, 4 and 5 were shown in Table 1. It is obvious that the error is about 3%, although the error of model 5 is 6.5%, this is because as the increase of internal pressure, the geometric axis of the ellipse is changed and FEA result is deviated. Figure 4 shows the FEA result of other models.

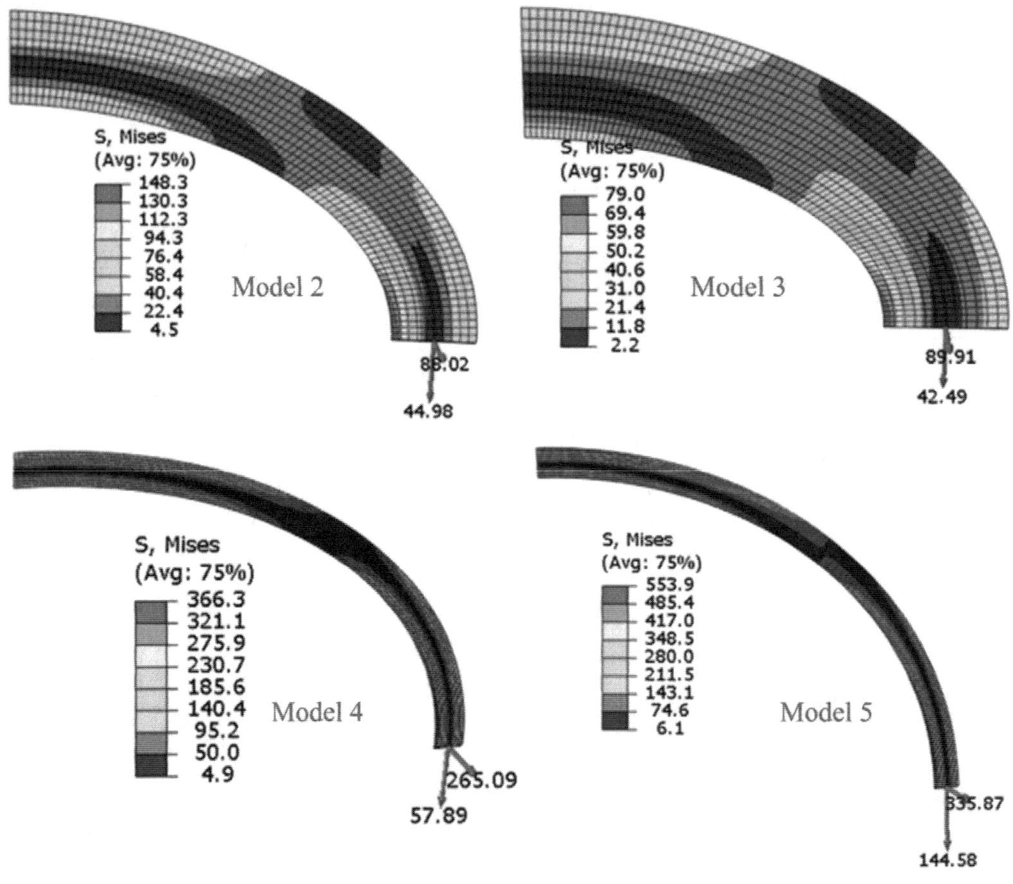

Figure 4. FEA results of models 2, 3, 4 and 5.

Figure 5 shows the variation curves of moment M_A changed with the internal pressure p. It can be seen that the moment M_A is proportional to the pressure p, and with the increasing of internal pressure, the simulation error is increasing. Moreover, the FEA result of Young modulus E=5000GPa is more close to the theory result.

Figure 6 shows the additional bending moment M_A by theory changed with the ellipticity ω, where the ellipticity is defined as $\omega=(a_o-b_o)/r$ and the nominal radius r=25m; the thickness t=2mm and the internal pressure p=1MPa. It can be seen that additional bending moment has a significant linear correlation with the ellipticity ω.

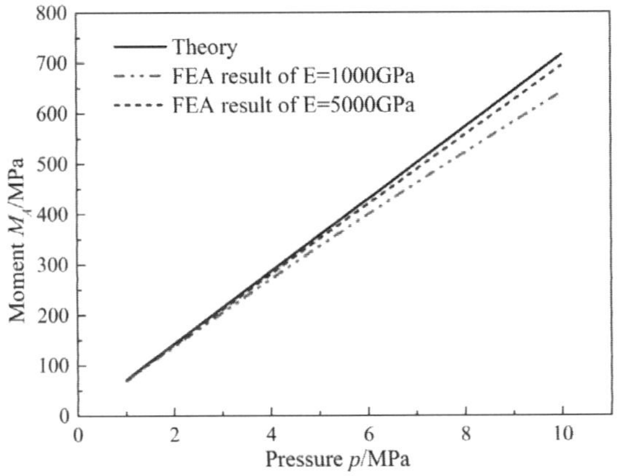

Figure 5. The additional moment M_A vs. pressure p.

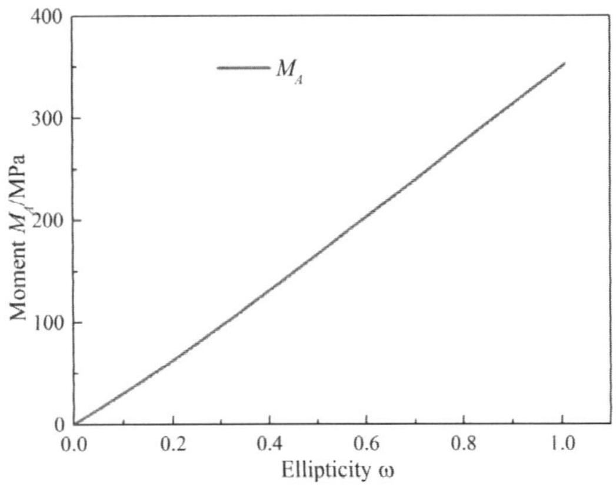

Figure 6. The moment M_A vs. ellipticity.

4. Conclusions

A mathematical model was developed successfully to calculate the additional bending moment of elliptic-sectional tube under the internal pressure, based on the structural static force method. This analytical model provides an accurate method to determine the additional moment and the stress distribution. The effects of internal pressure and ellipticity on the additional moment were discussed. At last, finite element analyses were conducted and the results were compared with the theoretical results to verify the correctness of this theoretical model. This model can provide a guideline for determining the maximum additional bending moment of elliptic, and help to design the boiler pipeline, and predict the failure and service life of the pipeline with the creep theory.

5. References

[1] Naphon P and Wongwises S 2006 *Renew Sust Energ Rev* **10(5)** pp 463-90
[2] Kanninen MF, Zahoor A, Wilkoski G, Abousayed I and Marschall 1982 *EPRI NP-2347 Electric Power Research Institute, Palo Alto*
[3] Xuan FZ, Li PN and Tu ST 2006 *Int J Mech Sci*, **48** pp 460-467
[4] Chen H F and Shu D W 2000 *Int J Pres Ves Pip* **77(1)** pp 17-25
[5] Lei Y 2008 *Int J Pres Ves Pip* **85(12)** pp 825-850
[6] Kanninen MF. 1976 *EPRI NP-192*
[7] Kurihara R, Ueda S and Sturm D 1988 *Nucl Eng Des* **106(2)** pp 265-273
[8] Kim Y J, Lee K H and Park C Y 2008 *Int J Pres Ves Pip* **85(6)** pp 360-367
[9] Kim Y J, Shim D J, Nikbin K, Kin Y J, Hwang S S and Kin J S 2003 *Int J Pres Ves Pip* **80** pp 527-40

Predictive modeling of surface roughness in the inner thread grinding considering the effects elastic deformation

Cui Fang#, Yongsheng Zhao, Congbin Yang, Zhifeng Liu

Beijing University of Technology, No.100, Pingleyuan, Chaoyang District, Beijing, China

fangcuihappy@163.com

Abstract. Surface roughness is an important factor to consider in the field of precision manufacturing. In this study, a model of the surface roughness in inner thread grinding that takes into account the thread helix angle and effect of elastic deflection between the wheel and the workpiece is presented. The surface roughness is analyzed under different spindle speeds, workpiece speeds, grinding depths, and thread helix angles. The proposed model provides a theoretical basis for the optimization of high-precision inner thread grinding.

1. Introduction

Surface roughness is an important indicator of the grinding quality and performance. Therefore, predicting the grinding force is essential to optimizing the grinding process. A number of grinding force prediction models were recently introduced in the literature. Feng et al. [1] developed an empirical model of roughness for steel turning, and the model could evaluate the roughness under certain cutting parameters. Agarwal et al. [2] developed a predictive model of roughness by considering the effect of grains overlapping. The experimental results showed that the model with considering the grains overlapping can improve the accuracy of surface roughness prediction. However, the model ignored the effect of material removal mechanism. Shao et al. [3] established a physics-based model of roughness considering the combined effect of brittle and ductile material removal mechanism. Zhang et al. [4] built surface roughness incorporating the effect of overlapping of triangular-shaped grooves, the predicted surface roughness showed a good agreement with experimental data. Sun et al. [5] developed a predictive model of the surface roughness considering the effects of abrasive grain size, processing parameters and grinding grooves distribution and discussed effects of various processing parameters on roughness. To date, none of the research conducted on the surface roughness in inner thread grinding has considered the effect of elastic deflection between the wheel and the workpiece.

The objective of this study was to develop a surface roughness model based on Wheel-workpiece contact length and the elastic deflection between the wheel and the workpiece in inner thread grinding. Then, the effects of spindle speed, workpiece speed, and grinding depth on the surface roughness were discussed.

2. Modeling of surface roughness in inner thread grinding

In contrast to normal cylindrical grinding processes, an angle between the wheel axis and workpiece axis exists in inner thread grinding, the effective equivalent wheel diameter can be defined as

$$d_e = \frac{d_s}{1 - \frac{d_s}{d_w}\cos\alpha} \tag{1}$$

Where, d_s is the diameter of the grinding wheel, d_w is the diameter of the workpiece, and α is the thread helix angle.

2.1. Wheel-workpiece contact length (l_c)

Contact length (l_c) is important to both grinding efficiency and workpiece surface integrity. Therefore, accurately calculating the contact length is crucial. As the effect of uneven wheel wear and other random factors, the wheel imbalance will occurs in the actual grinding process. In this paper, suppose the relative vibration between wheel and workpiece is a sine wave, the vibration equation of the grinding wheel center can be expressed as

$$M\ddot{y}+C\dot{y}+Ky=me\omega\sin(\omega t) \tag{2}$$

$$y(t) = A\sin(\omega t+\varphi) \tag{3}$$

Where M , K and C are the total mass, rigidity, and damping of the grinding wheel system respectively. ω is the angular speed of the grinding wheel, e is the eccentricity. A is the amplitude, $A = \frac{mer^2}{M\omega_n} g\omega^2 = \alpha\omega^2$, $\alpha = 2\times10^{-11} m\cdot s^2/rad^2$. φ is the initial phase, $\varphi=\arctan\frac{2\zeta\lambda}{1-\lambda^2}$, $\zeta=\frac{C}{2(MK)^{1/2}}$, $\lambda = \frac{\omega}{\omega_n} \ll 1$, ω_n is the natural frequency.

The motion of a single abrasive grain in the x, y and z directions can be expressed as follows

$$\begin{cases} x = v_t t \\ y = \frac{d_e}{2}\left(1-\cos(\omega_r t)\right) + A\sin(\omega t+\varphi) \\ z = \frac{d_e}{2}\sin(\omega_r t) \end{cases} \tag{4}$$

Where, ω_r is the relative angular speed of the grinding wheel to workpiece, $\omega_r = \frac{2v_s}{d_e} + \frac{2v_w}{d_w}$, v_t is the feed speed of the workpiece.

Then, the undeformed chip length can be expressed as

$$\begin{aligned} l_c &= \int_0^{\frac{(a_p d_e)^{1/2}}{v_s}} \left((\frac{dx}{dt})^2 + (\frac{dy}{dt})^2 + \left(\frac{dz}{dt}\right)^2 \right)^{1/2} dt \\ &= \int_0^{\frac{(a_p d_e)^{1/2}}{v_s}} \left(v_t^2 + (\frac{\omega_r d_e}{2}\sin(\omega_r t) + \omega A\cos(\omega t+\varphi))^2 + (\frac{\omega_r d_e}{2}\cos(\omega_r t))^2 \right)^{1/2} dt \end{aligned} \tag{5}$$

Where, a_p is the cutting depth.

2.2. The elastic deflection analysis between the wheel and the workpiece

The elastic deflection analysis between the wheel and the workpiece in inner thread grinding is assumed as a rigid grain impressing into an elastic binder, as shown in Figure.1

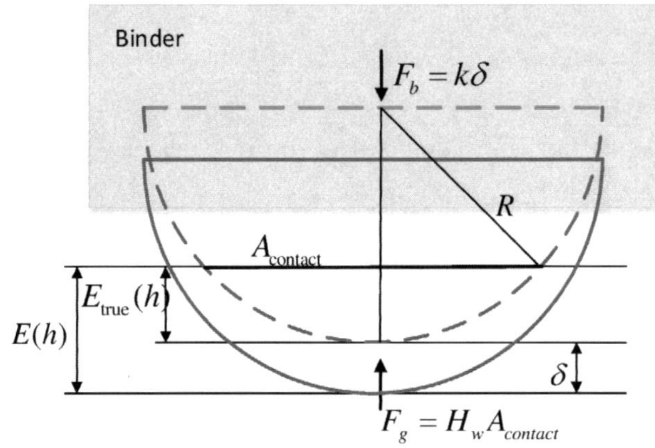

Figure 1. The geometry for grain deflection into the binder

The internal force of the binder to the grain can be written as

$$F_b = k\delta \tag{6}$$

Where, δ is the grain deflection, $\delta = E(h) - E_{true}(h)$, $E(h)$ is the average undeformed chip thickness of a single grain, $E(\mathrm{h}) = \left(\frac{a_p v_w}{2\phi\eta v_s l_c} \right)^{1/2}$, v_s and v_w are the grinding wheel speed and workpiece speed, respectively, ϕ is the effective coefficient of grains overlapping, $\phi = 0.79$, η is the number of active grains per unit area of the wheel periphery. $E_{true}(h)$ is the average undeformed chip thickness of a single grain with deflection. k is the stiffness, $k = 2E_b R(1-\upsilon^2)^{-1}$, E_b is the modulus of the binder, υ is the poisson ratio of the binder. R is the radius of the spherical grinding grain.

The normal grinding force acting on the grain can be expressed as

$$\begin{aligned} F_g &= H_w A_{contact} = H_w \pi R^2 \sin^2\theta \\ &= H_w \pi R^2 [1 - (1 - \frac{E_{true}(h)}{R})^2] \\ &\cong 2H_w \pi R E_{true}(h) \end{aligned} \tag{7}$$

Where, H_w is the workpiece hardness.

Since the internal force of binder is equal to the normal grinding force acting on the grain, $F_b = F_g$, $E_{true}(h)$ can be expressed as

$$E_{true}(h) = \frac{E(h)}{1 + \frac{2H_w \pi R}{k}} = \frac{E(h)}{1 + \pi(1-\upsilon^2)\frac{H_w}{E_b}} \tag{8}$$

2.3. Establishment of surface roughness model

The surface roughness Ra can be written as

$$R_a = \frac{1}{l}\int_0^l |y - y_{cl}| dl \tag{9}$$

Where, y_{cl} is the distance of the centerline , $y_{cl} = \beta\left(\frac{\pi}{2}\right)^{1/2}$, y is the random profile height.

The grooves can be divided into two types, either above or below the centre line y_{cl} , the total expected value of surface roughness considered overlapping factor can be expressed as

$$E(R_a) = P_1 E(R_{a1}) + P_2 E(R_{a2}) \tag{10}$$

According to Rayleigh probability distribution, the probability p_1 , p_2 of the undeformed chip thickness smaller and bigger than the centerline position , y_{cl} , can be calculated as

$$\begin{cases} p_1 = \int_0^{y_{cl}} f(\mathrm{h})\mathrm{dh} = 1 - \mathrm{e}^{-y_{cl}^2/2\beta^2} = 1 - e^{-\pi/4} & h < y_{cl} \\ p_2 = 1 - p_1 = e^{-\pi/4} & h > y_{cl} \end{cases} \tag{11}$$

$E(\mathrm{R}_{a1})$ and $E(\mathrm{R}_{a2})$ can be written as

$$\begin{cases} E(\mathrm{R}_{a1}) = E(\frac{A_1}{2h_1}) = \mathrm{y}_{cl} - (\frac{\pi}{4})\mathrm{E}(\mathrm{h}_1) = 0.656\beta \\ E(\mathrm{R}_{a2}) = E(\frac{A_2^{upper} + A_2^{lower}}{2h_2}) = 0.362\beta \end{cases} \tag{12}$$

The total expected value of surface roughness can be given as

$$E(R_a) = 0.522\beta = 0.417E_{\mathrm{true}}(h) = \frac{0.417}{1 + \pi(1-\upsilon^2)\frac{H_w}{E_b}}\left(\frac{a_p v_w}{2\phi\eta v_s l_c}\right)^{1/2} \tag{13}$$

3. Analysis of the model

The workpiece hardness H_w is HRC55. An inner thread $M42 \times 3$ was chosen, and the helix angle of the inner thread was $\alpha = 2^0 48'$. The grinding wheel was composed of metal-bonded diamond abrasives with diameter $d_s = 30mm$, the modulus of the binder is E_b=882GPa , the poisson ratio of the binder is υ=0.2 . The effects of spindle speed, workpiece speed, grinding depth and the thread helix angle on the surface roughness are presented in Figure.2.

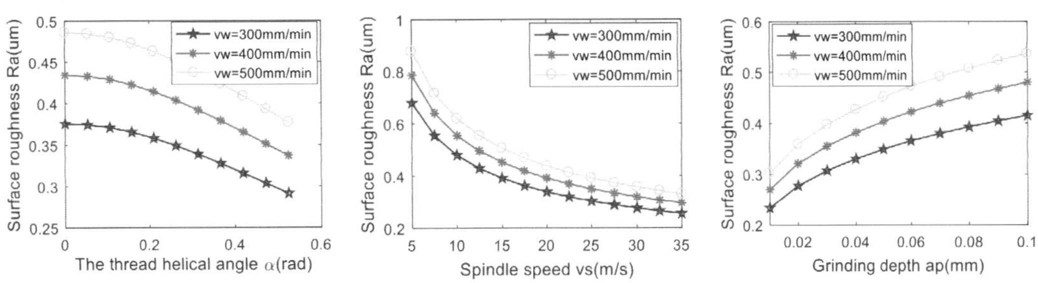

(a) $v_s = 5m/s$ $a_p = 0.05mm$ (b) $a_p = 0.05mm$ (c) $v_s = 18.84m/s$

Figure 2. Effects of spindle speed, workpiece speed, grinding depth and the thread helix angle on the surface roughness

As illustrated in Figure. 2 (a), higher workpiece speeds contribute to an improved surface roughness. Moreover, the higher the thread helix angle, the lower the surface roughness. The surface roughness varies from 0.4344 to 0.3372 as the thread helix angle varies from 0 to $\pi/6$, when $v_w = 400mm/\min$, $v_s = 5m/s$, and $a_p = 0.05mm$. The surface roughness is reduced by 22.4%, therefore, the thread helix angle should be taken into consideration to ensure an accurate prediction of the surface roughness in inner thread grinding. The surface roughness decreases as the spindle speed increases, as shown in Figure.2 (b). Furthermore, the larger the grinding depth, the higher the surface roughness under the same workpiece speed, as illustrated in Figure. 2 (c).

4. Conclusions

In this study, a model of surface roughness based on the effective equivalent wheel diameter and the elastic deflection between wheel and workpiece was presented for inner thread grinding. A higher workpiece speed and grinding depth, along with a lower spindle speed and a lower thread helix angle contribute to an improved surface roughness. The results suggest that surface roughness due to the thread helix angle and the elastic deflection between wheel and workpiece should not be ignored.

5. References

[1] Feng C X and Wang X 2002 *Int. J. Adv. Manuf. Technol* **20** 348
[2] Agarwal S and Rao PV 2005 *Proc. Inst. Mech. Eng. Part. B. J. Eng. Manuf* **219** 811
[3] Shao Y M, Li B Z and Steven Y L 2015 *Mach. Sci. Technol* **19** 325
[4] Zhang J H and Wang LY 2016 *Int. J. Adv. Manuf. Technol* **86** 407
[5] Sun J L and Chen P 2018 *Precision. Engineering* **51** 625

Acknowledgments

The authors would like to thank the Beijing Natural Science Foundation No.3162003 and the National Natural Science Foundation of China No. 51805012, for supporting the research.

Structure Design of Detachable Knotter mouth and Its Fatigue Performance Analysis

Hai-Tang Cen, Yu-Wei Song, Rui-Tao Wei, Ling Li, Jian-Guo Qin

College of Mechanical Engineering,
Inner Mongolia University of Technology,
Hohhot, Inner Mongolia, 010051, P.R. China

author's e-mail address: 841980521@qq.com

Abstract. Knotter mouth is the key part of the knotting on the knotter. Under the action of variable amplitude cyclic load, such as extrusion, shear, torsion and so on, the surface wear of knotter pliers and the fracture of clamp hooks are easy to occur. It takes a long time to disassemble and assemble the knotter mouth, which seriously affects the field operation efficiency of the square baler. In view of the present situation that the knotter mouth of D-type knotter is easy to fail and difficult to replace, on the premise of not affecting the normal knotting function of the knotter, the detachable knotter mouth is designed and completed by reasonably designing the thread structure and rotation direction of the knotter mouth. The disassembly and assembly time is significantly reduced, which is beneficial to the continuous operation of the square baler. ANSYS simulation analysis shows that, compared with the original knotter mouth, the maximum equivalent stress and maximum equivalent strain of the detachable knotter mouth are increased by less than 3%, and the maximum impact fatigue life of the detachable knotter mouth is 60 000 times. The fatigue life of the knotter mouth is much lower than that of the original knotter mouth, but it can still meet the rated life requirements of 40, 000 knotting times of the knotter. The research content makes full use of the safety margin of the knotter mouth structure, which has reference value for facilitating the maintenance and replacement of the knotter mouth and improving the operation efficiency of the whole machine.
Keywords: the mouth of knotter, detachable, structural design, fatigue performance

1. Introduction

The assembly of the D-type knotter on the square baler is shown in figure 1. The process of removing the knotter is troublesome As shown in figure 2, the gear and the rod of knotter mouth are connected by a pin. It is observed that the pin type is a coil type elastic pin, and the outer diameter of the pin is measured to be 5 mm, which is slightly larger than the diameter of the pin hole, so, the link is tight. Cooperate. Therefore, if the disassembly is to be carried out directly, it is likely to make the frame damaged, as shown in figure 3.

The knotter mouth is the part of the D-type knotter on the square baler, which is responsible for knotting the rope. [1] The top-end is subjected to the variable amplitude cyclic load of the rope, such as pressing, shearing and torsion, and the rope is caught by the joint action of other mechanisms. A series of complicated movements, such as buckles, are the wearing parts on the knotter, so it has a very important influence on the overall performance and service life of the D-type knotter. [2]

Figure 1. The assembly drawing of D-type knotter.

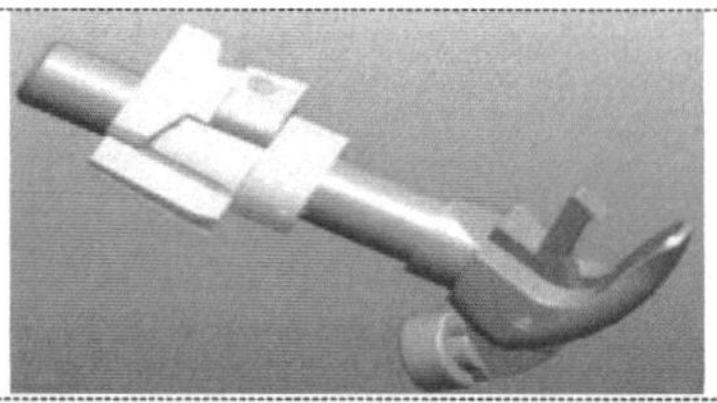

Figure 2. The schematic diagram of the pin connection with the knotter mouth of a gear.

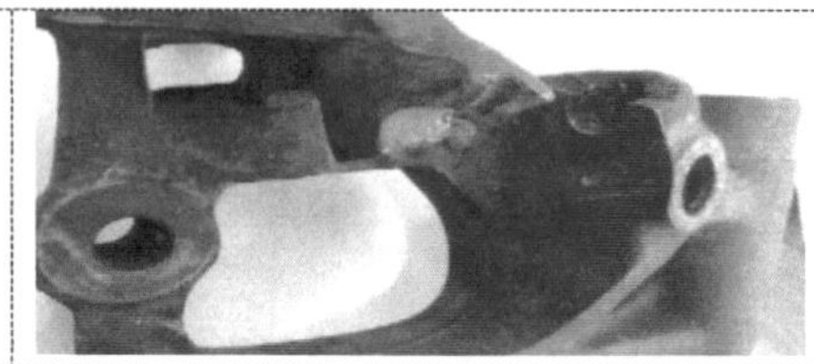

Figure 3. The fracture map of the shaft hole on the knotter mouth.

Figure 4 shows the damaged knotter mouth that is removed from the D-type knotter. It can be seen from the figure that there are two forms of damage to the knotter mouth: 1. The surface of the knotter mouth is seriously worn; 2. The joint of the roller on the bite bar is broken. No matter what happens in the knotter, the D-type knotter will not continue to work and must be replaced to continue working. [3]

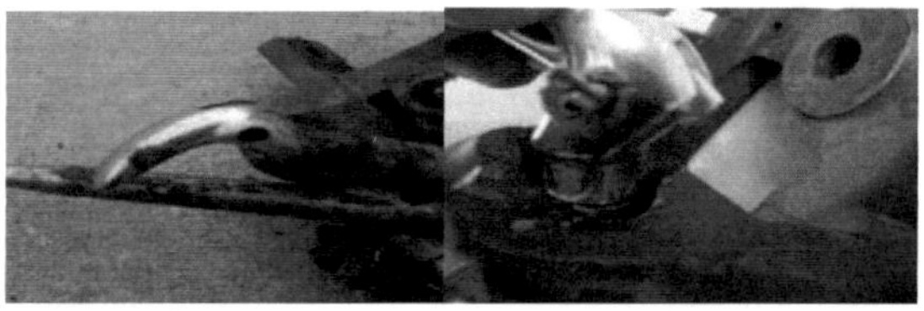

Figure 4. The damage map of the knotter mouth.

Based on the above, this paper has a detachable design for the knotter mouth, which is easy to damage and not easy to replace. The fatigue performance is analyzed and compared with the fatigue performance of the original knotter mouth.

2. knotter mouth detachable design

This article does not affect the normal knotting function of the D-type knotter. A detachable improvement was made. The improvement content is shown in figure 5: the knotter mouth bar and the knotter mouth are originally divided into two, and the two are threaded links. According to the power input direction of the D-type knotter and the rotation direction of the knotter mouth, the end of the knotter bar is designed as an external thread structure, and the end of the knotter mouth is designed as an internal threaded structure. [4]When the knotter mouth is broken, like the break of the bite bar and the invalidation of the knotter mouth, the incomplete gear on the rod does not need to be disassembled. The knotter mouth can be disassembled and replaced directly on the square baler. The specific data of the thread design is shown in Table 1. The relevant parameters of low carbon alloy steel are shown in Table 2. [5]

Table1. The parameters of tread design

Maximum torque	Knotter rod diameter	Knotter mouth material	Thread Typ
45 N·m	15 mm	Low carbon alloy steel	Single thread

Table2. The parameters of low-carbon alloy steel

Yield stress	Elastic modulus	Poisson's ratio
620.4Mpa	210Gpa	0.28

Since the thread is subjected to the preload force and the preload force is controlled, the safety factor n=1.2 is selected. The strength is calculated according to the tight bolt connection that only bears the preload. [6]

$$T \approx 0.2F_0 d \qquad (1)$$

The tensile stress of the dangerous section is:

$$\sigma = \frac{F_0}{\frac{\pi}{4} d_1^2} \tag{2}$$

The torsional shear stress of the dangerous section is:

$$\tau = \frac{\tan\lambda + \tan\varphi_v}{1 - \tan\lambda \tan\varphi_v} \cdot \frac{2d_2}{d_1} \cdot \frac{F_0}{\frac{\pi}{4} d_1^2} \tag{3}$$

According to the fourth strength theory, the calculated stress is:

$$\sigma_{ca} = \sqrt{\sigma^2 + 3\tau^2} \approx 1.3\sigma \tag{4}$$

That is, the threaded connection subjected to the pre-tightening force is subjected to the composite stress of the torsion caused by the frictional moment between the thread pair, and the strength can be calculated according to the pure tension, but the tensile force is increased by 30%， because of the effects of torsional moments.
Therefore, the strength condition of the dangerous section is:

$$\sigma = \frac{1.3F_0}{\frac{\pi}{4} d_1^2} \le [\sigma] \tag{5}$$

At the same time, it is necessary to consider the strength of the thin-walled round shaft which is caused by the thread of the knotter mouth. If the thread diameter is too large, the threaded hole of the knotter mouth may be damaged due to insufficient thin-wall strength. The maximum diameter allowed is calculated by twisting the circular axis. [7]
The maximum shear stress for the torsion of a circular shaft is:

$$\tau_{\max} = \frac{TR}{I_p} \tag{6}$$

$$W_p = \frac{I_P}{R} \tag{7}$$

The torsion section factor is:

$$W_P = \frac{2I_P}{D} = \frac{\pi D^3}{16}\left(1 - \alpha^4\right) \tag{8}$$

Of which, $\alpha = d/D$,which represents the ratio of the inner and outer diameters of the hollow circular section. [8]
It can be seen from the figure 5,that the detachable knotter mouth is assembled, and there is no change in the shape and the original knotting mouth, and other sizes and materials are consistent with the original. Because when the D-type knotter works, the driving toothed disc drives the incomplete gear on the knotter mouth to rotate clockwise (from the knotter mouth rod to the direction of the knotter mouth), the binding rope will apply a torque in the opposite direction to the knotter mouth, so If the pre-tightening force is not applied, the mouth will be tightened more tightly during the roping process. [9]

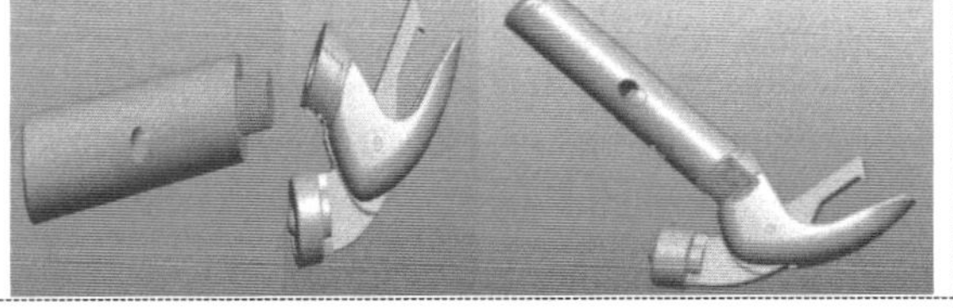

Figure 5. The schematic diagram of a detachable knotter mouth

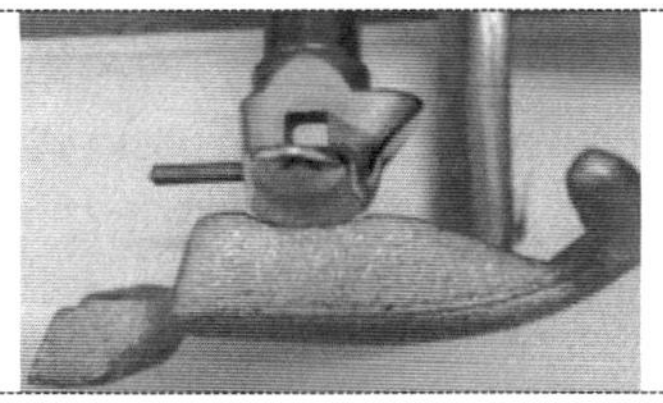

Figure 6. The contact diagram of the knotter mouth and the disengaging bar

IOP Conf. Series: Materials Science and Engineering **784** (2020) 012019 doi:10.1088/1757-899X/784/1/012019

Once the rotational displacement is generated, the timing of the knotter mouth and the disengaging rod will change. As shown in figure 6, it is the relative position of the disengaging bar and the knotter mouth when the tripping action is about to be performed. The preload force can be determined by referring to the maximum torque applied by the rope to the knotter mouth, t is 45 N·m . When disassembling the broken knotter mouth, only need to fix the incomplete gear, and then use the vise to clamp the end of the knotter mouth to counterclockwise (from the knotter mouth to the direction of the knotter mouth rod) to rotate.

3. Strength check and fatigue verification

The thread structure of the detachable knotter mouth weakens the strength of the knotter mouth, so it is checked for strength, and it can be verified by simulation calculation of fatigue life. Design life of D-type knotter: knotted 40,000 bundles. [10]

3.1. Finite element analysis

The main material of the knotter mouth is low carbon alloy steel. After multiple heat treatment processes and surface degree chromium treatment, the yield stress and other parameters are shown in Table 2. Define materials for more than one parameter in the workbench's engineering material library.

Through the analysis of the force of the knotter mouth, it is mainly affected by two kinds of loads: the torque of the tied rope to the knotter mouth and the collision of the disengaging bar against the knotter mouth. [11]This part of the collision cannot be avoided. In order to obtain the magnitude of the load during the collision, the entire D-type knotter model was introduced into the ADAMS for motion simulation, and the force curve of the contact portion at two cycles was acquired. The result is shown in figure 7 below:

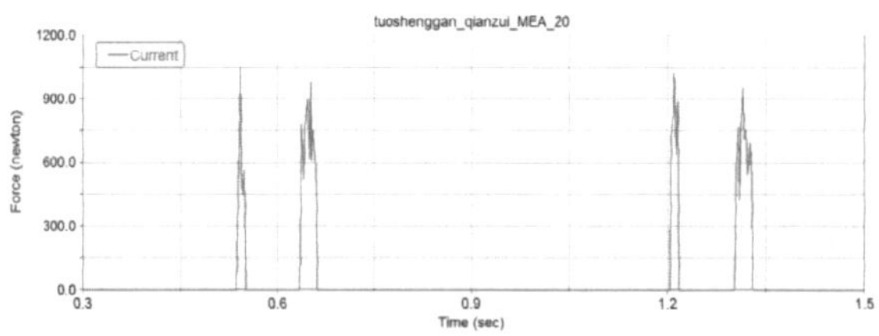

Figure 7. The force curve of the contact part of the knotter mouth and the disengaging bar

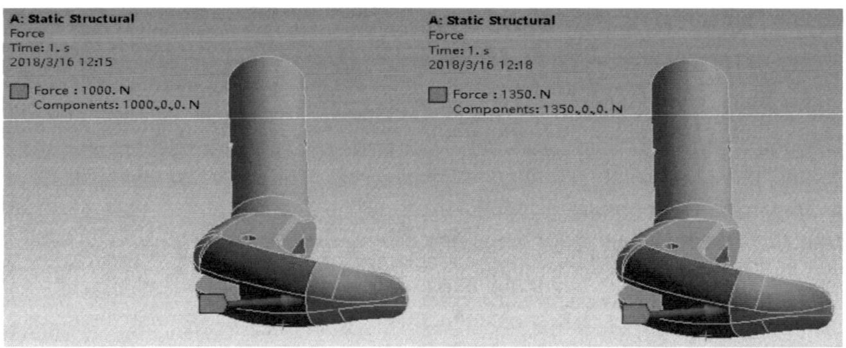

Figure 8. The schematic diagram of the load on the knotter mouth

It can be seen from figure 7 that the maximum contact force between the knotter mouth and the disengaging bar is close to 1000N. The maximum force found in the literature is 1320N. In summary, the original knotter mouth and the detachable knotter mouth are respectively applied with loads of 1000N and 1350N in the position that shown in figure 8 below. After the static solution is obtained, ncode is started to calculate the stress fatigue life that is alternating load under constant amplitude.

3.2. Simulation rsults of transverse load on the knotter mouth

When performing stress simulation on the detachable knotter mouth, the threaded connection of the knotter mouth bar to the knotter mouth is set as the binding contact. The simulation results take the 1000N lateral load of the knotter mouth as an example, and the maximum equivalent stress and equivalent strain of the original knotter mouth are obtained as shown in figure 9. The maximum equivalent stress and equivalent strain of the detachable knotter mouth are as follows figure 10 shows.

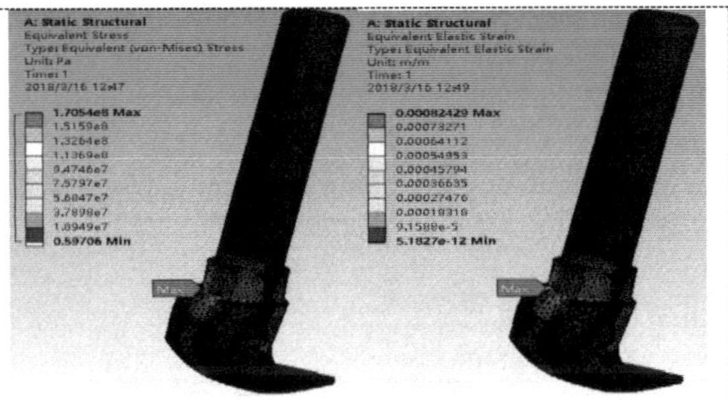

Figure 9. The equivalent stress and strain cloud chart of the original knotter mouth

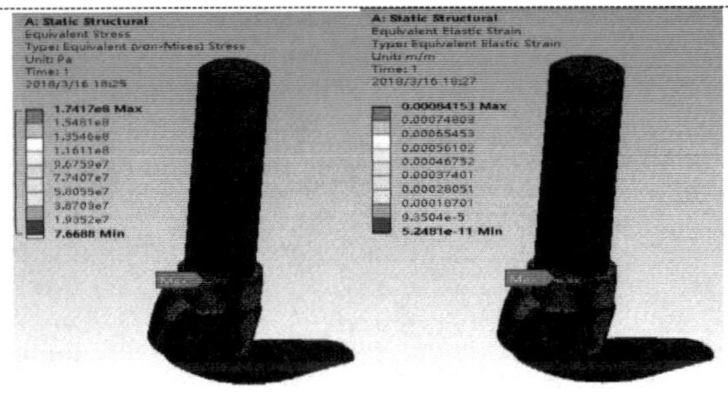

Figure 10. the equivalent stress and strain cloud chart of dismountable knitter mouth

It can be seen from the above two figures that the maximum equivalent stress of the two is much smaller than the material yield stress of 620 MPa. Under the same boundary conditions and equal load, the maximum equivalent stress and the equivalent strain of the detachable knotter mouth is 2.16% and 2.18% larger than the original knotter mouth. It can be seen that the threaded connection still has an effect on the strength of the knitter mouth, but it is completely within an acceptable range. [12]

Therefore, when calculating the fatigue life in ncode, the load map should be set as shown in figure 11 below: Right-click the stress fatigue analysis engine block Perform load map editing, setting the maximum factor and the minimum factor to 1 and -1 respectively, which means that the load loading form is the amplitude round-trip loading. In this loading form, the fatigue life of the two knotters is shown in figure 12. The left side of the picture is the original knotter mouth, and the right side of the picture is the detachable knotter mouth.

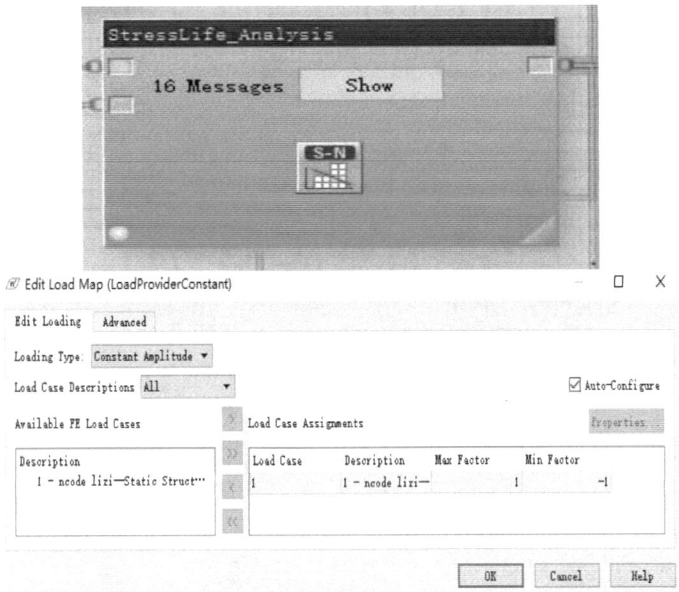

Figure 11. The ncode load mapping setting diagram

In order to ensure the reliability of the simulation results, for the maximum collision load of 1320N on the knotter mouth found in the literature, the simulation calculation is carried out with the same condition under the load of 1350N amplitude, and the result of the cloud diagram shown in figure 13 is obtained:

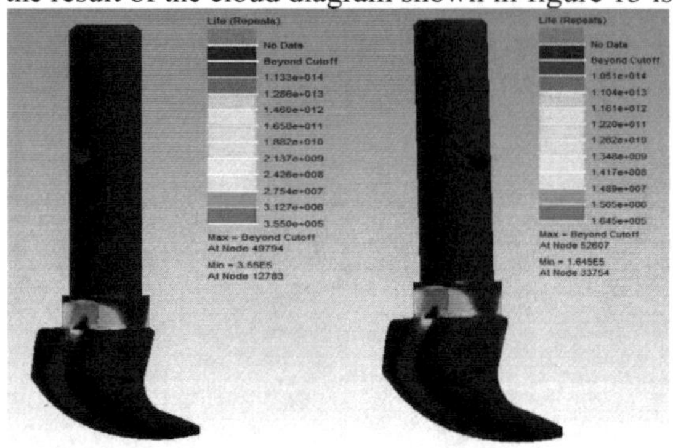

Figure 12. Fatigue life comparison diagram of knotter mouth

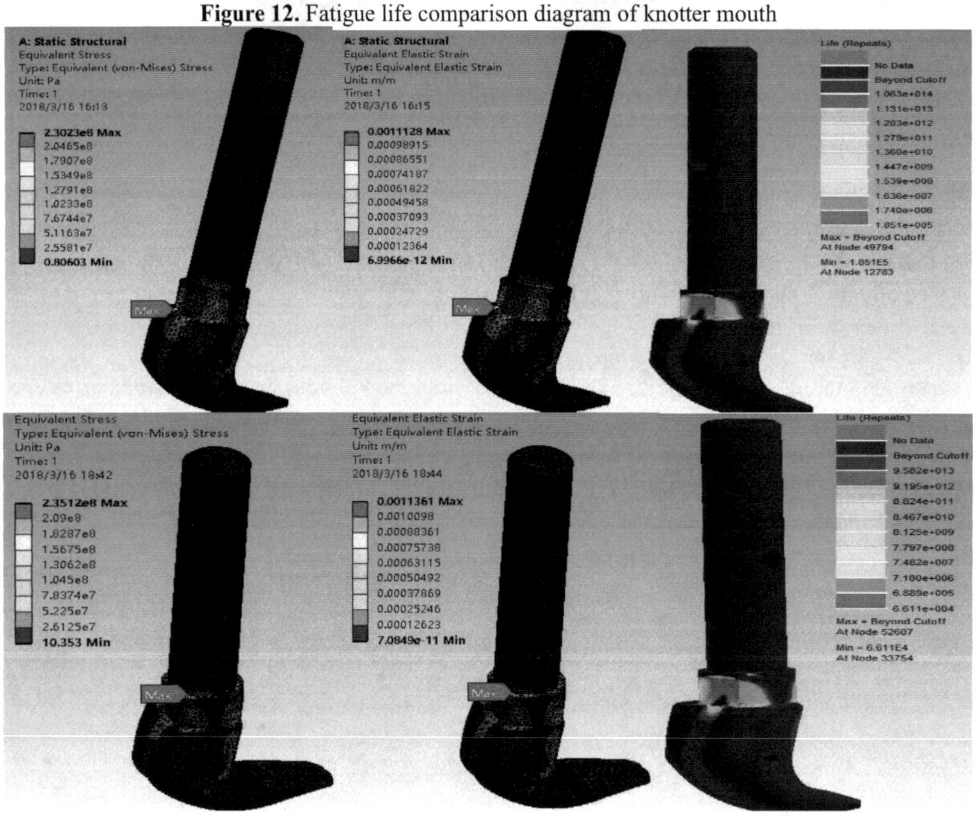

Figure 13. The cloud chart of the simulation results of the 1350N load

Figure 13 (top) is the result of the original knotter mouth, and figure 13 (bottom) is the result of the detachable knotter mouth. The above results are summarized to obtain the results shown in Table 3 below:

Table3. The simulation results of fatigue life of knot mouth

		Original knotter mouth	Detachable knotter mouth
1000N	Maximum equivalent stress(Mpa)	171	174
	Maximum equivalent strain(mm)	0.82	0.84
	Minimum fatigue life(times)	3.5×105	1.6×105
1350N	Maximum equivalent stress(Mpa)	230	235
	Maximum equivalent strain(mm)	1.11	1.14
	Minimum fatigue life(times)	1.8×105	6.0×104

It can be seen from the above results that under the load of 1350N, the fatigue life of the knotter mouth has also reached 60,000 times, which still exceeds the design life of the D-type knotter. Therefore, the detachable design of the knotter mouth fully satisfies the requirements of the D-type knotter in terms of stress damage. [13]

3.3. Mulation results of torque load on the knotter mouth

The maximum torque of the tied rope to the knotter mouth is 45 N·m, which is rotated around the knotter mouth rod and applied to the knotter mouth to obtain a stress cloud diagram as shown in figure 14 below:

The left side of figure 14 is the stress cloud diagram of the original knotter mouth, and the right is the stress cloud diagram of the detachable knotter mouth. It can be seen from the figure that the maximum equivalent stress of the detachable knotter mouth is about 25Mpa larger than the original knotter mouth, but the maximum equivalent stress of both is less than 200Mpa than the yield stress of the material, so it is safe. [14]

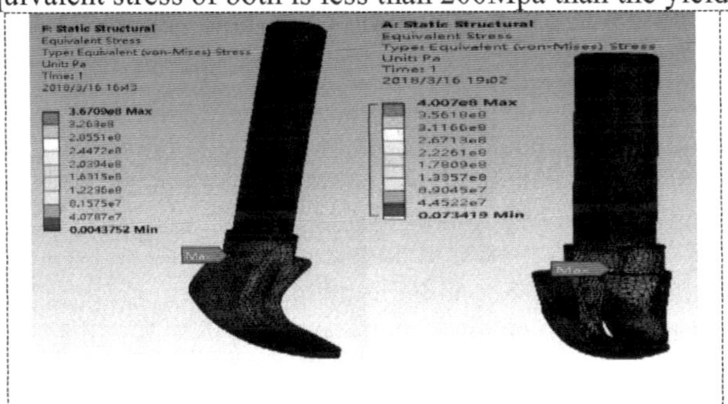

Figure 14. The contrast diagram of the torque simulation results of the knotter mouth

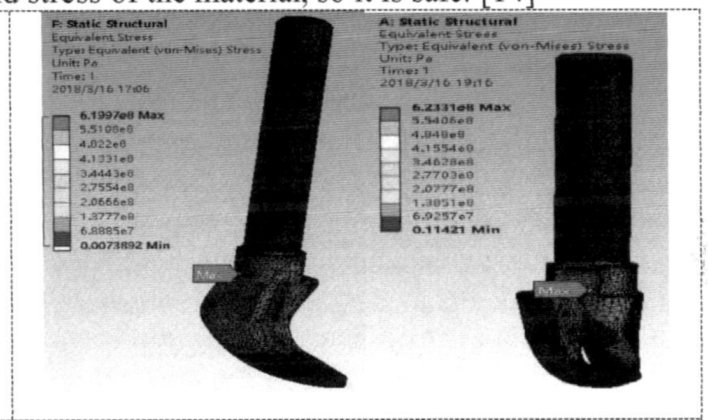

Figure 15. The simulation results of critical torque in a knotter mouth

In order to obtain the critical value of the torque damage, after multiple simulations, the result shown in figure 15. That is, when the maximum equivalent stress value is close to the yield stress value of the material, the maximum torque that the original knotter mouth can withstand the bundle rope is 76N.m, the detachable knotter mouth is 70N.m, which is nearly double the torque value that the rope is generally applied to the knotter mouth, so it is proved that the detachable design of the knotter mouth is also satisfactory in terms of resistance to torque damage. [15]

4. Conclusions

The traditional knotter mouth is prone to damage such as surface wear and nipper rod breakage, but due to the small disassembly space, the disassembly and assembly is not easy, which makes the replacement of the knotter mouth extremely complicated; if it is directly disassembled, it is extremely easy to damage the frame. Caused irreversible damage and further expanded losses. Therefore, a detachable design of the knotter mouth is more advantageous for quick replacement when the knotter mouth is damaged, and does not affect other components, and improves the overall knotting efficiency.

The ANSYS is used to compare the detachable knotter mouth with the original knotter mouth. The results show that after the detachable design of the knotter mouth, although it has been affected in terms of strength and fatigue life, it still has good reliability under the condition that the strength and fatigue life meet the requirements of use.

Considering the interaction and cooperation between the knotter mouth and the incomplete bevel gear, it should be noticed that the relative position between the knotter mouth and the rod of the knotter mouth during processing and assembly. The detachable knotter mouth is identical in size with the original knotter mouth, and does not require re-matching design of other components, thereby solving the problem of difficulty in disassembly and replacement after the knotter mouth is damaged.

Acknowledgements

This research was supported by Major special project of Inner Mongolia Autonomous Region (20190021) and National Natural Science Foundation of China (51365031).

References

[1] Chen Longjian, Li Cheng, Zhang Anqi, et al. Reconstruction and motion simulation of D-type knotter based on reverse engineering [J]. Journal of Agricultural Machinery, 2014, 45(13): 104-108.
[2] Li Haitao, Xiong Ya, Chen Longjian, et al. Wear analysis and improved design of knotter cutting rope tripping mechanism [J]. Journal of Agricultural Machinery, 2014.46(3): 118-124.
[3] Xiong Ya, Li Haitao, Zhang Shaoying, et al. Research on the movement law and design basis of knotter cutting rope tripping mechanism [J]. Agricultural Mechanization Research, 2015. (7): 113-118.
[4] Zhang Anqi, Chen Longjian, Dong Hao. Analysis of spatial structure parameters of D-type knotter cutting rope tripping mechanism. Agricultural Machinery: 2017, 48(1): 73-81.
[5] Wan Qi, Bu Ku, Li Danting. Analysis of Static Stress and Displacement of Vulnerable Parts of D-type Knotted Frames[J]. Xinjiang Agricultural Mechanization, 2009,04(34):18～22.
[6] Wang Yu, Gao Guixian, editor. Mechanical design basis [M]. Beijing: Beijing Institute of Technology Press. 2011.
[7] Shan Huizu, ed. Material Mechanics 1 [M]. Beijing: Higher Education Press. 2016.
[8] Manson SS. Behaviour of materials under condition of thermal stress [M]. NACATN- 2933. 1954.
[9] A. Pramanik; A. R. Dixit;S. Chattopadhyaya;M. S. Uddin;Yu Dong;A. K. Basak;G. Littlefair.Fatigue life of machined components[J].Advances in Manufacturing.2017,Vol.5(No.1):59-76.
[10] Manik Mia;Anjan K. Bhowmick.A finite element based approach for fatigue life prediction of headed shear studs[J].Structures.2019:161-172.
[11] Kihm, Frédéric, Halfpenny A, Ferguson N S. Fatigue Life from Sine-on-random Excitations[J]. Procedia Engineering, 2015, 101:235-242.
[12] Lei X U, Cao H, Liu H, et al. Assessment of fatigue life of remanufactured impeller based on FEA[J]. Frontiers of Mechanical Engineering, 2016, 11(3).
[13] Zhang W Q, Yin J J, Shahhosseini A M. Mechanics analysis of the knot-biting mechanism of D-knotter and bundling-straw tests[C]// International Conference on Design. 2015.
[14] Gang W, De-Qing S, Zhong X, et al. Design and analysis on compression mechanism of small square bales of sugarcane leaf baler[J]. Guangdong Agricultural Sciences, 2013, 40(21):182-185.
[15] Laber K B, Dyja H S, Kawalek A M, et al. Influence of the temperature and strain rate on the deformability of low-alloy carbon steel[J]. Steel in Translation, 2016, 46(9):620-623..

The study on Reliability Design of Knotter Cam Gear of D-knotter

Hai-Tang Cen, Yu-Wei Song, Rui-Tao Wei, Ling Li, Jian-Guo Qin

College of Mechanical Engineering,
Inner Mongolia University of Technology,
Hohhot, Inner Mongolia, 010051, P.R. China

author's e-mail address:doctorcen@126.com

Abstract. The outer gear of knotter cam gear of D-knotter and pinion of knotter hook to achieve coordination and cooperation is important guarantee for the movement of "winding rope - biting rope". This paper analyzes the damage mechanism of the outer gear of knotter cam gear. By application the probability theory for equal effect accuracy synthesis, the outer gear of knotter cam gear and pinion of knotter hook selected from 8 grade precision, and modification of teeth top along the normal direction of the measurement of 50.04 μm for pinion of knotter hook, can control the additional impact dynamic load less than 0.085 times of the calculated load. Two methods of traditional fatigue strength calculation and finite element analysis are used to check the strength of the incomplete outer teeth. And the results obtained by the finite element simulation described in this paper are closer to the real value. This study is of practical value in improving the reliability of the knotter cam gear and promoting the localization of D-knotter.
Keywords: D-knotter; knotter cam gear; reliability design; finite element analysis

1. Introduction

Knotter is the key component of square baler. Its space structure is complex, and the manufacture of key parts is difficult [1,2]. The outer gear of knotter cam gear of D-knotter and pinion of knotter hook to achieve coordination and cooperation is important guarantee for the movement of "winding rope - biting rope". If the gear pair's gear wear is serious or the gear precision is low, it will result in the serious consequences of breakage of the knotter hook and the knife arm or even the breaking of the lock surface of the pinion of knotter hook. According to the precision of the outer tooth of the knotter cam gear higher and easy to wear characteristics, this paper analyzes the damage mechanism of the gear tooth plate. By application the probability theory for equal effect accuracy synthesis, the design of precision and gear top modification of the outer gear and pinion of knotter hook are calculated. Finally, two methods of traditional fatigue strength calculation and finite element analysis are used to check the strength of the incomplete outer teeth.

2. Failure mechanism of the outer gear of knotter cam gear

The movement of the clamps is driven by a pair of by a pair of incomplete straight bevel gear. If the gear pair's gear wear is serious or the gear precision is low, it will result in two serious consequences. One is that it is unable to complete the "winding rope" correctly , because of the shock of gear too big, causing the rope to be entangled on the knotter hook. It can result in the breaking of the knotter hook and the knife arm(see figure 1). Two is the pinion of knotter hook no correctly return, resulting in crushing the locking surface of the pinion of knotter hook [2] (see figure 2). Figure 3 is a magnified picture of the wear of the outer gear of knotter cam gear. It can be seen from the figure that the 2 short teeth wear are very serious at the end of the outer gear.

Figure 1. Fracture of knife arm.

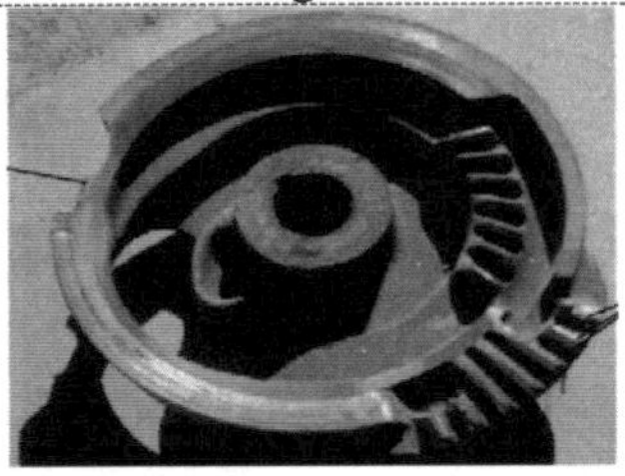

Figure 2. Damage of knotter cam gear.

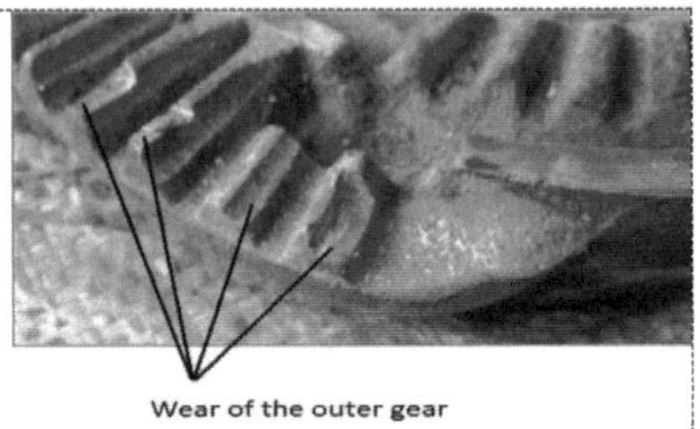

Figure 3. Wear of the outer gear.

There are three main reasons for this phenomenon. One is that the four teeth at both ends of the outer teeth are incomplete because of realizing the motion characteristics of the pinion of knotter hook. It makes the tooth stress concentrated and aggravates the wear of the outer teeth. Two is that the design space of the knotter frame is limited. In order to avoid knotter cam gear and the frame body interference, the outer gear design of teeth width is also limited , and the teeth width can not be increased. It makes the additional impact dynamic load of the unit tooth width on the outer gear difficult to be reduced. Three is that the knotter cam gear is obtained by precision casting. The casting parts have the disadvantages of low dimensional precision, loose casting organization and poor bearing capacity inevitably. As shown in figure 4, sand inclusion can be seen on the outer gear because of improper casting.

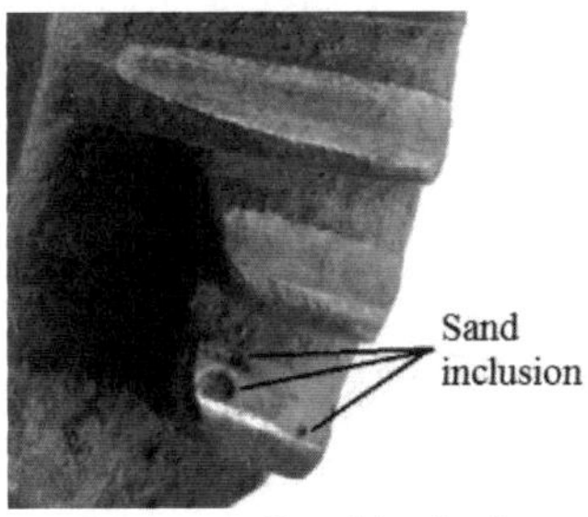

Figure 4. Sand inclusion.

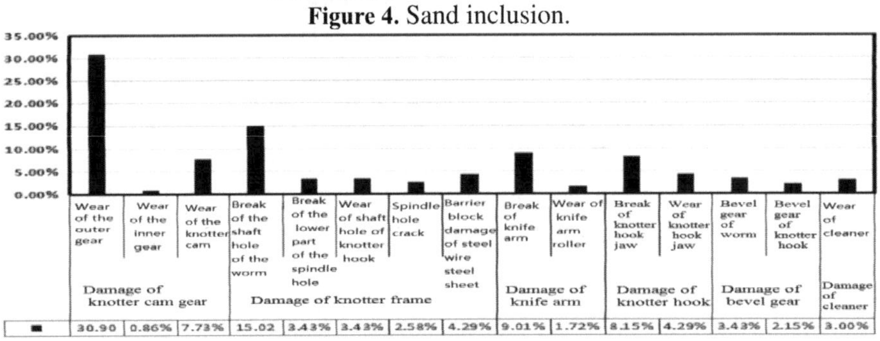

Figure 5. Damage of D-knotter.

The data is the statistics by Inner Mongolia Huade forage Machinery Co. Ltd in figure 5. It is the statistical data of the number of maintenance of imported RS3770 D-knotter in the last two years. There are 233 pieces in all. The fault caused by the wear of the outer gear occupies 30.9% of the total failure rate. According to the statistics obtained by domestic factories, the D-knotter obtained by reverse engineering is not only difficult to reach the life requirement of the imported knotter, but also the outer gear fault rate accounts for 70% of the total fault rate. This difference is mainly due to the low precision

of the outer gear obtained through reverse engineering, and it is impossible to reach its original design standard.
In order to reduce the wear of the outer gear, we should redesign the outer gear of knotter cam gear and pinion of knotter hook.

3. Determine the outer gear and pinion of knotter hook parameters

At the time of design, the gear material is 40Cr. This material has good impact resistance and mechanical properties. The speed of the knotter cam gear is set "n_1=90r/min". The simulation of the D-knotter model obtained by reverse engineering can be obtained, the maximum torque of the outer gear of knotter cam gear T_1=55N·m (see figure 6), the speed of pinion of knotter hook n_2=605r/min (see figure 7). Table 1 is the basic dimension obtained by reverse engineering of the outer gear and pinion of knotter hook, and the basic parameters obtained by the gear design formula in reference [3]. At the time of calculation, set $h_a^*=1$ and $c^* = 0.25$.

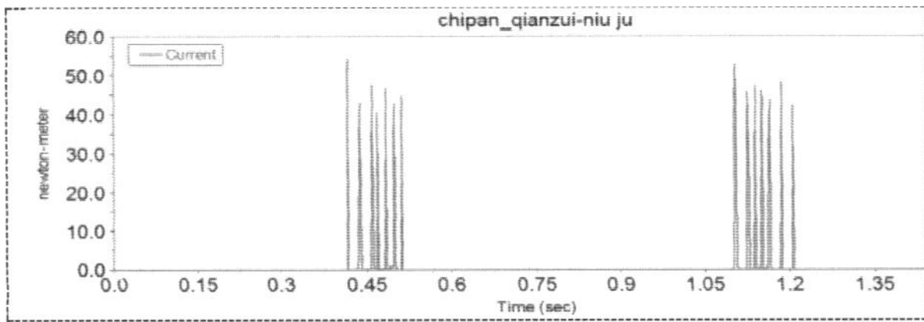

Figure 6. The outer gear - pinion of knotter hook torque diagram.

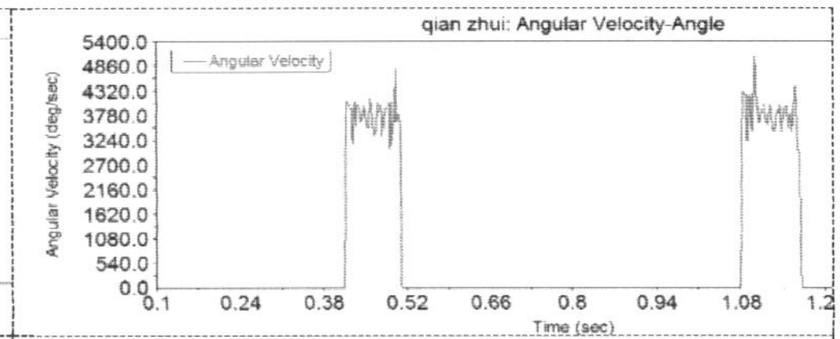

Figure 7. The speed of pinion of knotter hook

Table 1. Gear basic parameters.

Parameter name	The outer gear of knotter cam gear	The pinion of knotter hook
Theoretical teeth number	$z_1 = 54$	$z_2 = 8$
Tooth width	$b_1 = b_2 = 26mm$	
Modulus	$m_1 = m_2 = 4mm$	
Pressure angle	$\alpha = 25°$	
Shaft angle	$\Sigma = 98°$	
Reference cone angle	$\delta_1 = 89.48°$	$\delta_2 = 8.52°$
Addendum modification coefficient	$x_1 = -0.4594$	$\boldsymbol{x_2 = 0.4594}$
Outer cone distance	$R_e = 107.9953mm$	
Main aspects pitch diameter	$d_1 = 216mm$	$d_2 = 32mm$
Midpoint pitch diameter	$d_{m1} = 190mm$	$d_{m2} = 28.15mm$
Tooth addendum	$h_{a1} = 2.16mm$	$h_{a2} = 5.84mm$
Tooth dedendum	$h_{f1} = 6.64mm$	$h_{f2} = 2.96mm$
Top clearance	$c_1 = c_2 = 1mm$	
Main aspects pitch diameter tooth thickness	$s_1 = 4.57mm$	$s_2 = 7.99mm$
Space width	$e_1 = 7.99mm$	$e_2 = 4.57mm$
Equivalent gear rolling circle teeth distance	$p_{vb} = 10.02mm$	
Length of line of action	$g_{vb} = 14.44mm$	
Contact ratio and overlap ratio	$\varepsilon_{v\alpha} = 1.44$	

4. The outer gear and pinion of knotter hook reliability design

The impact load will affect the transmission accuracy of the external gear meshing pair of the knotted hook. In order to reduce the impact load, it is necessary to gear top modification for pinion of knotter hook. The influence of the additional dynamic load on the gear transmission precision is the dynamic load coefficient $\boldsymbol{k_v}$. In order to ensure the lower transmission error and reach the "precise transmission", the dynamic load coefficient is usually 1 ~ 1.1 in the bevel gear transmission [4]. But the designed gear are the nonstandard bevel gear. So the reliability design will be based on the impact dynamic load theory, using probability method to calculate the pitch deviation and gear top modification amount of bevel gear, determining the precision range of the meshing gear. This way can reduce the additional dynamic load, decrease its vibration and enhance the service life of the outer gear.

4.1. Determine the dynamic load coefficient of bevel gear

The general expression of the dynamic load coefficient is [5]

$$k_v = \frac{w_t + w_d}{w_t} = 1 + \frac{w_d b}{F_{mt} K_A} \qquad (1)$$

In the formula: w_d —The additional dynamic load on the tooth width of the gear tooth caused by the impact, N/mm; w_t —The calculated load on the unit tooth width of the bevel gear, N/mm, $w_t = F_{mt}K_A/b$;F_{mt} —The circumferential force at the middle point of the tooth width of the active bevel gear, N, $F_{mt} = 2000T_1/d_{m1}$; T_1 —Torque transmission of active bevel gear, N • m; d_{m1} —The pitch diameter at the middle point of the tooth width of the active bevel gear, mm; b —Tooth width of bevel gear, mm; K_A —Application coefficient, because there is a slight vibration during the meshing of the knob gear, taking$K_A = 1.35$.

According to the theory of impact, the derivation of reference [6] can obtain:

$$k_v = 1 + \mathrm{N}\sqrt{(1.8f_{pt\Sigma} - C_a)} \qquad (2)$$

In the formula:

$$\mathrm{N} = \psi V(1+i)b\sqrt{C_\gamma m_{red}} \times 10^2$$

$$\sqrt{\frac{2(1.8f_{pt\Sigma} - C_a)d_{2a}sin\alpha_{vt}}{a_v d_{1a}}} \Big/ F_{mt}K_A$$

By querying the relevant formulas of reference [5] and [7], can know Impact attenuation coefficient ψ=0.54,The linear velocity of the equivalent gear pitch diameter of the driven bevel gear, $V = 0.9014$m/s, Gear ratio,i=6.75, Curvature radius of the circular tooth profile of the small bevel gear equivalent gear tooth top $d_{2a} = 15.37mm$, Curvature radius of the circular tooth profile of the large bevel gear equivalent gear tooth root $d_{1a} = 264.59mm$, End face pressure angle of equivalent gear $\alpha_{vt} = 25°$, Center distance of the equivalent gears drive $a_v = 662.49mm$, Meshing stiffness of bevel gear teeth $C_\gamma = 6.012N/(mm \cdot \mu m)$, Induction quality of straight bevel gear pair $m_{red} = 2.91kg\ /\ mm$.Finally , we can get it, $N = 0.01429$.

4.2 Estimation of the pitch deviation and gear top modification by probability theory

The so-called "equal effect accuracy synthesis" is a method to determine the offset of every variable quantities in the function according to the allowable output deviation of a function when all variables in the function have the same influence on the output deviation of the function. Suppose that there is an output function $\boldsymbol{y = f}$ ($\boldsymbol{x_1}$, $\boldsymbol{x_2}$, $\cdots, \boldsymbol{x_n}$) , $\boldsymbol{x_1}$, $\boldsymbol{x_2}$, $\cdots, \boldsymbol{x_n}$ of which are independent random variables. Then it is known by probability theory that the relation between the deviations of the variables in the function and the output deviation value of the function is shown in the following formula [6]:

$$\Delta x_i = \frac{\Delta y}{\sqrt{2}|\partial_y/\partial_x|} \qquad (3)$$

Where $\boldsymbol{\partial_y/\partial_x}$- Error transfer function

First, the function of the Y is constructed. The formula can be obtained by the formula (2).

$$(k_v - 1)^2 = N^2(1.8f_{pt\Sigma} - C_a)$$

Therefore,

$$y = (k_v - 1)^2 = N^2(1.8f_{pt\Sigma} - C_a) \qquad (4)$$

Then, it is determined that the allowable deviation of y is Δy. According to the literature [8] and the probability theory, the coefficient of variation of the dynamic load coefficient is:

$$\tau_{k_v} = \frac{(k_v - 1)}{3k_v} \quad (5)$$

Then the standard deviation of the dynamic load coefficient k_v is:

$$\sigma_{k_v} = k_v \tau_{k_v}$$

$$\Delta k_v = 3\sigma_{k_v} = k_v - 1 \quad (6)$$

It is also known by the formula (4):

$$\frac{dy}{dk_v} = 2(k_v - 1)$$

$$dy = 2\ (k_v - 1)\ dk_v$$

Hypothesis,

$$dy = \Delta y,\ dk_v = \Delta k_v = k_v - 1$$

Therefore,

$$\Delta y = 2(k_v - 1)\Delta k_v = 2(k_v - 1)^2 \quad (7)$$

It is assumed that the pitch deviation $f_{pt\Sigma}$ and modification of teeth top along the normal direction C_a of the straight bevel gear are random variables. Then the dynamic load coefficient k_v and the function y are also random variables, and the variables are independent of each other. So y is a function of $f_{pt\Sigma}$ and C_a, that is to say, $y = f\ (f_{pt\Sigma},\ C_a)$.

According to the formula of equal influence precision (3), we know that:

$$\Delta f_{pt\Sigma} = \frac{\Delta y}{\sqrt{2}\left|\partial_y / \partial_{f_{pt\Sigma}}\right|} \quad (8)$$

$$\Delta C_a = \frac{\Delta y}{\sqrt{2}\left|\partial_y / \partial_{C_a}\right|} \quad (9)$$

The transfer functions of each deviation are:

$$\frac{\partial_y}{\partial_{f_{pt\Sigma}}} = 1.8N^2$$

$$\frac{\partial_y}{\partial_{C_a}} = -N^2$$

By the formula (8), (9) to calculate the $\Delta f_{pt\Sigma}$ and ΔC_a, make $C_a \approx \Delta C_a$, $f_{pt\Sigma} \approx \Delta f_{pt\Sigma}$. It is known from the literature [3] that the difference of the limit deviation of the tooth distance between the large bevel gear and the small bevel gear is 1～3 u m. According to this formula [6],

$$f_{pt\Sigma} = \sqrt{{f_{pt1}}^2 + {f_{pt2}}^2} \quad (10)$$

The f_{pt1} of the outer gear and the f_{pt2} of pinion of knotter hook can be determined.

Under guaranteeing the transmission precision, the machining cost is reduced as much as possible, and the dynamic load coefficient of the calculation is k_v=1.085.

By making the known parameters into the formula, we can know, $\partial_y / \partial_{f_{pt\Sigma}}$=3.6757× 10^{-4}, $\partial_y / \partial_{C_a}$=2.0421× 10^{-4}, Δy=0.01445, $f_{pt\Sigma}$ ≈27.79μm, C_a≈50.04μm。

Through formula （10）, we can see that the value of f_{pt1} for the limit deviation of the tooth distance of the outer gear is between (20~21) μm, and the value of the f_{pt2} of the pinion of knotter hook is between (18~19) μm. Reference documentation [3] shows that the bevel gear corresponding to the precision level between 7 grade and 8 grade. To reduce processing costs, will use 8 grade precision. And modified of teeth top along the normal direction of the measurement of 50.04 μm for pinion of knotter hook. Reference documentation [3] shows that the modification amount of the normal direction of the 8 level precision is 60 μm, which is different from the calculated value. The amount of modification obtained by calculation is used in the gear design.

4.3. Precision grade calculation of bevel gear considering vibration factor

In this section, the method of considering the vibration factors will be adopted to calculate the dynamic load coefficient $\boldsymbol{k_v}$, and the precision grade of the gear will be obtained.

The simplified formula for calculating k_v is [3]:

$$K_v = \left[\frac{A}{A+\sqrt{200v_{et}}}\right]^{-B} \qquad (11)$$

Where $A = 50 + 56 \times (1 - B)$ $B = 0.25 \times (C - 5)^{0.667}$

In the formula, C-Transmission precision coefficient, it can be used as a bevel gear designed precision grade ; v_{et} - The linear velocity of bevel gear Main aspects node, m/s , $v_{et} = \pi d_1 n_1/60/1000$

When K_v is 1.085, C is 7.74. Therefore, the precision of the bevel gear is 8 grade.Therefore, when the the dynamic load coefficient k_v is determined, considering the impact factor or the vibration factor, the precision grade of the bevel gear is basically the same. At the same time, it is proved that the way is correct and feasible by probability theory to calculate the precision grade of bevel gear.

4.4. Traditional theory checking of the outer gear

This check is mainly to check the 4 incomplete teeth of the outer gear which are usually serious wear. Through the reverse engineering survey, the tooth width of 4 incomplete teeth is 14mm, that is to say, $\boldsymbol{b_q = 14mm}$.

The length of meshing line in the middle of the tooth [3] :

$$l_{bm} = \frac{2b_q\sqrt{\varepsilon_{v\alpha}-1}}{\varepsilon_{v\alpha}} = 12.9mm$$

According to reference [4], the basic value of contact stress on the tooth surface σ_{H0} is known.

$$\sigma_{H0} = Z_M Z_H Z_E Z_{LS} Z_\beta Z_K \sqrt{\frac{F_{mt}}{d_{m2} l_{bm}} \times \frac{\sqrt{i^2+1}}{i}} \qquad (12)$$

The formula of contact stress of tooth surface is known.

$$\sigma_H = \sigma_{H0}\sqrt{K_A K_V K_{H\beta} K_{H\alpha}} \qquad (13)$$

For open gear drive, the formula of permissible contact stress:

$$\sigma_{Hp} = 1.05\sigma_{Hlimmin} \qquad (14)$$

By querying the relevant formulas of reference [3] and [7], can know Application coefficient $K_A = 1.35$, Dynamic load coefficient $K_V = 1.085$, Longitudinal load distribution $K_{H\beta} = 1.65$, Transverse load coefficient $K_{H\alpha} = 1.2$, Coefficient of node region $Z_H = 2.28$, Coefficient of middle point region$Z_M = 1.0502$, Coefficient of elasticity$Z_E = 189.8\sqrt{MPa}$, Helix angle coefficient $Z_\beta = 1$, Bevel gear coefficient $Z_k = 0.8$, Load distribution coefficient$Z_{LS} = 1$, Contact fatigue limit of test gear $\sigma_{Hlimmin} = 1000MPa$.

Finally, we can know $\sigma_{H0} = 441.32MPa$, $\sigma_H = 751.14MPa < \sigma_{Hp} = 1050MPa$. The bending fatigue strength of the tooth root can also be known by the formula of reference [3]. The basic value of the bending stress of the root of the outer gear tooth is $\sigma_{F0} = 45.17MPa$. The calculation value of the bending stress of the outer gear tooth root is $\sigma_F = 130.5MPa$. The wear coefficient is K_m =1.25. The checking value of the tooth root bending fatigue strength is $K_m\sigma_F = 163.1MPa < \sigma_{Fp} = 210MPa$.

5. Finite element contact analysis of the outer gear

In order to see more clearly the stress change of the designed outer gear, ANAYS Workbench finite element analysis software is applied to conduct finite element contact analysis of the outer gear in this paper. The process includes three major steps [9].

5.1. The preprocessing stage of contact analysis

Set up the three-dimensional model of gear. In the three-dimensional modeling software, the two gear shaft angle of 98 degrees finish assembling. Then the file is saved as "*.STEP" and directly into the ANAYS Workbench.

Set the basic parameters of the model. Young's modulus of material $E = 2.06 \times 10^{11} Pa$, Density of materialsρ $= 7.86 \times 10^{3}\ kg/m^{3}$, Poisson ratio $\mu = 0.29$.

Mesh division of gear model. An automatic mesh is used for the whole model and the mesh is refined at the contact area, as shown in figure 8.

Create contact pairs. The potential contact surface of the outer gear is the contact surface, and the potential contact surface of pinion of knotter hook is the target surface. And set the relevant parameters, Then the contact type of the tooth surface is frictional, Frictional coefficient of gear tooth surface $f = 0.2$, Normal contact stiffness factor $FKN = 2$.

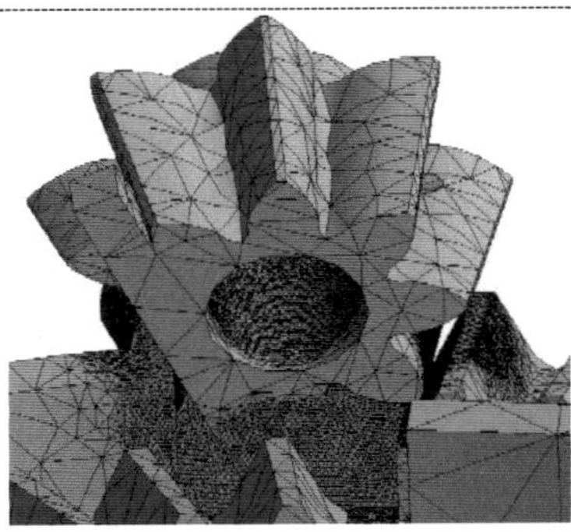

Figure 8. Mesh division.

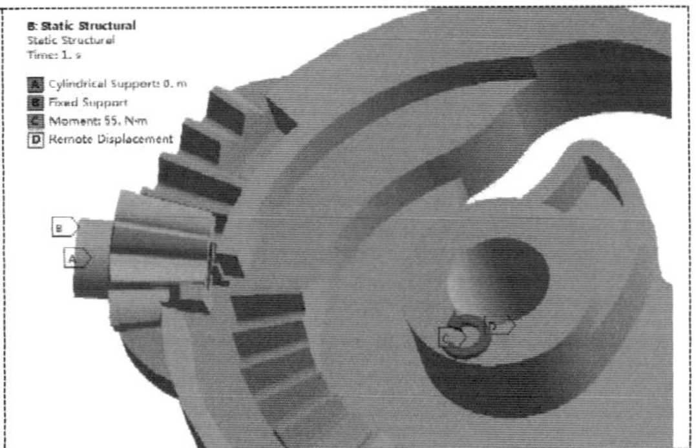

Figure 9. Applied loads and constraints.

5.2. Loading and solving phase of contact analysis

The gear contact analysis is a statics analysis. The pinion of knotter hook is driven wheel. Therefore, the inner surface of the shaft of the gear and the end face of the gear are all constrained. The outer gear is the driving wheel. So, the teeth are defined as X and Y direction constraints, and the Z circumference direction is free. And the torque of T1=55N • m is applied to the rotation center of the active wheel. As is shown in figure 9, is the applied loads and constraints.

5.3. Post-processing stage

Post processing stage is mainly based on graphic display for structural analysis. figure 10 is contact stress cloud chart of "the outer teeth-the bevel pinion". figure 11 is equivalent stress cloud chart of gear pair. figure 12 is equivalent stress cloud chart of outer teeth.

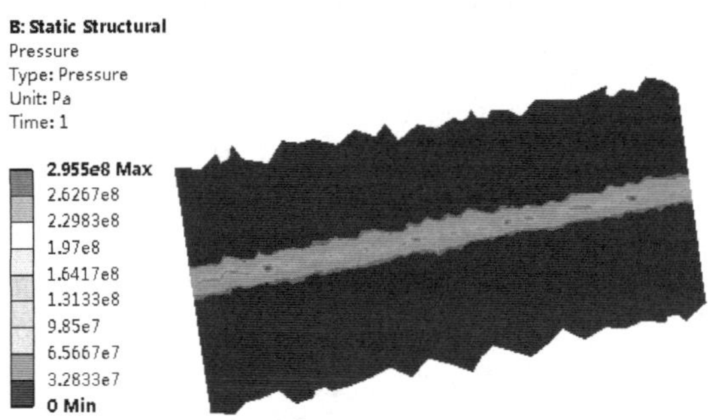

Figure 10. Contact stress cloud chart of gear pair

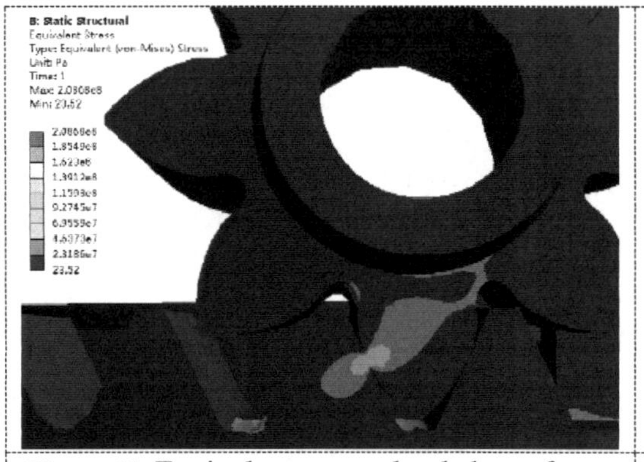

Figure 11. Equivalent stress cloud chart of gear pair.

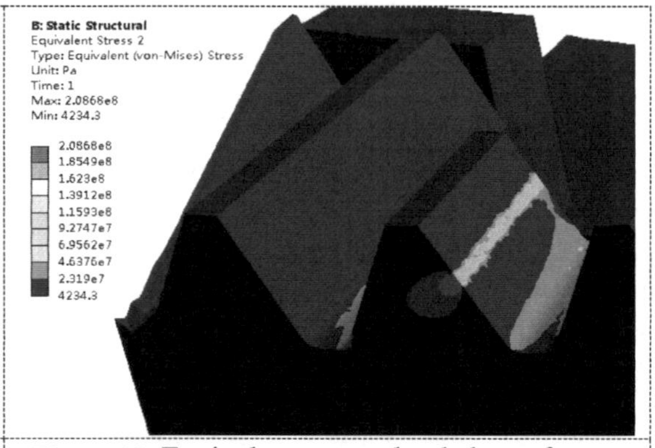

Figure 12. Equivalent stress cloud chart of outer teeth.

Table 2 shows the contact stress of the engagement pair and the error values compared with the traditional checking results. Table 3 shows the bending stress of the outer gear tooth root and the error values compared with the traditional checking results. Among them, the formula of error value is: error value (%) = (traditional calculation value -simulation value) / traditional calculation value.

Table 2. The result and comparison of finite element calculation of contact stress of teeth surfaces

Gear pair name	Traditional Calculation Value （Mpa）	Simulation Value（Mpa）	Error Value	Load coefficient	Check value （Mpa）	Allowable Stress （Mpa）
The outer teeth-The bevel pinion	441.32	295.5	33.04%	1.71	505.31	1050

Table 3. The result and comparison of the finite element calculation of bending stress of the root of teeth

Gear name	Traditional Calculation Value （Mpa）	Simulation Value （Mpa）	Error Value	Load Coefficient	Wear Coefficient	Check value （Mpa）	Allowable Stress （Mpa）
The incomplete outer teeth	45.17	41.36	8.4%	2.9	1.25	149.93	210

From the data of table 2 and table 3, we can see that the strength checking of the incomplete outer teeth is qualified. The stress values obtained by finite element simulation are generally less than those obtained by traditional calculation. This is mainly to improve the safety level of checking, and the various calculation coefficients selected by traditional checking methods are generally large. It is also found that the simulation values of gear contact stress deviate greatly from the calculated values. This is mainly due to the larger experimental value of node area coefficient Z_H. Therefore, the results obtained by the finite element simulation described in this paper are closer to the real value.

6. Conclusions

According to the impact dynamic load theory and application the probability theory for equal effect accuracy synthesis, the outer gear of knotter cam gear and pinion of knotter hook selected from 8 grade precision, and modification of teeth top along the normal direction of the measurement of 50.04 μm for pinion of knotter hook, can control the additional impact dynamic load less than 0.085 times of the calculated load.

According to the impact dynamic load theory and application the probability theory for equal effect accuracy synthesis, the approximate precision grade of bevel gear is in agreement with the precision grade (GB standard) which is now used in China to consider the vibration factors.

Two methods of traditional fatigue strength calculation and finite element analysis are used to check the strength of the " the outer teeth-the bevel pinion " obtained by the design. Finally, we know that the fatigue strength of the incomplete outer teeth is all qualified. And the results obtained by the finite element simulation described in this paper are closer to the real value.

Acknowledgments

This research was supported by Major special project of Inner Mongolia Autonomous Region (20190021) and National Natural Science Foundation of China (51365031).

References

[1] J. J. Yin, W. Q. Zhang, Y. M. Chen, et al, “Parameters analysis of rope-holding motion, knot-winding motion, rope-biting motion of knotter and knotting test,” Transaction of the Chinese Society for Agricultural Machine, vol. 46, no. 9, pp. 135-143, 2015.

[2] R. T. Wei, H. T. Cen, P. W Li, et al, “Failure analysis and research progress of D type knotting device,” Journal of Chinese Agricultural Mechanization, vol. 38, no. 3, pp. 14-18, 2017.

[3] C. B. Wen. Mechanical Design Manual: Gear Drive. Beijing: Machinery Industry Press, 2014.

[4] X. Z. Deng, B. Y. Wei. A new method of bevel gear design [M]. Beijing: Science Press, 2012.

[5] X. L. Zhu, Z. K. E. The super gear bearing capacity analysis [M]. Beijing: Higher Education Press, 1992.

[6] Z. Y. Shao,P. Gou. Probability design of precision and modification of spiral bevel gears [J]. Journal of Wuhan University of Science and Technology (Natural Science Edition), 2002, 25 (3), 263-265.

[7] Bevel gear bearing capacity calculation method [s]. Beijing: China Standard Press, 2003.

[8] Y. M. Lu. Reliability design for mechanical parts [M]. Beijing: Higher Education Press, 1989.

[9] W. X.Wu, W. Liu, F. Zhou, et al. The Analysis of Static Contact Stressin Cone Gear Based on Workbench.Group technology and production modernization, 2011,28 (1), 24-29.

IOP Conf. Series: Materials Science and Engineering **784** (2020) 012021 doi:10.1088/1757-899X/784/1/012021

Stress State Measurement and Result Analysis of Car Wheels

Liu Cheng, Ren Nianzu, Yang Qingsong, Chen Chao, Tian Cheng, Liu Zhengguo
China Automotive Technology & Research Center Co.Ltd, Tianjing,300300

corresponding author's e-mail:363944446@qq.com

Abstract. The stress state of the car wheel is one of the guiding design indicators for the wheel structure. This paper uses the combination of finite element method and bench test to study the stress state of the wheel when driving straight. Through the finite element simulation of bending fatigue and radial fatigue, the position of dangerous points during the use of the wheel is obtained. The location of the dangerous point is used as the strain measuring point, and the test bench is built. The strain gauge is used to measure the strain of these points in the straight running condition, and the test results are analyzed. The results show that when the vehicle is traveling straight, the tire will have a greater impact on the load of the wheel. The squeezing deformation between the tire and the wheel, as well as the positive and negative rotation, will have an effect on the stress state of the wheel. The maximum strain of the wheel appears on the inside of the wheel rim, the rim is in contact with the tire and along the circumference of the wheel. The strain near the center of the wheel is small. Different loads have no effect on the variation of wheel strain, but have an effect on the maximum and minimum peaks.
Key words: car wheel, tress state, dangerous points, strain measure

1. Introduction

Car wheel is one of the most important safety components in the vehicle. If the vehicle has fatigue and fracture damage during driving, it directly affects the safety of life and property of the driver and passenger. Therefore, the reliability design of the wheel is the most important content of the car design[1]. In recent years, with the stricter regulations on energy conservation and emission reduction of vehicles, higher requirements have been placed on the weight reduction of wheels. Many new wheel structures have emerged, and more new materials have been applied, such as composite wheels, new processes, etc[2]. The current wheel reliability design method is mainly carried out from two aspects: structural optimization design based on simulation calculation and product optimization design based on experiment [3][4][5]. In the early stages of wheel design, since the rim of the wheel is in direct contact with the tire, its structural form is determined by the standardized tire size, and the space for design optimization is small. The finite element structure optimization method is generally used to optimize the design of the spoke structure. The finite element can be used to calculate the stress state of the wheel under a specific load [6], but the final fatigue strength of the wheel is not only related to the basic structure, but also to the processing of the wheel. The process is related to the state of use of the wheel, etc., and the final state of force of the wheel is the result of the combination of these factors [7]. These factors are not fully considered in the simulation calculation process [8][9][10][11]. The other method is the bench reliability test. The bench test is used to evaluate the fatigue life under typical working conditions of the wheel. The test sample is the final product state, which can truly reflect the wheel load state. However, it is difficult to measure the load state of all positions of the wheel during the test, and

it is generally based on engineering experience to measure the unidirectional load of some special positions[12][13]. These locations are not necessarily the location of the wheel load hazard.
In response to the above problems, this paper establishes a method combining simulation and experiment to analyze the force state of the wheel. This article will first analyze the load state of the wheel during use. Then, the wheel rig test method is selected according to the load state, and the wheel finite element model is established to predict the position of the dangerous point of the wheel during the reliability test. In the third step, the strain flower is used to measure the stress state of the wheel at the dangerous point and under the straight running condition of the wheel, and comparative analysis is carried out. Through the above analysis, the load state of the wheel under driving conditions is obtained, which provides a reference for the optimization design of the fatigue reliability of the wheel.

2. Load distribution when using car wheels

The rim of a typical car wheel is used in conjunction with a tire, and the center of the wheel is connected to the hub. The load of the tire transmits the load to the hub through the rim, the spokes, and the wheel center during the running of the vehicle, and then transfers the load to the vehicle suspension and the vehicle body through the hub. At the same time, the body load is transmitted to the wheels through the hub. For this reason, the wheel will withstand the translation load in three directions and the three-direction rotational load during use, as shown in the figure 1. In this paper, the positive direction definition of the wheel load is consistent with the SAE standard, the forward direction is the forward direction of the vehicle, Fx is the driving force or braking force of the wheel, Fy is the lateral force direction of the wheel, and Fz is the vertical load aspect of the wheel. The moment around the x direction is the rolling moment Mx of the wheel, the moment around the y direction is the braking or driving torque My, and the moment around the z direction is the yaw moment Mz[1]. In a large number of engineering practices, wheel engineers have found that car wheels mainly include straight-line running conditions and cornering conditions during use. The contribution of various loads to the fatigue damage of the wheel under different driving conditions is also different. The following is a brief overview of the wheel load status under these two conditions.

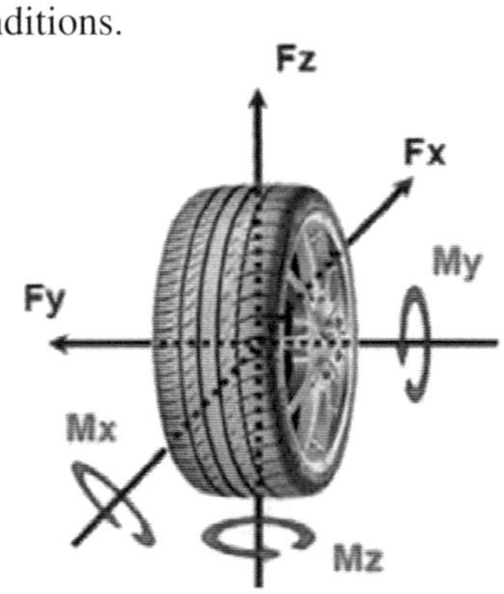

Figure 1. Wheel load direction definition

2.1. Straight driving condition

During the use of the vehicle, most of the driving conditions are straight-line driving. The vertical force analysis is shown in Figure 2 when the wheel is in a straight line. The driving torque transmitted by the hub and the friction torque generated by the tire are balanced with each other, and the wheel is mainly subjected to a vertical load. The wheel bears the vertical load F_g transmitted by the body to the hub, and the vertical load F_υ that the tire transmits to the wheel, which can be expressed as:

$$F_\upsilon = F_T k \qquad (1)$$

F_T For the static vertical load of the wheel, k the dynamic load factor of the wheel.

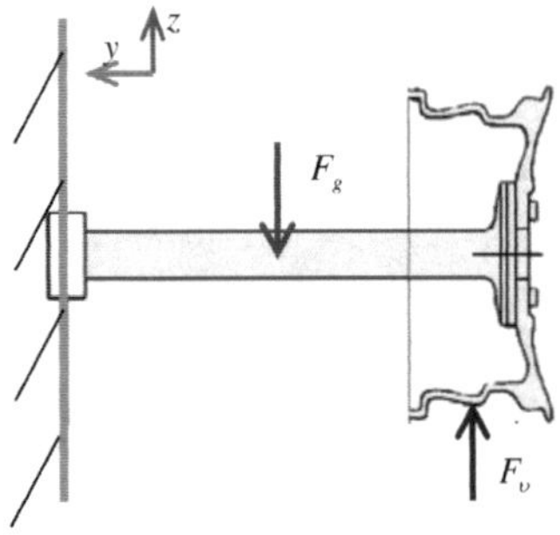

Figure 2. Force analysis of wheel running condition

2.2. Turn driving conditions

Another more working condition in the use of the wheel is the turning driving condition. The force of the wheel under this condition is shown in Figure 3. At this time, the wheel not only needs to bear the vertical load, but also needs to bear the lateral load. Assume that the integrated action point between the tire and the wheel is point C. At this point the wheel is subjected to a vertical load F_g and a lateral load F_y. The equivalent bending moment generated at the center of the wheel is:

$$M_x = F_\upsilon d + F_f R \quad (2)$$

Where: F_υ is the dynamic vertical load transmitted to the wheel for the tire, F_f is the lateral load transmitted to the wheel by the tire, d is the wheel offset, and R is the wheel rolling radius.

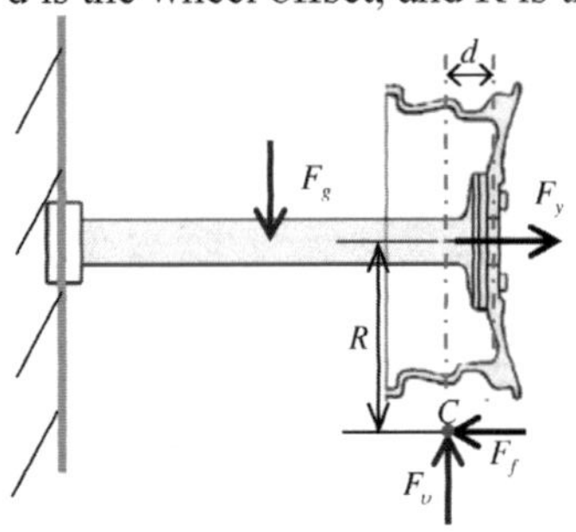

Figure 3. Stress analysis of wheel turning conditions

3. Finite element calculation method for car wheel reliability

Through the analysis of the wheel fatigue test principle and process in the previous section, it can be seen that the load state of different parts of the wheel is investigated by different test. For example, the radial fatigue test of the wheel mainly investigates the ability of the wheel rim to withstand the load, while the bending fatigue test mainly investigates the ability of the wheel spoke to withstand the bending load. And the above fatigue test is only used as the final reliability assessment of the wheel, and cannot directly guide the design. Only knowing the stress distribution under these test conditions can provide a basis for further optimization design. Therefore, this paper proposes a bench test to measure the stress distribution of the wheel under different working conditions, based on which the wheel fatigue life is predicted. However, the engineer does not know the position of the wheel stress concentration point. To solve this problem, this paper proposes a finite element method to predict the stress distribution during wheel fatigue test. Thereby, the position of the stress measurement point of the wheel fatigue bench test can be determined. The following is a brief description of the finite element simulation method for radial fatigue and bending fatigue.

3.1. Radial fatigue finite element simulation

*3.1.1. Finite element model establishment.*According to the principle of radial fatigue test, during the bench test, the drum transmits the load to the wheel through the tire. In order to reduce the complexity of the simulation, the interaction force between the tire and the wheel is generally simplified to the contact force of the tire acting on the wheel rim position. However, it is difficult to accurately express the load distribution of the tire on the rim. According to the literature [1] , the radial load can be approximated as a cosine distribution at the position where the tire is pressed into contact with the wheel rim. Assume that the loading pressure in the wheel plane is a cosine function distribution in the 2θ angle range. The axial direction of the bead seat on both sides of the wheel rim is evenly distributed, as shown in Figure 4. Assume that the load at the contact position of the wheel with the tire bead is symmetric, that is, the load on each side is

$$F' = \frac{F_\gamma}{2} \tag{3}$$

Where: F_γ is the radial load corresponding to the bending fatigue bench test.

On the other hand, depending on the load distribution definition between the tire and the wheel, the load on each side of the tire bead position can be expressed as:

$$F' = B\int_{-\theta_0}^{\theta_0} A r_b d\theta \tag{4}$$

Where, B is the bead width, $2\theta_0$ is the angle range of the tire and the wheel load, and r_b is the radius of the tire bead.

A is the pressure distribution amplitude of the tire bead position at any position, which can be defined as:

$$A = A_0 \cos\left(\frac{\pi\theta}{2\theta_0}\right) \tag{5}$$

According to the above formula, the maximum amplitude of the pressure distribution of the tire bead position can be obtained:

$$A_0 = \frac{\pi F_r}{8\theta_0 B r_b} \tag{6}$$

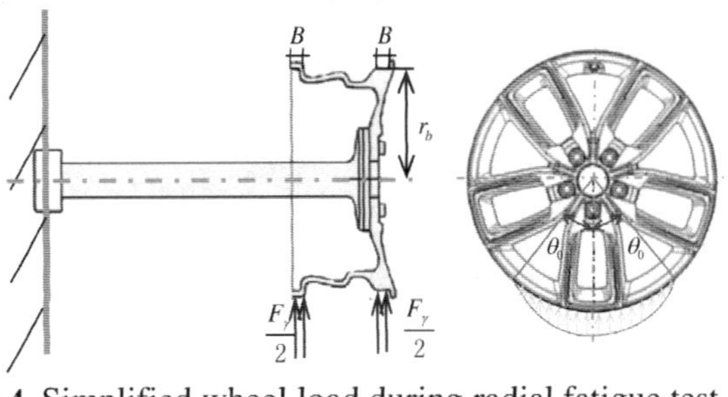

Figure 4. Simplified wheel load during radial fatigue test

According to Stearns J et al., the general assumption is $\theta_0 = \frac{2\pi}{9}$. According to the size of the wheel studied in this paper, the maximum amplitude A0 of the pressure is 1.4MPa.

Figure 5. Radial fatigue finite element model

IOP Conf. Series: Materials Science and Engineering **784** (2020) 012021 doi:10.1088/1757-899X/784/1/012021

After determining the load, the wheel finite element mesh is then divided. Due to the complex structure of the wheel, geometric cleaning is required before meshing. Clean up the small fillets and chamfers in the geometry, and finally use the hypermesh software for tetrahedral meshing. The size of the grid is 5mm, and the minimum size is controlled above 2mm, and the resulting grid is shown in the figure 5. The material of the wheel studied in this paper is ZL101A-1 aluminium alloy wheel. According to the structural characteristics of the wheel studied in this paper, the maximum radial hole of the wheel under radial load is the weakest and the stress concentration is easy. Therefore, the position of the wheel rim bead bearing the maximum pressure is as shown in the figure A point. It is more complicated to load continuous cosine pressure on the wheel finite element model. It can be discretized into finite segments during simulation. In this paper, it is set to 20 segments, and each segment is loaded with uniform pressure. The pressure amplitude is the average amplitude in this segment. At the same time, in order to simulate the effect of the inflation pressure of the tire on the surface of the wheel, a uniform pressure distribution load is applied at a position where the surface of the rim contacts the gas in the tire. In this paper, the applied inflation pressure is defined as 0.45 MPa according to the requirements of [12] the wheel under study. In order to be consistent with the radial fatigue reliability test state, the degree of freedom of all bolt holes at the position of the spokes is constrained during the test.

3.1.2. Analysis of finite element calculation results. According to the method described in the previous section, the finite element model is established and submitted to the static solver of the solution software. Finally, the stress distribution state of the wheel under this load can be obtained, as shown in Figure 10 below.

Figure 6. Radial fatigue stress simulation analysis results

It can be seen that there is a large stress concentration point at the rim position of the wheel, and the deformation amount at the spoke position is small. The stress of the wheel is mainly concentrated on the position where the wheel rim is in contact with the tire, and the stress on the inner side of the corresponding wheel rim is larger than the outer side of the corresponding position.

3.2. Finite element simulation of bending fatigue

3.2.1. Finite element model establishment. With reference to the wheel bending fatigue bench test standard and the test equipment principle, a 0.9m long loading shaft is assembled in the wheel finite element model. Import the model into Hypermesh software for finite element preprocessing. The assembly model is meshed using a tetrahedral element. The wheel model has a cell size of 5 mm and a minimum size of 2 mm. According to the method requirements of test standard [12], a fixed constraint is imposed on the inner edge to constrain all degrees of freedom. The bolt connection between the flange face of the load shaft and the wheel mounting surface is simulated with the rbe2 unit. A loading force perpendicular to the loading axis is applied at the end of the loading shaft. The direction of the force is the centerline pointing to the position of the largest through hole of the spoke, as shown in Figure 7.

Figure 7. Bending fatigue simulation load loading method

3.2.2. Analysis of finite element calculation results. Set up the analysis step and finally submit the static analysis model to the software solver. The wheel stress distribution corresponding to the analysis results is shown in the figure 8(a). It can be seen that in the radial fatigue test, the main deformation position of the wheel is located on the spoke. In the bending fatigue simulation, the wheel spokes and the wheel center position are greatly deformed. The deformation of the rim position is small.

The wheel center position is partially enlarged, as shown in figure 8(b). It can be seen that stress concentration occurs at the connection position between the spoke and the wheel center. At the same time, there is stress concentration at the wheel bolt mounting position.

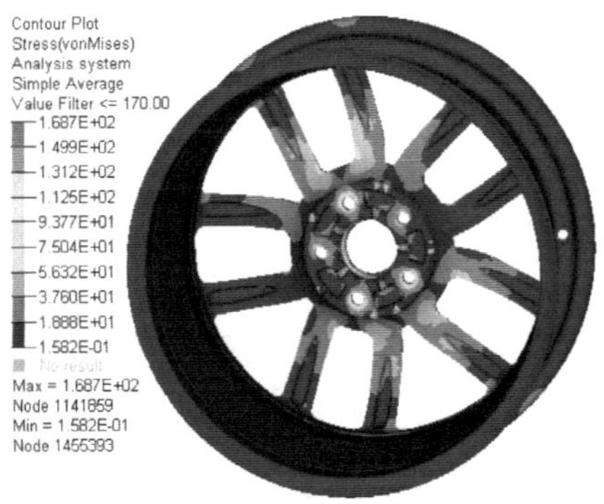

(a) stress sate distribution of wheel

(b) Partial magnification of stress sate

Figure 8. Bending fatigue stress analysis

4. Measurement and analysis of position stress of dangerous points on car wheels

4.1. Strain measurement

*4.1.1. Measuring point arrangement position.*Through the above finite element analysis, as the picture 9 shows, it can be concluded that in the bending fatigue test, the maximum stress point may occur at the chamfering position 4 of the spoke and the position of the wheel center position 5. In the radial fatigue simulation, the stress concentration is likely to occur at points 1, 2, and 3 of the spoke position, and at

IOP Conf. Series: Materials Science and Engineering **784** (2020) 012021 doi:10.1088/1757-899X/784/1/012021

the position where the tire is in contact with the tire bead seat, and the stress on the inner side of the rim is large. However, since the strain gauges cannot be pasted at this position, the inside of the wheel facing the bead seat serves as the stress measurement point 6.

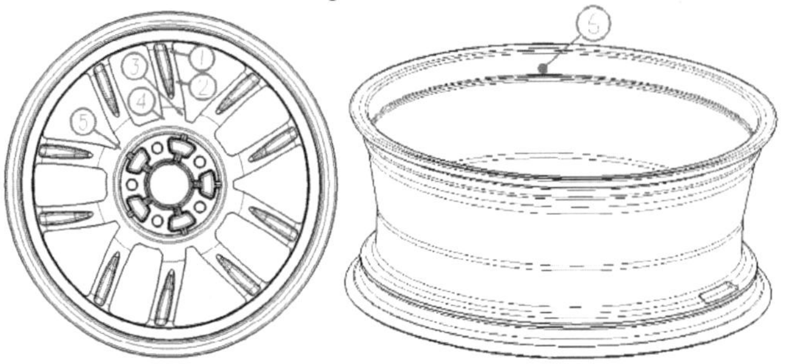

Figure 9. Strain measurement point location

According to the determined stress point concentration in the above figure 9, the measurement strain gauge is pasted. As shown in figure 10, in order to study the change of the plane stress at the measuring point position, the plane stress is measured by using 0-45-90 strain flower in this paper. At the same time, when the strain gauge flower is pasted, the distribution of the strain gauges at 0, 45, and 90 degrees is defined as the horizontal direction, the middle direction, and the vertical direction. The direction of the strain in the horizontal direction is the direction of the contour parallel to the corresponding position. For example, the horizontal direction of the strain gauges 1, 2, and 3 is parallel to the longitudinal direction of the spoke, and the vertical direction is perpendicular to the length of the spoke. The horizontal directions of the 4th and 5th strain flowers are chamfered directions parallel to the corresponding positions. The horizontal direction of the strain gauge No. 6 is parallel to the wheel extension, and the corresponding vertical direction is parallel to the axis direction of the wheel. Due to the small temperature variation of the wheels during the test, in order to simplify the test process, a quarter bridge was used in the test.

Figure 10. Test wheel strain paste position

4.1.2. Test bench construction. The stress measurement in the radial fatigue of the wheel is shown in Figure 11. The wheel strain sensor is directly connected to the data collector during the test, and the data acquisition parameters are controlled by the computer. The wheel is loaded with radial loads by means of a drum. In the process of wheel radial fatigue test, there is no special regulation on the wheel rotation speed in the standard. In order to reduce the influence of the wheel dynamic response on the test results, only the wheel load change during the quasi-static rolling process is considered in the test, and the wheel rotation speed is considered as small as possible.

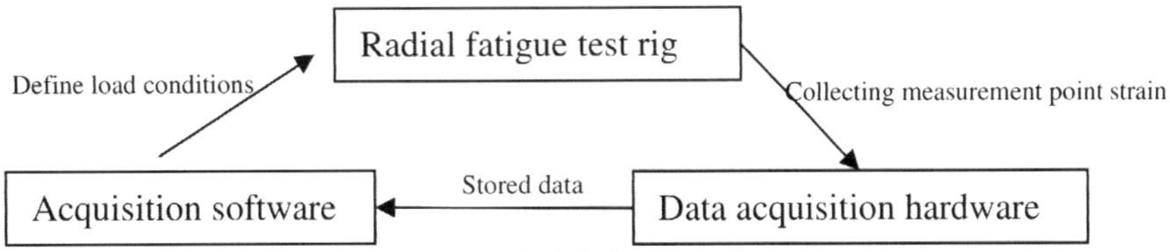

Figure 11. Radial fatigue test bench

4.2. Analysis of measurement results

Due to space limitations, only the strain variation laws of several typical locations are shown. The following is a comparison of the horizontal and vertical strains of measuring points 1 with the change of the wheel angle, as shown in figure 12. This measuring points are located at the spoke.In the figure 12, the C curve indicates that the wheel rotates clockwise, and the A curve indicates that the wheel rotates counterclockwise. It can be seen that during the rolling of the wheel, two change cycles appear at one measuring point, and the two positions are exactly 180 degrees apart. This is because the position of the maximum peak-to-peak corresponds to the position where the measuring point is closest to the ground. In another position, the measuring point is located farthest from the ground when the wheel is rotated 180 degrees from the first position. At this time, the pressure of the tire and the bead position of wheel is increased due to the wheel extrusion deformation, so that the deformation at the position of the spoke of the wheel is increased. It can be seen that the peak occurring in the change of the horizontal and vertical stress amplitudes at the first point position is the largest. The law of load change in the forward and reverse directions is basically the similar. However, the peak parameters vary, mainly due to the nonlinearity of the tire. At the same time, it can be seen that at the measuring point 1, the peak in the horizontal direction is larger than the vertical direction. From the horizontal and vertical phases, it can be seen that the phase of the horizontal phase changes exactly opposite to the phase of the vertical direction.

The strain at the position of the measuring point 5 is also the rolling periodic variation of the wheel, as shown in figure 13. The change in amplitude is relatively small relative to the position of the measuring point 1. This is because the measuring point 5 is located at the center of the wheel and bears less load under straight driving conditions. As shown in figure 14,it can be seen from the strain law at the position of the measuring point 6 that the strain change process is basically the same as that of the measuring point 1, but the maximum peak value changes rapidly. It indicates that the load on the position of the measuring point 6 changes rapidly when the tire is in contact with the ground during the rolling process of the wheel. There is a certain difference in the load change in the vertical direction of the measuring point 6 when the wheel rotates in the forward and reverse directions. The horizontal strain value is 10 times larger than the vertical direction.

Comparing the above three measuring points, it can be seen that the rim of the wheel bears the maximum load during the radial load test, and the load in the horizontal direction has an absolute advantage with respect to the vertical direction. At the same time, there is also a large load change in the horizontal direction of the spokes and rim. The load change at the center of the wheel is small.

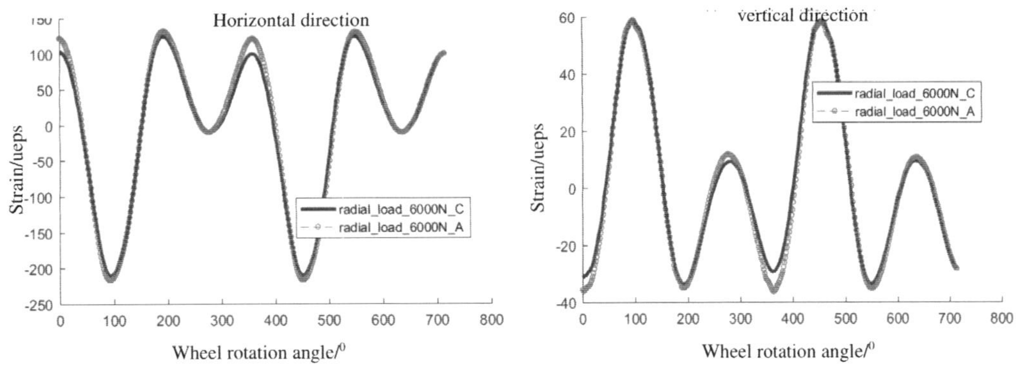

Figure 12. Measuring point 1 position strain change history

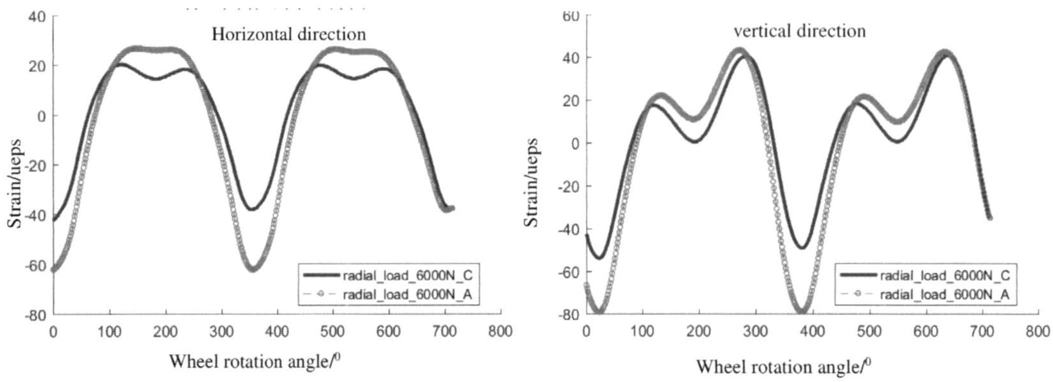

Figure 13. Measuring point 5 position strain change history

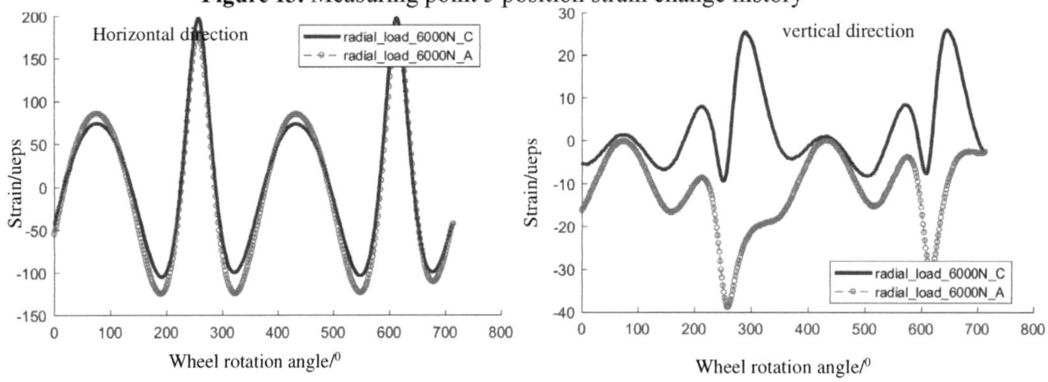

Figure 14. Measuring point 6 position strain change history

5. Conclusion

Based on the specific conditions of the wheel, this paper analyzes two typical operating conditions of wheel reliability. Based on this, the finite element simulation and test method is established to analyze the stress variation law during the radial fatigue test of the wheel. The following conclusions can be drawn:

- During the use of the wheel, the tire will have a large influence on the load of the wheel. The deformation of the tire and the ground, as well as the positive and negative rotation, will have an impact on the stress state of the wheel.
- When the wheel is running straight, the maximum stress of the wheel appears at the position where the tire is in contact with the ground. The stress near the center of the wheel is small.

Acknowledgements

During the project development process, the project was supported by the Center for Key Projects of China Automotive Technology and Research Center Co., Ltd. The project name is the bi-axis fatigue life test of the passenger car wheel, project number is 18200117

References

[1]. Zhang Shuai. Research on multi-disciplinary lightweight Optimization Design Method for Fatigue-Impact-Aerodynamic Performance of the Wheel [D]. *Jilin University*,2018.

[2]. Wang Yiliu. Multi-objective Optimization Design for the Lightweight Wheel of Structure-Materials Integration [D]. *Jilin University*,2016.

[3]. Grubisic V , Fischer G. *SAE Technical Paper Series [SAE International SAE International Congress and Exposition -, (FEB. 28, 1983)] SAE Technical Paper Series* - Automotive Wheels, Method and Procedure for Optimal Design and Testing[C]// SAE International Congress & Exposition. 1983.

[4]. Fischer G , Grubisic V. *SAE Technical Paper Series [SAE International SAE International Truck and Bus Meeting and Exposition -, (DEC. 03, 1984)] SAE Technical Paper Series* - Cast Aluminum Wheels for Trucks and Buses — Testing and Evaluation[J]. Fahrzeugrad, 1984, 1.
[5]. Grubisic V. Criteria and methodology for lightweight design of vehicle components subjected to random loads[J]. *SAE Transactions*, 1985: 71-85.
[6]. Rupp A N, Heinrietz A, Ehl O. Simulation of the experimental proof out of wheels and hubs[R]. *SAE Technical Paper*, 2002.
[7]. Neugebauer J R, Grubisic V, Fischer G. Procedure for design optimization and durability life approval of truck axles and axle assemblies[R]. *SAE Technical Paper*, 1989.
[8]. Heinrietz A, Lehrke H P, Rupp A N, et al. Identification of parametric tire models for the fatigue evaluation of suspension components[J]. *SAE transactions*, 2003: 1279-1286.
[9]. Rupp A N, Heinrietz A, Ehl O. Simulation of the experimental proof out of wheels and hubs[R]. *SAE Technical Paper*, 2002.
[10]. Fischer G , Grubisic V V . *SAE Technical Paper Series [SAE International International Truck & Bus Meeting & Exposition -, (NOV. 16, 1998)] SAE Technical Paper Series* - Design Criteria and Durability Approval of Wheel Hubs[J]. 1998, 1.
[11]. Herbert A, Fischer G, Ehl O. Load program development and testing of super single wheels in the biaxial wheel test rig and numerical pre-design[R]. *SAE Technical Paper,* 2004.
[12]. GB/T 5334-2005, Performance requirements and test methods of passenger car wheels [S].
[13]. Stearns J, Srivatsan T S, Prakash A, et al. Modeling the mechanical response of an aluminum alloy automotive rim[J]. *Materials Science and Engineering: A*, 2004, **366**(2): 262-268.

IOP Conf. Series: Materials Science and Engineering **784** (2020) 012022 doi:10.1088/1757-899X/784/1/012022

Development of Decision Support System for Solving the Machine Selection Problem in an Intermittent Manufacturing System Design

P Buathong, A Vilasdaechanont

Department of Industrial Engineering, Faculty of Engineering, Chulalongkorn University, Bangkok, Thailand

phinlada.bua@gmail.com

Abstract. Capacity planning, which is integral to resource selection and allocation, is an important task in manufacturing system design. As budget is limited, investment in resources must be done carefully and appropriately in order to achieve desired capacity and system efficiency. One of the most expensive and valuable resources in the manufacturing system is machinery. Therefore, selecting an appropriate machine is of high significance. Among many types of manufacturing systems, intermittent manufacturing system is complex, yet flexible. The intermittent manufacturing system enables manufacturers to produce multiple products using multiple processes and multiple machines. The system comprises a network of multiple types of multi-process machines. A machine selection problem in the context of the intermittent manufacturing system is a delicate process, as it must correspond appropriately with production complexities. To alleviate this problem, we develop a decision support system for solving the machine selection problem in an intermittent manufacturing system design. We study several manufacturing systems producing multiple types of products with various multi-process machines to lay out the framework. This decision support system consists of four main parts, namely; data entry, data conversion, optimization model and report. To solve the machine selection problem, an optimization technique is used to identify the optimal number of machines and the allocation of production process of each machine. The objective of this optimization problem is to minimize total production costs (machine cost and operating cost) while maintaining production capacity and other production constraints. The decision support system is developed for and validated with the manufacturing company.

1. Introduction

The intermittent manufacturing system allows companies to make various types of product, process and order quantities. Machinery should be flexible enough to produce a wide variety of operations. Machinery and equipment are usually grouped as stated by function they perform. Moreover, the flow of production is varied and depend on product features. Then, the operation can be made more than one time for each workstation.

Due to the increasing of alternative machines in the market, it leads to the complexity of machine selection decision. Most of the time, selecting an appropriate machine is time consuming, costly and it also requires expert knowledge. Therefore, we develop the decision support system of machine selection for help non-specialist to determine appropriate machine to achieve both of product and process of capacity planning. Existing capacity in the manufacturing system effects overall production performance such as capacity utilization, quality, on-time delivery, operating efficiency and production costs.

In this research, we propose the decision support system model that consists of four main parts because the machine selection process requires various data types and many steps. We need to clarify step-by-step processes that starts from data entry, data conversion, optimization model is used to determine both number and type of machines including the allocation of production process of each machine. The objective of optimization model aims to minimize total costs both machine cost and operating cost by consider the general manufacturing constraints of production time, number of machines and production constraints in order to satisfy customer's need. The final part is selected machine report.

2. Literature Review

Researches about machine selection process can be divided into two aspects, which are quality-based method and quantity-based method. The quality-based method is the selection method that uses quality data (criteria lists) to select the suitable alternative machine such as [1] provides the multi-criteria decision method and selects the highest score of machine lists. The quantity-based method is selection method that uses quantity data (capacity requirement, costs) such as [2] pays attention to the decision about product mix and capacity involve choosing between dedicated and multi-product facilities that concern both machine cost and operating cost. [3] proposes 0-1 linear programming and genetic algorithm by considering multi-objective to minimize various production costs; machine cost, material-handling cost, setup cost as well as maximize machine workload time. The model was developed to solve machine-tool selection and operation allocation problem. Moreover, [4] presents both a heuristic method named Five Simple Procedure (FSP) heuristic method and branch and bound method to assign operations to appropriate machine-tool combinations. The objective of model is to minimize some production costs. The results demonstrate that the FSP was efficiency to find good solutions and it helps a decision maker to select machine. [5] develop a decision support system for high speed milling machine tool selection. Since machine selection process is time-consuming and required advanced knowledge, [6] the decision support system is developed for the selection of machine tools. Multi-criteria weight average method by considering different criteria is used in process to rank the machine list.

3. Methodology

Due to the complexity of machine selection problems in the context, we develop the decision support system for supporting machine selection process. The machine selection process requires various data types involved in the production system and many steps to solve the problem. To simplify the complexity of decision process, we propose the methodology in four main parts Figure 1 that starting from import data process to acquire the solution. First part is about the collected data in three types of data. Second part is about data conversion that the collected data from previous part will be converted into the specific model template. The converted data is used to find the optimal solution in third part. Then, the report shows the optimum number and type of machine including allocation of production process of each machine type in fourth part.

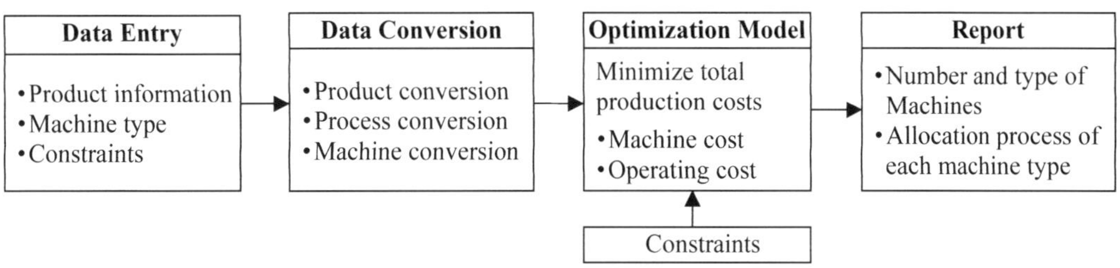

Figure 1. Flow process of the decision support system.

3.1. Data Entry

The machine selection process requires various types of data that can be collect from the person who involved in the manufacturing system design. The data entry is first part of the decision support system that consists of two data types namely; main data, which are basic product information, machine type, and constraints of the production system shows in Table 1 and additional data, which are changeover time and operating cost that was made machine selection model more realistic.

Table 1. Input data

Type of data	Components
Product information	Product type, Production process, Ratio and type of input and output part
Machine type	Machine type (Machine, Tool, Labor), Machine capability, Machine cost, Setting time
Constraints	Capacity requirement (Each product type), Production time (Working time per day), Number of machines (Maximum or minimum machine in system), Maximum budget

3.2. Data Conversion

The collected data are converted into the specific model template before inputting into the optimization model. The procedure of data conversion consists of three steps. First step, the product information and machine type are converted into specific abbreviation which the raw material of product type X is RMX, the process number Y is PY, the finished goods X is FGX and the machine type Z is MCZ. Second step, the converted data both product information and machine type are linked to define the capability of each machine type such as machine type Z can produce process Y of product X. Third step, the from-to table is generating for input changeover time by using both product and process capability of each machine.

3.3. Optimization Model

The converted data is used in the optimization model. We propose a mixed-integer linear programming (MILP) for solving machine selection problems. Both Visual Basic for Applications (VBA) and OpenSolver are develop for generating and solving the optimization model. We generate small problems to verify the proposed model could provide the correct solution.

3.3.1. Problem description. The optimization method is used to find the appropriate machine both number and type of machine and allocate the process to each machine. The objective of optimization model is to minimize total production costs both machine cost and operating cost. We also consider limitation of each manufacturing system such as production time, number of machine and production constraints. The assumptions of optimization model that the machine lifetime is five years.

3.3.2. Notations

Indexes

P	Process type index, P = 1,2, …, p
I	Part type index, I = 1,2, …, i
M	Machine type index, M = 1, 2, …, m

Parameters

$Input_{pim}$	Number of input part type i of process type p at machine type m
$Output_{pim}$	Number of output part type i of process type p at machine type m
$Capreq_i$	Capacity requirement of part type i
$Ratio_{pi}$	Ratio of process p of part type i
$MCost_m$	Machine cost of machine type m
$OCost_{pim}$	Operation cost of process type p of product i part type at machine type m
$TIME$	Production time per period
U_m	Percent allowance of capacity at machine type m
T_{pim}	Operating time of process type p of part type i produce at machine type m
$Budget$	Maximum budget
$MaxNum_m$	Maximum number of machine type m

IOP Conf. Series: Materials Science and Engineering **784** (2020) 012022 doi:10.1088/1757-899X/784/1/012022

Decision Variables

Num_m	Number of machine type m
$NumProcess_{pim}$	Number of cycles of process type p of part type i produce at machine type m

3.3.3. Model

Objective function

$$Minimize\ Z = \sum_{m\in M}(MCost_m)Num_m + \sum_{p\in P,i\in I,m\in M}(OCost_{pim})NumProcess_{pim} \quad (1)$$

Subject to constraints

$$\sum_{m\in M} Output_{pim}NumProcess_{pim} \geq Ratio_{pi}Capreq_i \quad ;\forall p\in P, \forall i \in I \quad (2)$$

$$\sum_{m\in M} Output_{pim}NumProcess_{pim} \geq \sum_{m\in M} Input_{p+1,i,m}\, NumProcess_{p+1,i,m} \quad ;\forall i \in I, \forall p \in P \quad (3)$$

$$Num_m U_m TIME \geq \sum_{p\in P,i\in I} T_{pim}NumProcess_{pim} \quad ;\forall m \in M \quad (4)$$

$$Num_m \leq MaxNum_m \quad ;\ \forall m \in M \quad (5)$$

$$\sum_{m\in M} MCost_m Num_m \leq Budget \quad ;\ \forall m \in M \quad (6)$$

$$Num_m, NumProcess_{pim} \geq\ 0, integer \quad ;\forall i \in I, \forall p \in P, \forall m \in M \quad (7)$$

The objective (1) is to minimize total costs consists of machine cost and operating cost. Constraint (2) is the capacity requirement each process of product. Constraint (3) is balancing constraint to be built finished goods. Constraint (4) is the available time of machine. Constraint (5) is the maximum number of each machine type. Constraint (6) limits the investment budget for the machine selection. Finally, decision variables are imposed to be positive integer in constraint (7).

3.3.4. Report. The output from the optimization model provides the optimal number and type of machine and process allocation to each machine type at the minimum total production costs. The selected machine is able to produce multiple products and multiple processes. One process can be produced by multi-machine types by considering constraints. Table 2 shows the selected machine solution that three MC1 machines allocate to produce process P1 of product A and P1, P2, P3 of product B.

Table 2. Selected machine report

Selected machine report			
Machine type	Number of machines	Product type	Process type
MC1	3	A	P1
		B	P1, P2, P3

4. Numerical Example

The decision support system is validated with various types of the manufacturing system that have difference characteristic such as mouse key manufacturing has to produce and assembly process with one product and fruit and vegetable processing manufacturing is multiple products using multiple processes. Therefore, we present one example is clamps manufacturing.

A decision maker has to prepare capacity for clamps manufacturing system. The production system consists of two product types that is characterized by size such as Large(A) and Small(B) clamps. Both of two product has seven steps. The processes have difference input and output part ratio and five difference types of machines. They operate eight hours per day and five days a week to produce clamps 100,000 units per month with a ratio 1:1. The report of example is shown in Table 3.

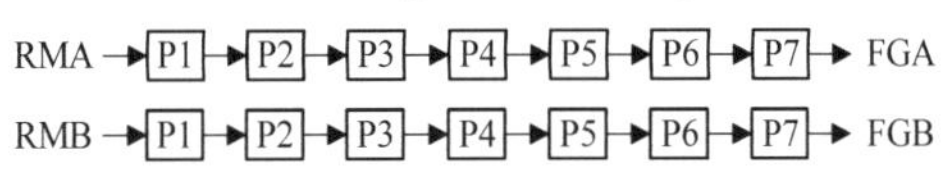

Figure 2. Clamps production process.

Figure 2 shows the clamps production process. To achieve finished goods A (FGA), the raw material of product A have to produce in seven steps (P1, P2, P3, P4, P5, P6, P7). The product B also have the production process as product A

Table 3. Report of example

Selected machine report			
Machine type	Number of machines	Product type	Process type
MC1	7	A, B	P1, P2, P3, P4, P5
MC3	1	A, B	P4, P6
MC4	2	A, B	P6, P7

Table 3 shows the optimal solutions of the clamps manufacturing system. The number of selected machine types MC1, MC3, and MC4 are seven, one and two respectively. Each process can be produced by multi-machine types such as machine MC3 and MC4 allocate to produce process P6 of product A and B. Moreover, only one machine type is able to produce multiple products and multiple processes such as machine MC1 is allocated to produce process P1, P2, P3, P4 and P5.

5. Results and Discussion

The results of the model evaluation indicate the ability of the decision support system that can be applied in various types of the intermittent manufacturing system. Then, the decision support system evaluated by users (engineers) who working with the manufacturing system design such as clamps manufacturing, fruit and vegetable processing manufacturing and car seat manufacturing.

Their responses show the program capabilities that user-friendly design, quick and easy way to identify a solution, able to handle complex and unexpected machine selection problems and selecting an appropriate machine at minimum costs. However, a user has to understand the machine selection process which the data affects the solution of selected machine and import data requirements correctly.

6. Conclusion

This research, we develop the decision support system that intends to help decision makers to select an appropriate machine in order to achieve desired capacity and system capability. The model attempts to minimize total production costs by considering constraints. The proposed model can be applied in other intermittent manufacturing systems that contains multiple products using multiple processes including multi-machine types. In this research, the numerical example used to validate the model.

The result shows that the features of model are easy to import data and get the report. Moreover, the model focuses on simplify both complexity and uncertainty of machine selection problem to help decision makers (engineers, managers and owners) to reduce time and resources in the decision-making process.

Acknowledgement

This research is funded by government budget of Chulalongkorn University, Bangkok, Thailand.

References

[1] Özgen A, Tuzkaya G, Tuzkaya UR and Özgen D 2011 A Multi-Criteria Decision Making Approach for Machine Tool Selection Problem in a Fuzzy Environment *Int. J. Comput. Int. Sys.* **4** 431-45

[2] Karmarkar U, Kekre S 1987 Manufacturing configuration, capacity and mix decisions considering operational costs *J. Manuf. Syst.* **6** 315-24

[3] Soolaki M, Zarrinpoor N 2014 A new 0-1 linear programming approach and genetic algorithm for solving assignment problem in flexible manufacturing system *Int. J. Adv.* **75** 385-94

[4] Jahromi MHMA, Tavakkoli-Moghaddam R 2012 A novel 0-1 linear integer programming model for dynamic machine-tool selection and operation allocation in a flexible manufacturing system *J. Manuf.* **31** 224-31

[5] Alberti M, Ciurana J, Rodriguez C and Özel T 2011 Design of a decision support system for machine tool selection based on machine characteristics and performance tests *J. Intell. Manuf.* **22** 263-77

[6] Arslan M, Çatay B and Budak E 2004 A decision support system for machine tool selection *J. Manuf. Technol. Manag.* **15** 101-9

Life Cycle Cost Analysis of Gas Turbine Air Inlet Filtration System

Yunshan Bai, Jiangang Hao, Xin Tian

HuaDian Electric Power Research Institute, Hangzhou, 210037, China

yunshan-bai@chder.com

Abstract. As an important component of gas turbine, air inlet filtration system is of great significance for ensuring the normal operation of gas turbine. However, it is often difficult to select a suitable air inlet filtration system according to the operating characteristics of the unit and the external environment due to the large number of variables. From the perspective of life cycle cost, the main factors to be considered in the selection of inlet filtration system are discussed, and the corresponding calculation model is proposed. Through the cost calculation of a gas-steam combined cycle unit's inlet filtration system during the whole life cycle, the influence of the initial cost, operation cost ,maintenance cost and gas turbine performance degradation cost on the total cost is analyzed. The results show that the cost of gas turbine performance degradation accounts for 90.44%, and the initial cost only accounts for 0.62% in the total cost. Therefore, in the selection of the inlet filtration system, the system economy should be evaluated from the perspective of full life cycle cost. The calculation model has important guiding significance for the selection and cost calculation of the gas turbine inlet filtration system.

1. Introduction

Gas-steam combined cycle unit has high efficiency and low pollution, and has been widely used in the word[1-4]. With the development of technology, the power output of gas turbine continues to increase, and the simple cycle power of GE 9HA.01 gas turbine reaches about 400 MW. With the increase of power output, the intake air flow of gas turbine also increases, and the influence of the intake air quality on the performance of gas turbine becomes more significant [5]. Due to the presence of a large amount of dust, oil and other particles in the air, if it is not efficiently filtered or inhaled by the gas turbine, it will cause problems such as corrosion and fouling of compressor blades, which will lead to a decrease in power output and efficiency, and even affect the safety of gas turbine [6-7]. In order to solve this problem, an air inlet filtration system is usually installed at the inlet of gas turbine to reduce the amount of dust entering the compressor in order to increase availability of the unit.

There are few researches on the selection of gas turbine inlet filtration system and the cost analysis of the whole life cycle in the world. Evaluating the cost of air inlet filtration system properly is of great guiding significance to the selection of the system. Literature [8] compares two inlet filtration systems with filtration accuracy of F9 and H11 respectively. Through economic analysis of the two systems, it is pointed out that improving the filtration accuracy of the inlet filtration system appropriately can significantly improve the economic benefits of the unit. Literature [9] preliminarily discussed factors affecting the life cycle cost of the inlet filtration system, and elaborated on how to select the inlet filtration system properly. In Literature [10], the effects of blade corrosion and fouling on the output and efficiency of the gas turbine were studied. Through theoretical derivation and data observation, the effects of different degrees of corrosion and fouling on the performance of gas turbine were obtained. Literature [11] studies the optimal replacement cycle of inlet filtration system under different operating conditions aiming at minimizing the unit power generation cost within the service cycle of inlet filtration system of a gas turbine. Literature [12] discussed in detail the principle, influence and control method of compressor blade fouling, and discussed the effects of online and offline washing on gas turbine performance recovery respectively, pointing out that the combination of two washing methods is more effective in alleviating the problem of compressor blade fouling.

At present, the life cycle cost calculation is rarely applied in the selection of inlet filtration system. Usually, only the initial cost of the system is considered, the operation and maintenance cost and cost due to efficiency decline caused by poor air filter are rarely be considered and evaluated. In order to solve this problem, this paper puts forward the calculation method of the life cycle cost of the gas turbine inlet filtration system, and discusses the various factors affecting the system cost, and establishes corresponding calculation model through theoretical derivation and practical experience.

2. Life Cycle Cost Analysis

The life cycle cost analysis of gas turbine inlet filtration system refers to the comprehensive evaluation of the costs of procurement, installation, operation, maintenance and treatment from the long-term economic benefits within the expected using time for decision-making and management. The cost of inlet filtration system is a cost incurred continuously over a long period of time, which should take into account the time value of the capital. In the cost analysis, the expenditures of each year in the life cycle of the system are first calculated, and then converted to the net present value of the initial equipment purchase according to the social discount rate and inflation rate. The net present value is calculated as shown in equation (1), where A is the expenditure for the current year, i is the discount rate, and n is the year of expenditure. The calculation of the net present value in equation (2) takes into account the effects of inflation, and e is the inflation rate. If the annual expenditure is a fixed value during the system life cycle, the sum of the net present values of each year can be calculated according to formula (3) (4). The National Development and Reform Commission's "Economic Evaluation Methods and Parameters for Construction Projects (3rd Edition)" stipulates that the discount rate can be 8% and the inflation rate can be 4% [13].

$$NPV = A(1+i)^{-n} \tag{1}$$

$$NPV = A(1+(i-e))^{-n} \tag{2}$$

$$NPV = \frac{A}{i}\left[1-(1+i)^{-n}\right] \tag{3}$$

$$NPV = \frac{A}{i-e}\left[1-(1+i-e)^{-n}\right] \tag{4}$$

3. Life Cycle Cost Calculation of Gas Turbine Inlet Filtration System

The life cycle cost of gas turbine inlet filtration system includes initial cost, operation and maintenance cost, downtime cost, gas turbine performance degradation cost, removal and disposal cost, as shown in formula (5), where NPV_{total} is the life cycle cost of the inlet filtration system, NPV_{init} is the initial cost of the system, NPV_{main} is the operating and maintenance cost of the system, NPV_{down} is the cost of downtime, NPV_{GT} is the cost of performance degradation of gas turbine due to the inlet filtration system, and NPV_{disp} is the removal and disposal cost of the system. The calculation methods for each part are discussed in detail below.

$$NPV_{total} = NPV_{init} + NPV_{main} + NPV_{down} + NPV_{GT} + NPV_{disp} \tag{5}$$

3.1. Initial cost

The initial cost NPV_{init} refers to the sum of all expenditures during the infrastructure period, including the purchase cost of the filter NPV_{filt}, the installation cost NPV_{inst}, and the commissioning cost NPV_{comm}, as calculated by equation (6). The calculation of filter purchase cost can be consulted with the air inlet filtration system supplier or the purchase contract can be viewed. The prices of different manufacturers and different types of filters are not consistent. Under normal circumstances, for static filters, the final filter element price is 500~1000 yuan, the second stage filter unit price is 150~300 yuan, the first stage filter unit price is 100~150 yuan, a pair of pulse filter is 1200~2000 yuan. The determination of the number of filter elements is related to the model of the gas turbine and the user's preference. In this paper, the number of filter elements of several common types of gas turbines in Jiangsu and Zhejiang provinces is counted, which is shown in table 1. The cost of installation and commissioning can be referred to the previous gas turbine generator set, and the costs are basically the same.

$$NPV_{init} = NPV_{filt} + NPV_{inst} + NPV_{comm} \tag{6}$$

3.2. Operation and maintenance cost

Operation and maintenance cost NPV_{main} runs through the entire life cycle of the inlet filtration system and is an important part of the system's life cycle cost. It mainly includes the replacement cost of the filter element NPV_{repl}, the periodic inspection cost of the inlet filtration system NPV_{insp} and maintenance cost NPV_{repair}. The calculation is shown in equation (7).

In this part, it is important to determine the replacement period of the filter element. Due to the different environmental conditions, unit operation mode, and filtration system configuration in each area, the replacement period of the filter elements is quite different. For newly built gas turbine generator set, the replacement cycle of filtration system can be referred to the previous one. Table 1 shows the filtration system replacement cycle of five gas turbine generator sets in Jiangsu and Zhejiang. It can be seen that the replacement period of these five filtration systems is basically between 3000h to 4000h. The inspection and maintenance costs of the intake housing are relatively low and can be given empirically.

$$NPV_{main} = NPV_{repl} + NPV_{insp} + NPV_{repair} \tag{7}$$

3.3. Downtime cost

The cost of downtime refers to the loss caused by the abnormal operation of the inlet filtration system or the poor filtration performance, which causes the power loss of gas turbine, including filter replacement and offline water washing. The replacement of the filter needs to stop the gas turbine. For a gas turbine generator set with basic load and long-time running or for winter heating, it will cause loss of power output, thus increase the life cycle cost of the inlet filtration system. This cost can be calculated by the filter replacement cycle and replacement time. Generally the filter replacement can be completed in one day. For the peaking unit, the filter replacement and offline water washing can be performed at night without any increase in cost.

3.4. Gas turbine performance degradation cost

The filter performance of inlet filtration system has a great influence on gas turbine performance, which is reflected in the following two aspects:

(1) The effect of compressor fouling.

It is impossible for the air inlet filtration system to intercept all the particles in the air, so some particles will penetrate the filter and enter inside of gas turbine, and attach to compressor blades to increase the surface roughness of blade, and even change the aerodynamic shape of the blade. As a result, the power output of gas turbine is reduced and gas consumption rate is increased, which increases the operating cost of gas turbine [14-15]. In areas with serious environmental pollution, the average concentration of suspended particulate matter of 0.5 um or above in the air is $3\times10^7/m^3$. For the filter of F8 class, about 6×10^6 suspended particulate matter per cubic meter are inhaled into internal flow channel of gas turbine [8]. Literature [16] points out that the performance degradation of the unit is most obvious within the first 1000 h of operation time. After 1000h, the change gradually tends to be stable, and the degree of performance degradation can be expressed by the exponential form of operation time. Equation (8) (9) gives the calculation method for the cost increase due to compressor fouling. In the formula, A^j_{foul} is the cost due to compressor fouling in the jth year, C_{kwh} is the on-grid price, Q_{std} is the gas consumption of the unit during daily operation, C_{scm} is the natural gas price of per cubic meter, P is the capacity of the power generation unit, P_{loss1} is the power loss due to compressor fouling, η_{q1} is the increase of gas consumption due to compressor fouling, and T is the operation time of the unit in the jth year.

$$NPV_{foul} = \sum_{j=1}^{n} A_{foul}^{j}\left[1+(i-e)\right]^{-j} \tag{8}$$

$$A_{foul}^{j} = \int_{0}^{T}\left[P_{loss1}C_{kwh} - P_{loss1}\left(1+\eta_{q1}\right)Q_{std}\times C_{scm}\right]dt + \int_{0}^{T}\left[P\times\eta_{q1}Q_{std}\times C_{scm}\right]dt \tag{9}$$

(2) Influence of pressure loss in the inlet filtration system

The inlet air pressure has a great influence on the performance of gas turbine generator set. Literature [17] points out that for every 250 Pa increase in inlet air pressure loss, the power output of gas turbine is reduced by 0.5% and the heat consumption is increased by 0.1%. The inlet air pressure loss generally increases non-linearly with the filter usage time, and the increase in the cost of the inlet filtration system is calculated as follows:

$$NPV_{\Delta p} = \frac{A_{\Delta p}}{i-e}\left[1-\left(1+i-e\right)^{-n}\right] \tag{10}$$

$$A_{\Delta p} = \int_{0}^{T}\left[P_{loss2}C_{kwh} - P_{loss2}(1+\eta_{q2})Q_{std}\times C_{scm}\right]dt + \int_{0}^{T}\left[P\times\eta_{q2}Q_{std}\times C_{scm}\right]dt \tag{11}$$

3.5. Removal and disposal cost

When the service time of filter reaches the requirement of replacement, the original filters need to be removed and processed. Literature [18] gives the estimated cost of removal and disposal of different types of filter elements. After the operation time of gas turbine reaches its life cycle, the residual value of the equipment exists in the steel structure such as the inlet housing shell and bracket, but the residual value is extremely low, which can be ignored in the cost calculation of the whole life cycle of the system.

Table 1. Filter number and replacement cycle of different type of gas turbine

Unit Type	9E	9F	M701F4	6F.03	V94.2
Filter type	Self-cleaning	Self-cleaning	Self-cleaning	Static	Self-cleaning
Number of filter elements	Fine filter:464	Fine filter:672	Course filter:153 Fine filter:900	Course filter:144 Fine filter:144	Course filter:504 Fine filter:616
Replacement cycle	3400 h	3600 h	3100 h	Course filter:2200 h Fine filter:2800 h	4300 h

Table 2. NPV of different items and the corresponding percentage

item	Initial cost	Operation and maintenance cost	Downtime cost	Performance degradation cost	Removal and treatment cost
NPV of each part (Ten thousand yuan)	102.00	1438.35	0.00	14941.56	38.01
percentage(%)	0.62	8.71	0.00	90.44	0.23

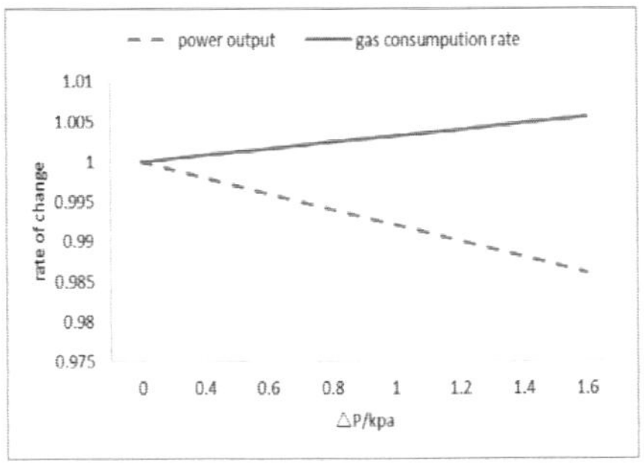

Fig 1. Influence of inlet pressure loss to power and heat efficiency

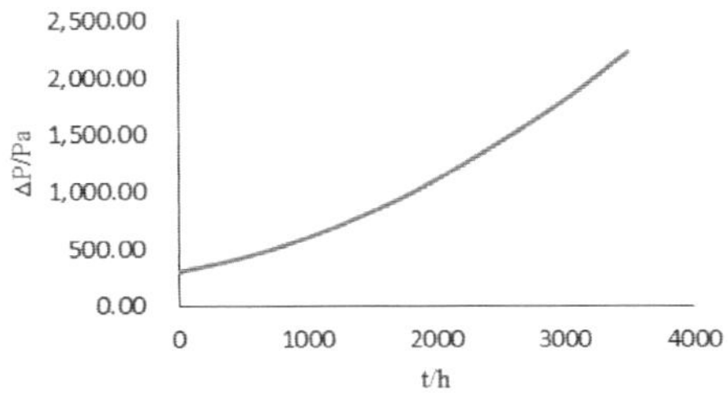

Fig 2. Curve of inlet pressure loss versus operating time

4. Examples and Analysis

The rated power generation of a 9FA gas-steam combined cycle generator set is 390 MW, and the gas turbine power is 255 MW. It is a single-shaft combined cycle generator set with gas consumption rate of 0.18 Nm^3/kWh. The inlet filtration system is equipped with self-cleaning filters, in total of 670 pairs of filter elements, and the average replacement cycle of filter elements is 3500h. The average annual running time of the unit is also 3500h, the on-grid price is 0.873 yuan/kWh, and the price of natural gas is 2.17 yuan/m^3. The cost of inlet filtration system during life cycle is calculated as follows:

(1) Initial cost: The purchase cost of the filter element of the air inlet filtration system is 1500 yuan/pair, plus installation and commissioning costs, the initial cost is 1.02 million yuan.

(2) Operation and maintenance cost: Assume that the life cycle of gas turbine is 20 years and the filter element is replaced once a year. In the life cycle of gas turbine, the net present value of the filter replacement cost is 13.71 million yuan, and the inspection and maintenance cost of the inlet housing is the 50000 yuan/year, the net present value of the operation and maintenance costs of the inlet filtration system during the life cycle of gas turbine is 14.33 million yuan.

(3) Downtime cost: Because the unit is a peaking unit, the replacement of the filter element and the washing of the compressor can be carried out at night, without affecting the normal power generation of the unit, so the cost of downtime is not considered.

(4) Performance degradation cost of gas turbine: In actual operation of the unit, the load rise and fall usually follow instructions of the dispatching. In order to facilitate calculation, assuming that the unit operates at base load, the rate of power loss of the unit is 1.5% per year, and the increase of gas consumption rate is 0.5% per year. According to the formula (8)(9), the net present value of performance degradation cost caused by compressor fouling is 88.378 million yuan. The curves of pressure loss variation with time and the power output and gas consumption variation curve of the unit with pressure loss are shown in figure (1) and (2) respectively. According to equation (10) and (11), the net present value of the performance degradation cost caused by the air pressure loss is 61,036,900 yuan.

(5) Removal and disposal cost. Assuming that the cost of removal and disposal of each pair of filters is 30 yuan, the net present value of filter removal and disposal costs during the life cycle of gas turbine is 380100 yuan.

In summary, the net present value of the total cost of the inlet filtration system during life cycle of gas turbine is 16.519.92 million yuan. The net present value and percentage of each item are shown in Table 2. Among the total costs, the performance degradation cost accounted for the highest proportion, which was 90.44%, followed by the operation and maintenance cost, accounting for 8.71%, and the initial cost ratio was only 0.62%. For this type of gas-steam combined cycle generator set, if conditions permit, the cost caused by inlet air pressure loss and blade fouling on the performance of gas turbine should be reduced by increasing the inlet filter area and improving the filtration accuracy, and the various costs of the system should be balanced reasonably.

5. Conclusion

(1) From the perspective of life cycle cost, this paper comprehensively considers the initial cost, operation and maintenance cost, downtime cost, gas turbine performance degradation cost, removal and disposal cost, and proposes a new method to calculate the life-cycle cost of gas turbine inlet air filtration system.
(2) Factors affecting the cost of whole life cycle of the inlet filtration system are discussed, and the calculation model of each part is established based on the actual use of the domestic gas turbine generator set, which solved the problem that the cost of each part cannot be quantified in the previous inlet filtration system selection.
(3) The calculation of life cycle cost of the inlet filtration system of a combined cycle generator set shows that gas turbine performance degradation cost accounted for the highest proportion, reaching 90.44%, and the corresponding initial cost and operation and maintenance cost accounted for 0.62% and 8.71% respectively, the degradation cost of gas turbine performance is a key factor affecting the economics of the inlet filtration system.
(4) In the initial design and selection of the inlet filtration system, in addition to the initial procurement cost, the cost of gas turbine performance degradation caused by the inlet filtration system should be paid more attention. According to the operating characteristics of the unit, it is possible to obtain the optimal air inlet filtration system and the maximum economic benefits by minimizing the total cost and balancing the costs comprehensively within the whole life cycle.

6. References

[1] Wright I G and Gibbons T B 2007. *Recent developments in gas turbine materials and technology and their implications for syngas firing. International Journal of Hydrogen Energy.***32(16)**:3610-3621.
[2] JIANG Hongde, REN Jing and LI Xueying 2014. *Status and Development Trend of the Heavy Duty Gas Turbine. In: Proceedings of the CSEE.* **34(29)**: 5096-5012.
[3] Xie Daxing, Shi Yongfeng and Zheng Jian 2016. *The Latest Development of H-class Gas Turbine. Gas Turbine Technology.***29(4)**:1-3.
[4] Poullikkas A 2005. *An overview of current and future sustainable gas turbine technologies. Renewable & Sustainable Energy Reviews.***9(5)**:409-443.
[5] Frumento A 2015. *Performance modeling of industrial gas turbine with inlet air filtration system filtration. Case Studies in Thermal Engineering.* **5(11-12)**:160-167.
[6] Heymanns E, Kazmeier F and Schild W 1992. *Gas Turbine Axial Compressor Fouling: A Unified Treatment of its Effects, Detection, and Control. International Journal of Turbo & Jet Engines.* **9(4)**:311-334.
[7] Mishra R K 2015. *Fouling and Corrosion in an Aero Gas Turbine Compressor. Journal of Failure Analysis & Prevention.***15(6)**:1-9.
[8] Schroth T and Cagna M 2008. *Economical Benefits of Highly Efficient Three-Stage Intake Air Filtration for Gas Turbines. In: ASME Turbo Expo 2008: Power for Land, Sea, and Air.* 5177-85.
[9] Wilcox M and Brun K 2011. *Gas Turbine Inlet Filtration System Life Cycle Cost Analysis. In: ASME 2011 Turbo Expo: Turbine Technical Conference and Exposition.* 675-682.
[10] Zwebek A and Pilidis P 2001. *Degradation Effects on Combined Cycle Power Plant Performance—Part I: Gas Turbine Cycle Component Degradation Effects. Journal of Engineering for Gas Turbines & Power.* **125(3)**:651-657.
[11] CHEN Jianhong, CHENG Yuan and SHENG Deren 2014. *Research on Update Interval Optimization Strategy Model of Gas Turbine Filtration[J]. In: Proceedings of the CSEE.* **34(20)**: 3302-3307.
[12] Kiakojoori S and Khorasani K 2016. *Dynamic neural networks for gas turbine engine degradation prediction, health monitoring and prognosis. Neural Computing & Applications.* **27(8)**:1-36.

[13] National Development and Reform 2008, *Commission Ministry of Construction. Economic Evaluation Method and Parameter Summay About Construction Projects.3rd edition.*(Beijing: China Planning Press)

[14] Kiakojoori S and Khorasani K 2016. *Dynamic neural networks for gas turbine engine degradation prediction, health monitoring and prognosis. Neural Computing & Applications.* **27(8)**:1-36.

[15] Li Shuming, Li Shidong and Zhang Ying 2015. *Quantitative Analysis of Impact Factors of Aircraft engine compressor performance degradation. Science Technology and Engineering.* **(32)**:74-78.

[16] Hepperle N, Therkorn D and Schneider E 2011. *Assessment of Gas Turbine and Combined Cycle Power Plant Performance Degradation. In: ASME 2011 Turbo Expo: Turbine Technical Conference and Exposition.* 569-577.

[17] Wilcox M, Kurz R and Klaus B 2011. *SUCCESSFUL SELECTION AND OPERATION OF GAS TURBINE INLET FILTRATION SYSTEMS. Mass Transport Phenomena in Ceramics. Springer US*, 149-154.

[18] Matela D M and Arnold B D 2005. *Life-Cycle Costing of Air Filtration. Ashrae Journal.* **47(11)**:30-32.

IOP Conf. Series: Materials Science and Engineering **784** (2020) 012024 doi:10.1088/1757-899X/784/1/012024

Projection Intensity Adjustment Method Based on Multi-threshold for Fringe Projection Technology

Changzhi Yu[1,*], Fang Ji[1] and Fang Li[1]

1.Institute of Mechanical Manufacturing Technology, China Academy of Engineering Physics, Mianyang 621000, China

*Corresponding author's e-mail: yuczcaep@163.com

Abstract. Three dimension measurement based on fringe projection technology is widely used in precision manufacturing. However, when measuring objects with reflective surfaces, the measurement accuracy is reduced due to image saturation. This paper presents a projection intensity adjustment method based on multi-threshold. The reflectivity information of fringe pixels and scale factor are used to calculate the multi- threshold. According to the multi-threshold, the optimal projection intensities are obtained, which reduces the image saturation and improves the measurement accuracy. The experimental results show that this method can improve the measurement accuracy effectively.

1. Introduction

Fringe projection technology (FPT) has been widely applied for 3D shape reconstruction in precision manufacturing, reverse engineering, etc. [1-2]. As we all known, the fringe projection system consists of several cameras and several projectors [3]. With the help of phase-shifting algorithm, fringe projection technology can easily achieve high-precision measurement. In recent years, with the progress of digital light processing (DLP) projector, fringe projection technology has been developed rapidly.

However, FPT is more suitable for diffuse reflective surfaces. If the object with high reflective surface was measured by FPT, saturation regions will appear on the fringe images. The phase-shifting algorithm could not calculate the phase values of saturation regions, which reduce the application of FPT[1-3]. To solve the limitations of FPT, many works have been studied. Zhang [4] developed a scanning technique to obtain fringe images under different exposures. Analogously, Waddington [5] obtained many fringe images with different intensities. Zhang [6] obtained several groups of fringe images with optimal light intensity. Combining with the advantages of binocular vision and monocular fringe projection, Yu [2] proposed an adaptive binocular fringe dynamic projection method. Lai [7] provided a Moire Fringe Patterns Sequence algorithm to measure the over-dark and over-bright surfaces. In this paper, we propose a projection intensity adjustment method (PIAM) based on multi-threshold. Firstly, we explain the phase error caused by image saturation. Then, the segmentation threshold of pixel reflectivity is calculated. Several reflectivity thresholds are obtained by introducing scale factor. Thus the optimal projection gray groups can be calculated. Experimental results show that the proposed method is effective.

2. Projection intensity adjustment method based on multi-threshold

For fringe projection technology, assuming A is the average intensity, B is the intensity modulation, φ is the phase to be solved for, the projection intensity with N-step phase-shifting algorithm can be represented as,

$$I(x,y)=A(x,y)+B(x,y)\cos\left[\varphi(x,y)+\delta_N\right],\ \delta_N=\frac{k*2\pi}{N},N=3,4,5\cdots,k=0,1,\cdots,N-1 \quad (1)$$

$$\varphi(x,y)=-\arctan\left[\frac{\sum_{n=1}^{N}I_n(x,y)\sin\delta_N}{\sum_{n=1}^{N}I_n(x,y)\cos\delta_N}\right]. \quad (2)$$

In order to calculate easily, the average intensity value is equal to the intensity modulation value, that is, A=B. Generally, three-step and four-step phase-shifting algorithm with equal phase-shifting are widely used in FPT [2]. For N-step phase-shifting algorithm, let $I_i^C(x,y)$ be the projection intensity, the gray value captured by camera can be written as,

$$\tilde{I}_i^C(x,y)=\begin{cases}I_i^C(x,y) & I_i^C(x,y)\le 255\\ 255 & I_i^C(x,y)>255\end{cases},\ i=1,2,\cdots,N. \quad (3)$$

Thus, the intensity error and phase error[8] caused by image saturation can be calculated as

$$\Delta I_i^C(x,y)=I_i^C(x,y)-\tilde{I}_i^C(x,y)=\begin{cases}0 & I_i^C(x,y)\le 255\\ I_i^C(x,y)-255 & I_i^C(x,y)>255\end{cases}, \quad (4)$$

$$\Delta\varphi(x,y)=\frac{4}{NI(x,y)}\sum_{i=0}^{N-1}\sin\left[\varphi(x,y)-\frac{2\pi}{N}\right]\cdot\Delta I_i^C(x,y). \quad (5)$$

From the equation (5), if N is large enough, the phase error can be negligible, which will result in low measurement efficiency. Let I_{opt}^C be the optimal intensity captured by camera. According to the intensity response function[2,6], the corresponding optimal projection intensity is presented as[6]

$$I_{opt}^P=\frac{I_{opt}^C-b}{a} \quad (6)$$

where a is the pixel reflectivity value, b is the maximum value of ambient light. Let T be the Otsu threshold of pixel reflectivity, and $k=(k_l,k_h)$ is scale factor. Thus, we can get three thresholds, that is, k_lT,T,k_hT. The corresponding projection intensities can be expressed as

$$I_{N1}^P=\frac{I_{opt}^C-b}{k_lT},I_{N2}^P=\frac{I_{opt}^C-b}{T},I_{N3}^P=\frac{I_{opt}^C-b}{k_hT} \quad (7)$$

According to equation (1) and (7), groups of projection intensities for 3-step and 4-step phase-shifting can be given respectively as,

$$G_i^3:\begin{cases}I_{3i1}^P=\frac{I_{3i}^P}{2}+\frac{I_{3i}^P}{2}\cos\left[\varphi(u,v)-\frac{2\pi}{3}\right]\\ I_{3i2}^P=\frac{I_{3i}^P}{2}+\frac{I_{3i}^P}{2}\cos\varphi(u,v)\\ I_{3i3}^P=\frac{I_{3i}^P}{2}+\frac{I_{3i}^P}{2}\cos\left[\varphi(u,v)+\frac{2\pi}{3}\right]\end{cases},\ i=1,2,3.\quad G_i^4:\begin{cases}I_{4i1}^P=\frac{I_{4i}^P}{2}+\frac{I_{4i}^P}{2}\cos\left[\varphi(u,v)\right]\\ I_{4i2}^P=\frac{I_{4i}^P}{2}+\frac{I_{4i}^P}{2}\cos\left[\varphi(u,v)+\frac{\pi}{2}\right]\\ I_{4i3}^P=\frac{I_{4i}^P}{2}+\frac{I_{4i}^P}{2}\cos\left[\varphi(u,v)+\pi\right]\\ I_{4i4}^P=\frac{I_{4i}^P}{2}+\frac{I_{4i}^P}{2}\cos\left[\varphi(u,v)+\frac{3\pi}{2}\right]\end{cases},\ i=1,2,3,4.$$

3. Experimental results

We analyze the phase error caused by image saturation for the three-step phase-shifting method and the four-step phase-shifting method, as shown in figure 1 and figure 2. To better illustrate the necessity of saturation modification, fringe images with different saturation degrees are simulated. Phase and phase error are also presented under different saturation degrees. The simulations are carried out with three-step and four-step phase-shifting under different conditions of saturation degree.

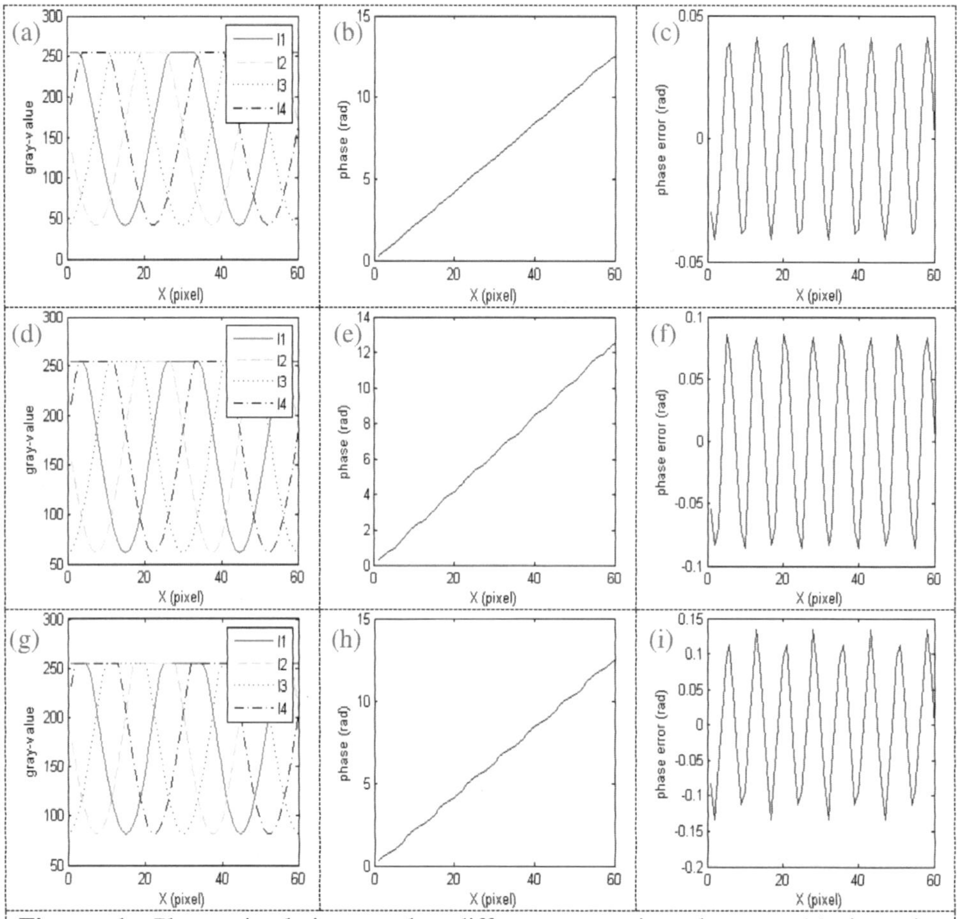

Figure 1. Phase simulations under different saturation degrees. (a) intensity distribution with low saturation; (b) absolute phase with (a); (c) phase error with (a); (d) intensity distribution with middle saturation; (e) absolute phase with (d); (f) phase error with (d); (g) intensity distribution with high saturation; (h) absolute phase with (g); (i) phase error with (g).

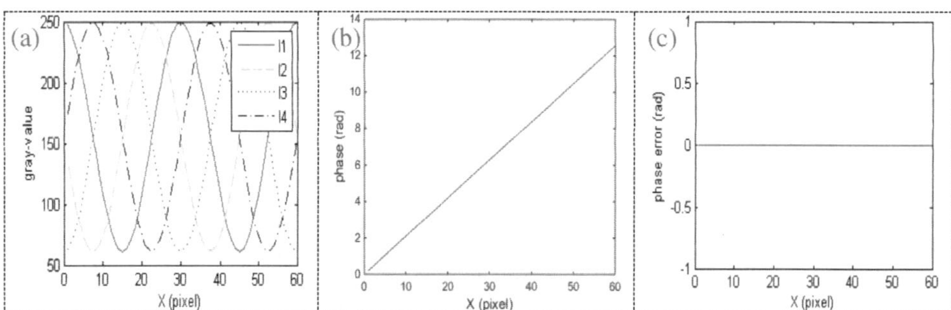

Figure 2. Phase simulation after modification. (a) intensity distribution after

modification, (b) absolute phase after modification, (c) phase error after modification.

From figure 1, the phase error raises with the increase of saturation degree. The maximum phase errors are 0.0411, 0.0859, and 0.1337 rad for figure 1(a), figure 1(d) and figure 1(g) respectively. Phase simulation after modification for figure 1(a) is shown in figure 2. Comparing with figure 1(a), the intensity distribution after modification presents sinusoidal variation, and the absolute phase is a straight line with a phase error about zero. Therefore, the simulation results show the effectiveness of the proposed method.

Several calibration balls with reflective surface are designed in figure 3, and their diameters are needed to be measured. To better verify the performance of PIAM method, compared with Coordinate Measuring Machine (CMM) and FPT method [4-5], the measurement results are shown in table 1.

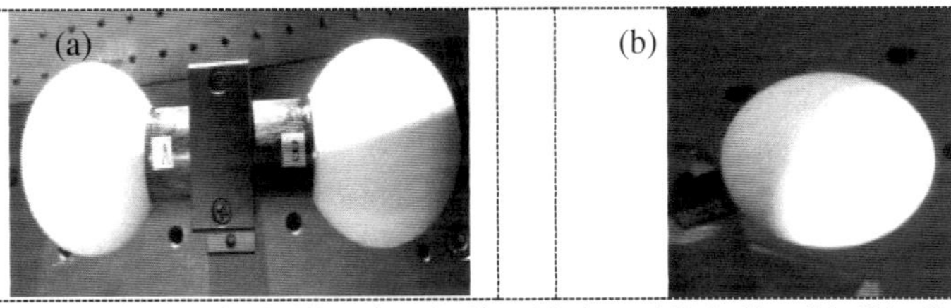

Figure 3. Calibration balls: (a) A and B balls; (b) C ball.

Table 1. Measurement results.

Balls	CMM /mm	FPT [4-5] /mm	PIAM/mm	Mean value/mm	Mean error/mm
			50.805		
			50.803		
A	50.799	50.782	50.807	50.804	0.005
			50.803		
			50.805		
			50.799		
			50.803		
B	50.797	50.784	50.803	50.802	0.005
			50.802		
			50.803		
			25.404		
			25.403		
C	25.399	25.421	25.404	25.403	0.004
			25.403		
			25.404		

Compared with FPT, the mean measurement error values with PIAM for three objects are 0.005 mm, 0.005 mm and 0.004 mm respectively, which shows that PIAM has a higher measurement accuracy. The above experiments demonstrate the effectiveness of the proposed PIAM method. The proposed PIAM method also extends the application range of fringe projection technology.

4. Conclusion

In this paper, we proposed a method to adjust the projection intensity for fringe projection technology. With the aid of intensity response function, we analyzed the segmentation threshold of reflectivity of image pixels. Through multi- threshold, we can obtain several groups of different optimal projection intensities. Thus, the accuracy of fringe projection measurement is improved. Experimental results on three objects show the effectiveness of the proposed method.

Acknowledgments

National Natural Science Foundation of China (51705485), and China Academy of Engineering Physics (K1204).

References

[1] Zhang, S. (2018) Absolute phase retrieval methods for digital fringe projection profilometry: a review. Opt. Lasers Eng., 107: 28-37.
[2] Yu C., Ji F., Xue J., Wang Y. (2019) Adaptive binocular fringe dynamic projection method for high dynamic range measurement, Sensors, 19(18), 4023.
[3] Sun, X., Liu, Y., Yu, X., Wu H., Zhang N. (2017) Three-dimensional measurement for specular reflection surface based on reflection component separation and priority region filling theory. Sensors, 17(1), 215.
[4] Zhang, S., Yau, S. T. (2009) High dynamic range scanning technique. Opt. Eng., 48(3), 033604.
[5] Waddington, C.; Kofman, J. (2014) Camera-independent saturation avoidance in measuring high-reflectivity-variation surfaces using pixel-wise composed images from projected patterns of different maximum gray level. Opt. Commun., 333, 32–37.
[6] Zhang, L., Chen, Q., Zuo, C., Feng, S. (2018) High dynamic range 3D shape measurement based on the intensity response function of a camera. Appl. Optics, 57(6), 1378-1386.
[7] Lai, J., Li, J., He, C., Liu, F. (2019). A robust and effective phase-shift fringe projection profilometry method for the extreme intensity. Optik, 179, 810-818.
[8] Sheng, H., Xu, J., Zhang, S. (2017) Dynamic projection theory for fringe projection profilometry. Appl. Optics, 56(30), 8452-8460.

Static Simulation of Electric Tilting Tricycle Frame Structure Using The Finite Element Analysis

Mutqin Abdul Muiz[1], Muhammad Kusuma Jati[1], Farras Prasrtyo Cahyono[1], and Danardono Agus Sumarsono[1,*]

[1]Mechanical Engineering Department, University of Indonesia, Depok 16424, Indonesia

*Corresponding author: danardon@eng.ui.ac.id

Abstract. Electric Tilting Tricycle is a three-wheeled vehicle that has tadpole configuration driven by two sources of motion, that motion by a manual pedal and power from an electric motor. The three-wheeled vehicle has better stability and the tilting system improves ergonomics at curves. This research aims to design and analyze the strengths frame with three different materials using the Autodesk inventor software. By using the Stress Analysis feature that comes with the method of finite element analysis (FEA), we can know the result of von misses stress, displacement, strain, and safety factor. The main consideration of structural strength was seen from the distribution of static loads using three types of materials (Aluminium 6061, Steel Mild and CFRP). The minimum factor of safety resulting from the design simulation was estimated to be 1.83 using steel mild material.
Keywords: Electric Tilting Tricycle, frames, Finite Element Analysis

1. Introduction

A three-wheeled vehicle can be alternative transportation that provides a higher level of safety because of its stability is equivalent to a four-wheeled vehicle with smaller dimensions. Three-wheeled vehicles with two-wheel configuration in front and one behind are also called tadpoles. The technology in this type of vehicle that can be implemented is tilting system. The tilting system on the tadpole increases the stability and comfort of driving, especially when around corners. When cornering using a tadpole that has a tilting system the driver does not feel the force caused by the force of gravity, centripetal acceleration, and torque from the tilting. When the tilting system works, the resultant direction of the force is perpendicular to the axis of the driver's body, that the lateral acceleration is less pronounced by the driver when turning. [1-2,7]
Engineers are currently developing energy sources that are friendly to the environment with renewable sources, one of the sources is electrical energy. The application of electrical energy to a bicycle can reduce the energy needed to pedal a bicycle and cut the time taken. [2-3]
The finite element method is a numerical approach using digital computing. With the advent of high-speed and large-capacity digital computers, FEA is increasingly being used throughout the industry as a practical analysis solution for designs with high complexity.[3-6] This means the design with an unusual shape, can be solved with this method. Also, FEA can resolve the conditions of heterogeneous materials. (Syehan, 2019)

2. Methods

2.1. Component Design and Assembly

The first step of this study was to make a design using Autodesk Inventor 2018 software to make the frame shape of the electric tilting tricycle. one of the things to consider in the design of the elements was the contacts of each tube of the frame. In the present study, the parameters considered were the design of the frame and the materials used without the effect of welding. Because of that reason, all of the tubes were assumed as single element without using weld as a joint.

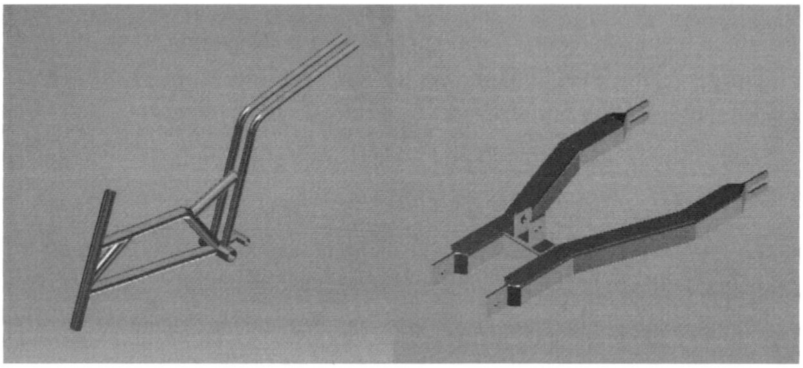

Figure 1. Electric tilting tricycle components.

The assembly process was a process of assembling the elements of the electric tilting frame that contain mainframe, suspension mounting, swingarm bushing, and swingarm. The important thing in the assembly process is the type of joint between the elements. Six types of joint provided by Autodesk Inventor namely rigid, rotational, slider, cylindrical, planar, and ball. In the present study, only two types of joints used to make the design represented the actual electric tilting tricycle frame. The elements to be assembled with the rigid joint was suspension mounting to mainframe and the rotational joint was swingarm to mainframe. Since it was a static simulation and to simplify the simulation, the suspension assumed as a rigid element.

2.2. Constarints and Load in Assembly

In present studies, the types of constraints used in the simulation of electric tilting tricycle design were fixed and pin. The use of fixed constraint is located at the rear of the swingarm which is used as the rear wheel shaft and also on the head tube that will be connected to the tilting system of the bicycle. These constraints were used to represent components not included in frame design. In the simulation, the fixed constraint of the rear axle shaft will give a reaction force from the load given by the wheels, as well as the head tube section which gives the reaction of the front wheel. Whereas the pin constraint is used at the joint between the swingarm and mainframe, this makes one degree of freedom at that point which limiting motion when the joint is given by reaction force of the rear wheel and the load of the driver on the seat tube.

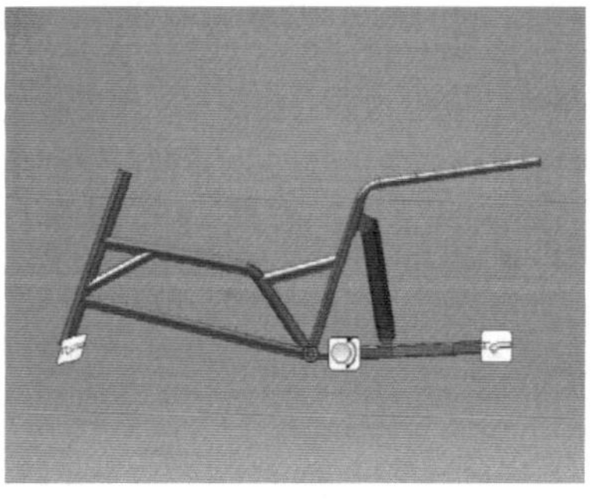

Figure 2. Constraints used on the electric tilting tricycle frame.

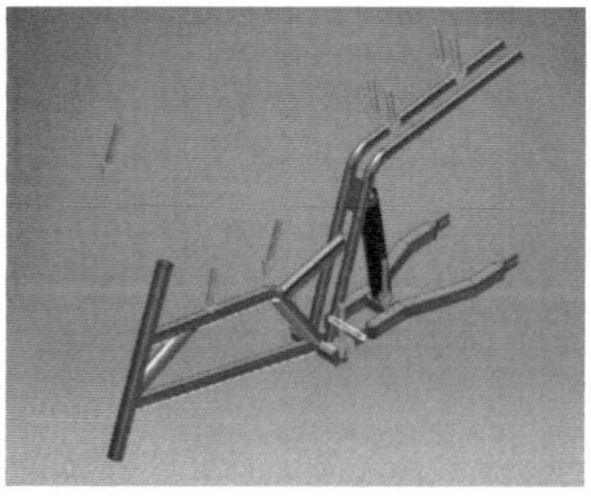

Figure 3. Load on the electric tilting tricycle frame.

In the electric tilting tricycle stress analysis study there were three loads (the rider's, electric motor, and battery). The electric tilting tricycle design can withstand rider'ss and luggage with a total mass of 90 kg (rider's load 82 kg and the 8 kg load luggage) located on the seat tube, pedal and also the handlebar. For the rider's, based on the present study, the rider's loads would be distributed to several ponits on the elements of the electric tilting tricycle, i.e. 50% on the seat tube, 100% on the pedals, and 44% on the handlebar.[1] The features used in the simulation of the tricycle frame were force, remote force, and gravity. The force feature was used to represent the driver's load on the seat tube and pedal. Besides, the force feature is also used to define the motor and battery located on the mainframe. Remote force was represented the load of the rider's on the handlebar. Since the handlebar was not included in the assembly frame components, so it defined using the remote force feature. The gravity force was used to define the direction of gravity on the tricycle with a magnitude of gravity was 9.81 m / s^2.

2.3. Mesh Process
In the present study, the size of elements was adjusted to the capabilities of inventor stress analysis for the best results of the simulation with an average size of 0.05 and the smallest size of the element 0.10. Next thing to arranged was the grading factor, with the amount set to the smallest size of 1.5. In addition, the maximum turning angle of the element is adjusted to the design of the bicycle which has curved tubes with the largest turning angle of 45^0 to maximize stress analysis on the tricycle frame.[6]

2.4. Material Selection
In the present study, there were 3 different materials selected as a comparison of the results of each stress analysis of the object. the material used is a material commonly used on bicycle frames. Following are the mechanical properties of the material used:

Table 1. Material properties.

	Al 6061	Steel mild	CFRP
Elastic Modulus	68.9 GPa	220 GPa	133 GPa
Poisson's Ratio	0.33	0.28	0.39
Shear Modulus	26 GPa	82 GPa	53 GPa
Mass Density	2.7 g/cm^3	7.85 g/cm^3	1.75 g/cm^3
Tensile Strength	310 MPa	367 MPa	577 MPa
Yield Strength	276 MPa	306 MPa	550 MPa

3. Results and discussion

3.1. Results of Frame Design Simulation Using the Steel Mild Material
The results of the static simulation were Von misses stress, deformation, safety factor, and strain. Based on the simulation results, the maximum value of Von misses stress is 113 MPa located in the middle of the seat tube. The maximum stress value that occurs due to the load given by the rider's on the seat tube with 0.814 mm deformation at the maximum Von misses stress value.

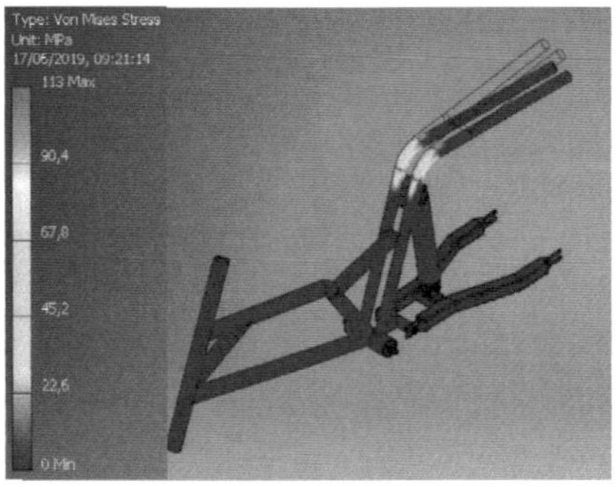

Figure 4. Von Mises stress simulation results of the tricycle frame using Steel Mild material.

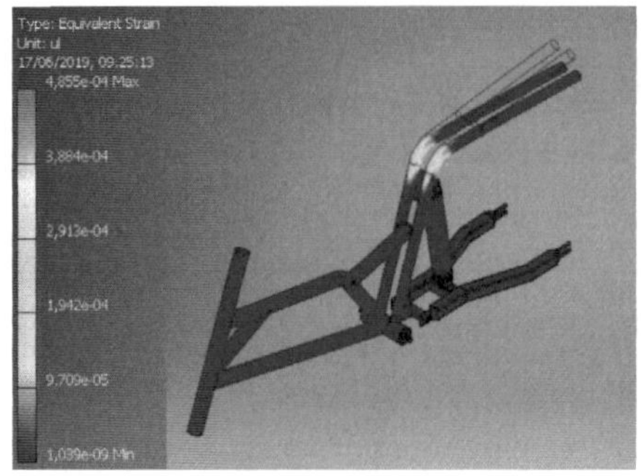

Figure 5. Strain simulation results of the tricycle frame using Steel Mild material.

The maximum deformation occurred on the frame with a mild steel material was 4.865 mm on the back of the seat tube that holds the load from the rider's with a load of 90 kg. The smallest safety factor value is 1.83 at the same point as the largest von misses stress value. The load generated from the prototype II prototype bicycle using mild steel material was 6.92 kg.

3.2. Results of Frame Design Simulation Using the Al 6061 Material

The results of static simulation with Aluminium 6061 material, the maximum Von misses stress value was 156.7 MPa which was located in the center of the seat tube which contact with the suspension mounting. As well as the other material, the maximum stress value that occurs due to the load given by the rider's with 0.532 mm deformation at the maximum Von misses value. The yield strength of the Aluminum 6061 material was 277 MPa, this means by using this material there was no plastic deformation on the object.

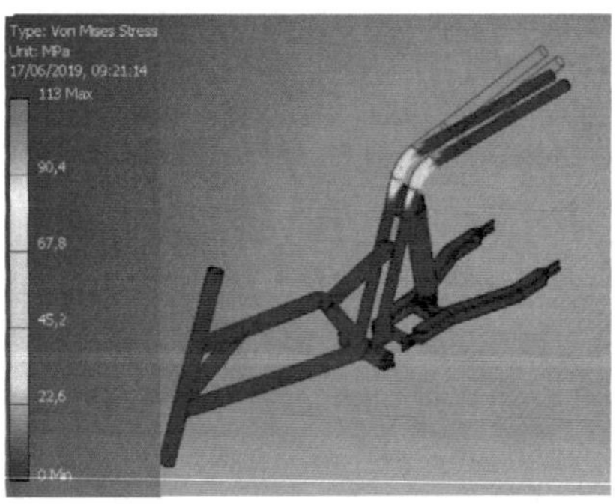

Figure 6. Von Mises stress simulation results of the tricycle frame using Al 6061 material.

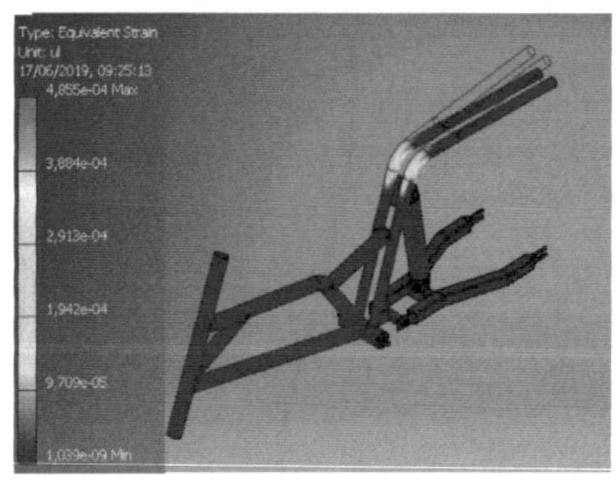

Figure 7. Strain simulation results of the tricycle frame using Al 6061 material.

The maximum deformation that occurred in the frame with aluminum 6061 material was 6 mm on the end of the seat tube that holds the load of the driver by 90 kg. While the smallest safety factor value of 2.41 lies at the same point as the maximum von misses stress value. The load generated from the prototype II prototype bicycle using Aluminum 6061 material is 4.19 kg. The tricycle with this material was safer than steel mild material, but it has bigger deformation on the end of the seat tube because of the young's modulus of the material.

3.3. Results of Frame Design Simulation Using the CFRP Material

Static simulation of the frame using CFRP material the maximum value of Von misses stress is 113.5 MPa which is located in the middle of the seat tube. The maximum stress value that occurs due to the load given by the rider's on the seat tube with 0.922 mm deformation at the maximum Von misses stress value.

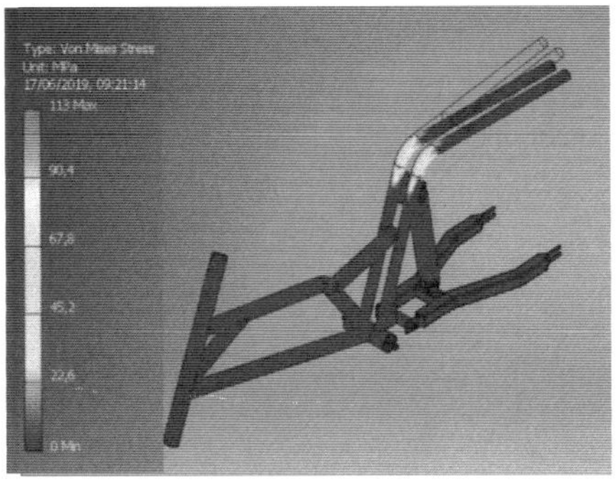

Figure 8. Von Mises stress simulation results of the tricycle frame using CFRP material.

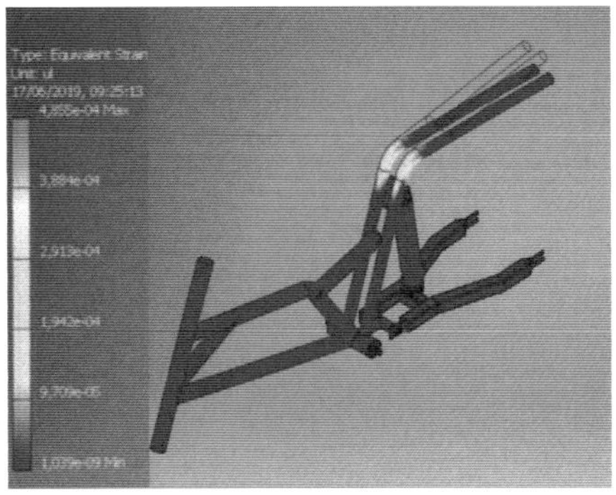

Figure 9. Strain simulation results of the tricycle frame using CFRP material.

The maximum deformation that occurs on the frame with CFRP material was 7.57 mm on the end of the seat tube that holds the load from the rider's with a load of 90 kg. While the smallest safety factor value of 2.64 is located at the same point as the largest von misses stress value. The load generated from the tricycle using CFRP material is 3.14 kg.

4. Conclusions

Based on the static simulation of the tricycle design, it can be concluded :

- CFRP material is the highest level of safety with a minimum safety factor value of 2.64, followed by Aluminum 6061 material with a safety factor value of 2.41, and Steel mild material with a safety factor value of 1.83.
- CFRP material is the lightest frame mass by 3.14 kg, followed by the aluminium 6061 mass by 4.19, and the steel mild mass by 6.92 kg.
- In the Static simulation results using several materials, since the maximum Von misses stress value below the yield point of the material, the design can be able to withstand the simulated load.

5. Acknowledgments

paper was supported by Hibah PITTA (Publikasi Internasional Terindeks untuk Tugas Akhir). Funded by DRPM Universitas Indonesia.

References

[1] Van Poelgeest A 2011 *The Dynamics and Control of A Three-wheeled Tilting Vehicle* (Doctoral thesis, Unversity of Bath, United Kingdom) p 1-8

[2] Syehan A, Pamungkas A T, and Sumarsono D A 2019 *IOP Conf. Series: Materials Science and Engineering* **494** (2019) 012088

[3] Cahyono S, Anwar M, Diharjo K, Triyono T, Hapid A, et. Al. 2017 AIP Conference Proceedings Vol. **1788** pp 30095

[4] Cook R D 1995 *Finite Element Modeling for Stress Analysis* (Canada: Jhon Wiley & Sons, Inc)

[5] Covill D, Begg S, Elton E, Milne M, Morris R, and Katz Tim 2014 Parametric Finite Element Analysis of Bicycle Frame Geometries. Procedia Engineering Vol. 72, p 441-446

[6] Bathe K J 2006 *Finite element procedures* (USA: Prentice Hall)

[7] Sharp A 1896 *Bicycles and Tricycles* Longmans Green (Reprinted M.I.T. Press, Cambridge, 1977)

Mechanical System Design and Simulation Research of a Demonstration Assembly Line

MA Ke[1*], CHEN Yu[1] and SUN Yi[1]
1. College of Mechanical and Electrical Engineering, Hohai University, Changzhou, 213022
2. Jiangsu Key Laboratory of Special Robot Technology, Changzhou, 213022
*Corresponding author's e-mail:bravery1978@163.com

Abstract. Most assembly lines use a bottom-up mechanical system design approach. In view of the problems of poor system interchangeability and low product versatility, a battery assembly line was designed based on the modular design method. The linear modules, cylinders and linear guides were used as the basic elements to build the production line modules. Different assembly requirements have improved the performance of the entire production line. The discrete-event system simulation method is used to simulate the quasi-running process of the production line, analyze the bottleneck of the production line, and improve the production efficiency of each module of the production line by changing the optimization method of the loading link and the production sequencing.
Key words: Non-standard automation; Plant Simulation simulation optimization; assembly production line; modular design

1. Introduction

In recent years, industrial automatic assembly technology has been widely used in manufacturing industry with its own advantages. The United States, Japan and Germany are leading the world in industrial automatic assembly technology [1].

At the same time, consumer demand presents personalized and diversified characteristics. There is less demand for a single product, and as companies order more, they order fewer products to deliver. The production mode of multiple varieties and small batch is a good solution to this problem. However, the following difficulties exist in the construction of automated assembly production line suitable for the production of multiple varieties and small batches [2-3]:(1) The product structure under this mode is complex and the complete set of products should meet the standard[4],which makes it difficult to use special equipment and reduces production efficiency;(2) When there is a large variety of products, controlling the production process is very complicated;(3) The production process of this mode makes the parts assembly process discontinuous;(4) It is very difficult to collect, process and transmit information about the size and specification of parts in the whole production process.

In this paper, aiming at the above difficulties and combining with the needs of enterprises, a modular design method [5] is adopted to design a battery assembly line. Production line Simulation optimization [6] was also conducted based on Plant Simulation to achieve the goal of improving both production line flexibility and production efficiency.

2. Product assembly process analysis

The designed battery assembly line in this paper produces a certain type of battery, whose main structure is shown in figure 1. The battery is mainly composed of battery box, A cell, B cell, battery box cover plate and some bolts, among which the battery box is composed of three parts: box body, bottom plate and cover plate. The box body and the bottom plate are connected by three pairs of bolts on the bottom plate before assembly. The difference between them is that the B cell has u-shaped sections on both sides, while the A cell does not, as shown in figure 1 (b). When A and B cells are together and the whole assembly is good, the battery can only supply power normally, as shown in figure 1 (a).

(a) Schematic diagram of battery assembly (b) Schematic diagram of each component of the battery

Figure 1. The battery mainly consists of structure

3.Modular design of production line

3.1. Production line module division

According to the modular design method, the mechanical system of the battery assembly line is divided into 3 subsystems (assembly subsystem, disassembly subsystem, personalized customization subsystem) and 15 modules, as shown in figure 2 and 3.

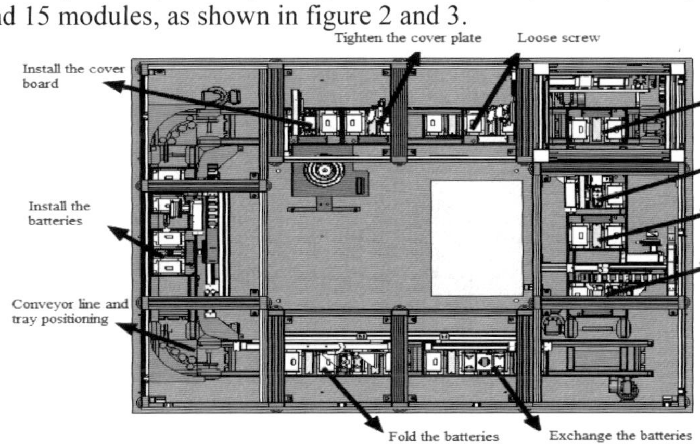

Figure 2.Division diagram of single product line mechanical module

In order to meet the second display demand of the production line -- production of personalized and customized demonstration battery, another three independent modules are introduced.The final model of the production line is shown in figure 3. The other three independent modules are: APAS robotic arm module, Active Assist manual assembly station module and Active Cockpit production management system module.

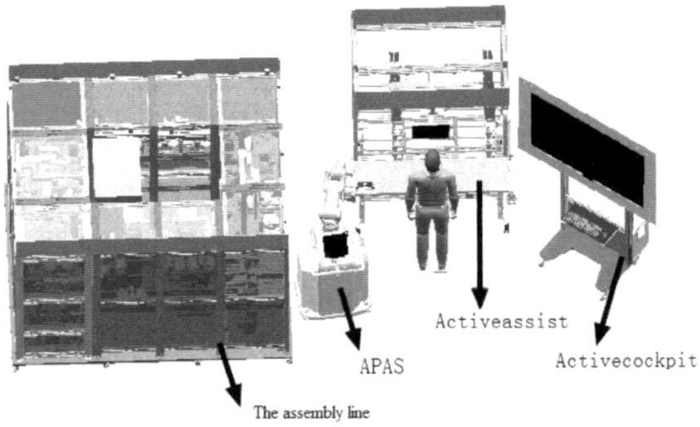

Figure 3. Assemble the final model of the production line

3.2. Assembly line running process

After dividing the modules, analyze the production process of the battery assembly line:

(1) Mass production (assembly and disassembly subsystem): The lower feeding—The tray arrives at the elevator and rises to the upper wire body—Remove the cover plate—Remove the batteries—The separation of batteries—Exchange of batteries—Superposition of batteries—Pack the cell into the battery box—Cover plate assembly—Tighten the cover plate—Loose screw—The tray arrives at the lift and descends to the lower line for a cycle.

(2) Personalized customization: Customers place personalized orders with the system using Active Cockpit—The system writes the order number to the RFID tag of the tray—The tray completes the assembly of the first half and waits after the battery module is installed—APAS grabs the battery which is located on the battery cell module tray and places it on the work table of Active Assist—The workers complete the assembly and show it to the customer—The staff place the demonstration battery on the battery tooling of the work table—The APAS grabs the battery and puts it back on the tray of the cell module.

4.Production line modeling simulation optimization and debugging operation

4.1. The simulation strategy

The simulation strategy of discrete event system is inseparable from the basic elements of the discrete event system, which is the control method of steadily advancing the simulation strategy along the simulation clock, and is the core of the control simulation system. The simulation strategy can be divided into: Event scheduling, Activity scanning and process interaction. Among them, event scheduling method analyzes the real system from the perspective of event, by defining the events and the changes in the state of the system caused by each event, the logical relations related to the occurrence of each event are executed in a chronological order. All events in the real system are stored in the event table. Preset a time controller, The controller selects the earliest event from the "event table", modifies the simulation clock to the time when the event occurs, and then calls the event handler corresponding to the event.After this event is processed, return to the time controller. Events are selected and processed over time until the simulation terminates .

In essence, event scheduling is a "scheduled event time" strategy. The simulation model must predetermine the ongoing events in the system. This strategy is more convenient for systems with more definite activity duration. In this paper, the event scheduling method is selected as the simulation strategy.

4.2. Simulation optimization evaluation index

The corresponding system evaluation indexes of the battery assembly line are as follows:(1) Equipment utilization;(2) The average waiting time of a product on the production line;(3) Production of the beat;(4) Maximum waiting time for production line products;(5) Number of demo products under maximum queue waiting time;(6) Total assembly time. These parameters can be clearly expressed in Plant Simulation.

4.3. Production line modeling

The battery assembly line is highly abstracted to analyze the production line system, find out the main parameters affecting the system changes and their mutual relations, and use Plant Simulation software to model.

(1) Modeling of feeding device

Complete the assembly task of demonstration battery and tooling tray, and simulate the feeding process. After the assembly is completed, it will be transported to the offline layer of transportation.

(2) Production line modeling of mass production mode

Under the mode of mass production, the production line has five working stations, which are divided into two layers of the same size line body.

The lower body has no assembly module and the upper and lower layers are connected by lifts. The equipment on the upper wire body is elevator module, remove cover plate module, remove cell module, separate cell module, exchange cell module, superposition cell module, install cell module, install cover plate module, install cover plate tightening module, remove screw module.

(3) Production line modeling of customization pattern

In the mode of personalized customization, all modules coordinate and match to complete the assembly and display tasks of the production line.

The motion logic of each device is controlled by Method and control options.

4.4. Simulation and optimization of production line

After modeling the battery assembly line, the assembly time of each module in the field was measured several times. The average assembly time of each module is taken as the input of simulation, and the assembly time of each module is shown in table 1.

Table 1. Assembly running time of each module

The module	Assembly running time /s	The module	Assembly running time /s
Elevator	10	Install batteries	20
Remove the cover plate	10	Install the cover board	10
Remove the batteries	20	Cover and tighten	25
The separation of batteries	10	Remove the screw	27
Exchange of batteries	15	APAS	12
Superposition of batteries	20	Active Assist	20

According to customer requirements, the demo line works 8 hours a day. Use the schedule in the resource to set up the battery assembly line for the rest of the week.

Set the number of demo batteries and pallets for initial feeding to 14 pairs. The production process of assembling 10 ordinary batteries and then 5 customized batteries is used to run the simulation model. The Event Controller is used to control the running process of the battery assembly line.

Simulation time of a working day, After the end of simulation, data analysis was conducted on the last module (screw removal). The battery assembly line circulated 990 demo batteries in eight hours of operation. The working rate of this station is 92.84% and the waiting rate is 7.16%.

After Bottleneck Analyzer analysis, the bottlenecks in the assembly module chart clearly expressed.

Generate resource information statistics of each module of the production line with Chart, including working time, setting time, waiting time, blocking time, failure time, stop time, pause time, unplanned time and power on/off time. The four main times are shown in table 2.

Table 2. Statistical table of resource information of each station (before optimization)

Station	Working time	Waiting time	Blocking time	The pause time
Elevator	69.44%	30.56%	0%	0%
Remove the cover plate	34.75%	65.25%	0%	0%
Remove the batteries	69.49%	30.51%	0%	0%
The separation of batteries	34.74%	65.26%	0%	0%
Exchange of batteries	52.10%	47.90%	0%	0%
Superposition of batteries	62.52%	37.48%	0%	0%
Install the batteries	69.44%	30.56%	0%	0%
Install the cover board	34.71%	65.29%	0%	0%
Tighten the screws	86.76%	174%	0%	0%
Remove the screw	9118%	6.32%	0%	0%
APAS	35.16%	64.84%	0%	0%
ActiveAssist	22.98%	65.69%	11.34%	0%

According to statistics, since the APAS robotic arm was in working state and could not put the dismantled demo battery back into the battery cell module in time, 11.34% of the blocking time was found in the Active Assist manual assembly station, which resulted in the unsmooth demonstration process of the entire production line.

Two optimizations were made to the battery assembly line by changing the optimization method of feeding link and production sequence:1. Before the optimization of the model, the original plan was to cycle the assembly according to the process of 10 ordinary batteries and 5 customized batteries. But considering the actual function of the battery assembly line, only one customized battery is displayed at one time during the operation. In this way, although the resource utilization ratio of APAS robotic arm is reduced, it ensures that the APAS robotic arm can act as the intermediate station for installing the battery module and Active Assist manual assembly station stably,and can grab and demonstrate the battery to flow between the two modules at any time; Meanwhile, in order to ensure that the corresponding modules of the personalized customization process can be fully utilized, the production process in the Method is modified into 1 ordinary battery and 1 customized battery, which is carried out in a continuous cycle.2. The original plan was to provide 14 demonstration batteries in the loading stage, which could not fully utilize the resources of each module. Now the number of batteries in the feeding stage is adjusted to make full use of the resources of each module in the production line. After simulation and field debugging, it is found that 22 demonstration batteries are the best.

After optimization, we performed data analysis on the last module (remove the screw). At this time, the daily assembly number of battery assembly line increases from 990 to 1054, and the working rate also increases slightly, rising from 92.84% to 98.86%, and the waiting rate decreases from 7.16% to 1.12%, blocking rate is 0.02%.

Table 3 is the statistical table of resource information of each station after optimization.

Table 3. Statistical table of resource information of each station (after optimization)

Station	Working time	Waiting time	Blocking time	The pause time
Elevator	7129%	26.21%	0%	0%
Remove the cover plate	55.76%	44.24%	0%	0%
Remove the batteries	74.29%	25.71%	0%	0%
The separation of batteries	55.68%	44.32%	0%	0%

Exchange of batteries	55.63%	44.38%	0%	0%
Superposition of batteries	66.69%	381%	0%	0%
Install the batteries	7148%	26.02%	0%	0%
Install the cover board	55.36%	44.64%	0%	0%
Tighten the screws	930%	7.80%	0%	0%
Remove the screw	98.86%	1.12%	0.02%	0%
APAS	52.16%	47.84%	0%	0%
Active Assist	497%	54.12%	51%	0%

According to the resource information statistics, the blocking time of each module was less than 2.5%, and the working time of each station was improved, making the working process of the battery assembly line more smooth. This shows that it is feasible to change the optimization method of feeding link and production sequence, and it can also be verified in the field debugging of production line.

5. Conclusion

(1) This paper USES the modular method to design the production line. Taking the linear module, cylinder and linear guide rail as basic elements, each module of the production line is built to meet different assembly requirements. It improves the interchangeability and universality of the whole system, and is also convenient for maintenance and repair. Build production line modules to meet different assembly requirements, and improve the performance of the entire production line.

(2) The bottleneck of the battery assembly line after the introduction of personalized subsystem was analyzed based on the Plant Simulation platform. The daily assembly number of demonstration batteries increased from 990 to 1054 by changing the optimization method of feeding link and production sequence. The working rate also increased slightly, from 92.84% at the beginning to 98.86% later, and the waiting rate decreased from 7.16% to 1.12%. Through simulation and optimization, the production line efficiency has been significantly improved.

6. References

[1] Sun M. Development status and trend of automatic assembly in foreign machinery industry [J]. Combination machine tool and automatic machining technology, 1985(04): 13-18+50.

[2] Zhu B. Study on layout optimization of multi - variety small batch production line [D]. Jinan university, 2013.

[3] Gun H. Research on optimization of multi - variety small batch production line based on simulation technology [D]. Tianjin university of technology, 2018.

[4] Zhou C. Analysis of complete set problem based on equipment manufacturing products [J]. Journal of tangshan university, 2012, 25(03): 90-92.

[5] Damayanti, D.D., Toha, I.S.. Model of spine configuration assembly line design for a Product family[P]. Industrial Engineering and Engineering Management (IEEM), 2011 IEEE International Conference on, 2011.

[6] Li M, Shi Y, Wang J, Cai B. Simulation study on packaging manufacturing system based on em-plant [J]. Manufacturing Automation, 2015, 37(23): 24-27.

Acknowledgments

This research was supported by the Key P&D Program of Jiangsu(BE2018004-1).

Multifunctional smart crib design

Honghui Xie [1], Runxi Sun [1], Zhaojin Yan [1], Zhilin Wu [1], Weichuan Ni [2]

1. School of Information Science， Xinhua College of Sun Yat-Sen University, Guangdong, Guangzhou 510520, China
2.School of Equipment and Laboratory Management, Xinhua College of Sun Yat-Sen University, Guangdong, Guangzhou 510520, China

Abstract. The environment for newborn babies is harsh, and it is often difficult for parents to understand whether the baby's current environment is comfortable. Taking care of babies is the most troublesome thing for young parents. In order to let parents know more about the environmental changes around the crib, when the environment changes greatly, timely treatment, such as the temperature difference between day and night when the seasons are alternating, it is difficult for parents to grasp whether the current temperature of the crib is comfortable, and the baby often has a fever due to sudden changes in temperature at night.Through the use of various sensors to collect the surrounding environment of the crib, real-time monitoring of the temperature, moderate, gas and other environmental data around the crib, through the infrared temperature measurement module to measure the baby's body temperature in real time, while adding music playback, shaking comfort, baby crying detection By uploading data to the cloud through the Internet of Things Internet of Things, parents can easily view the real-time status of the baby's growing environment, providing a healthy, scientific, and comfortable environment for babies while reducing the burden on busy parents.

1. Introduction

With the development of the Internet of Things, in recent years, various smart home products have gradually entered the family, and various smart products have greatly facilitated our lives. At present, most cribs on the market are traditional cribs. Traditional cribs cannot monitor the environment around the crib in real time. In reality, there are too many unforeseen factors affecting the healthy growth of infants and young children, such as environmental conditions. Parents still have to live and work normally on the basis of taking care of infants and young children, so it is inevitable that they will be negligent when taking care of infants and young children.In view of the above situation, we have designed a multifunctional smart crib. [1-3] The functions of the crib are: temperature and humidity collection in the crib, music playback, crib data abnormal reminder, automatic shaking, non-contact body temperature detection, sound detection, TFT Data display, air quality testing, and parents can view real-time data on cribs in the cloud at any time. Provide a comfortable environment for babies to grow.

About the Author:
Honghui Xie (1999-), male, Meizhou, Guangdong,China. Main research direction: embedded systems, artificial intelligence
Runxi Sun (1998-), male, Shantou,Guangdong,China. Main research direction: embedded system development and application, Internet of Things
Zhaojin Yan (1998-), male, Yunfu, Guangdong,China. Main research direction: Internet of things, artificial intelligence;
Zhilin Wu (1998-), male, Meizhou, Guangdong, China. Main research direction: single-chip application, Internet of Things
Corresponding Author:
Weichuan Ni (1990-), male, Jieyang, Guangdong, China. Assistant engineer, main research direction: image processing, machine vision; E-mail: linwc@xhsysu.edu.cn

2. Overall system design

The system uses multiple sensors to detect crib environment changes in real time. The MLX90614 non-contact infrared temperature sensor collects baby's body temperature data, DHT22 collects temperature and humidity changes around the crib, MQ135 detects air quality in real time, and realizes the automatic crib function of the crib by controlling the motor. The sound automatically detects music and shakes when the baby cries, uploads the data to the server through the ESP8266, [4-7]and displays the acquired data in real time through the TFT screen. The system uses ESP8266 Wi-FiSoC as the main control core, and the TFT screen is driven by Arduino nano.

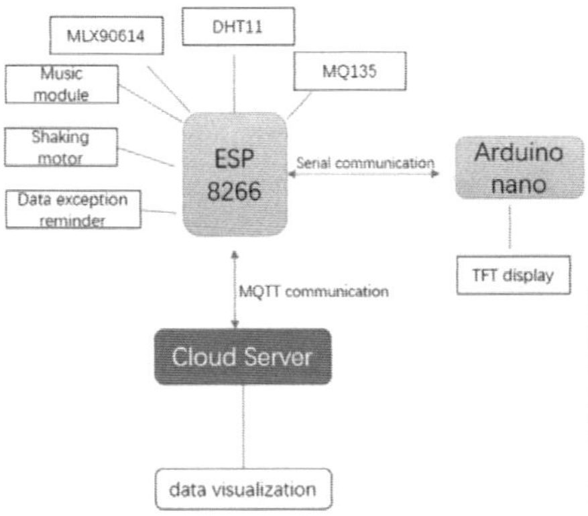

Figure 1 System framework diagram

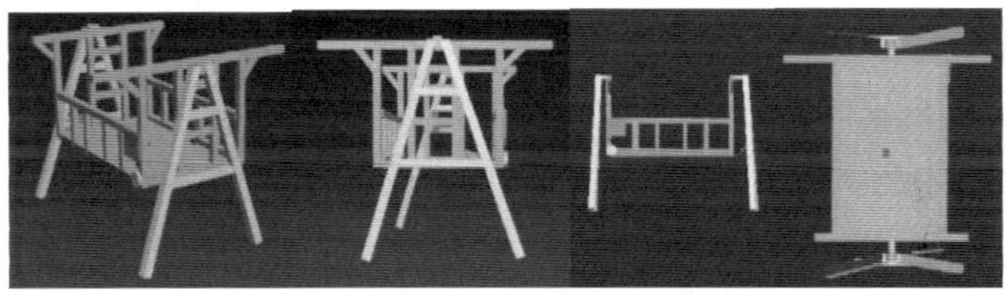

Figure 2 Crib renderings

2.1. System Master

The system adopts ESP8266 Wi-FiSoC as the system master. The ESP8266 has rich hardware interfaces and can support UART, IIC, PWM, GPIO, ADC, etc. It also has a built-in low-power 32-bit CPU, which can double as an application processor and is suitable for various Internet of Things applications. Occasion. ESP8266Wi-FiSoC is used as the main control of the system to process the data collected by the sensor and also realize data uploading. At the same time, the price has a large advantage and the overall cost performance is high. [5-6]

2.2. Non-contact temperature measurement

MLX90614 infrared temperature measurement module determines the temperature of the object according to the infrared radiation energy of the measured object, does not contact with the measured object, and has the temperature distribution field of the object to be measured. The temperature resolution is high, the response speed is fast, and the temperature measurement range is wide. It is not limited by the upper limit of temperature measurement and has good stability. The use of non-contact infrared sensors can reduce the interference of the environment on the sensor and the measurement accuracy is high, which can continuously monitor the body temperature of the baby.

The MLX90614 and the main control chip use IIC communication, just need to connect the corresponding VCC, GND, SCL, SDA to the corresponding pin of the main control chip.

2.3. Temperature and Humidity Measurement

The temperature and humidity measurement uses the DHT22 digital temperature and humidity sensor. The DHT22 digital temperature and humidity sensor is a temperature and humidity composite sensor with a calibrated digital signal output. The DHT22 is installed on the bottom of the crib to detect whether the temperature and humidity in the baby's bed are normal. When the baby wets the bed, the humidity will increase, and the corresponding bedtting humidity threshold is set. When the humidity exceeds the threshold, the bedwetting is determined, and the bed can be detected. Whether the internal temperature is comfortable, set the baby comfort temperature range, when the temperature is too high or too low, it is judged as uncomfortable temperature. When the temperature or humidity is abnormal, the system master control GSM module issues a warning reminder. At the same time, there is also a DHT22 digital temperature and humidity sensor around the crib for collecting the temperature and humidity around the crib so that parents can better understand the temperature and humidity of the current environment and prepare in advance.

2.4. Baby comfort

Baby comfort includes automatic shaking and music playback. Music playback uses JQ6500. JQ6500 is a serial port MP3 chip. It integrates hard decoding of MP3 and WMV. It adopts AD button control mode and controls the corresponding pins through the main control chip. The level change triggers the corresponding control. The automatic shaking adopts the design of DC motor and linkage rod. The rotation of the motor drives the linkage rod to realize the shaking function of the crib. The motor is driven by the uln2003 chip and directly controlled by the main control chip. At the same time, the button is also used to control the shaking, correspondingly realize the opening and closing of the shaking, and also can be timed off, and the parents can sleep on the baby by the shaking function and the music playing function.

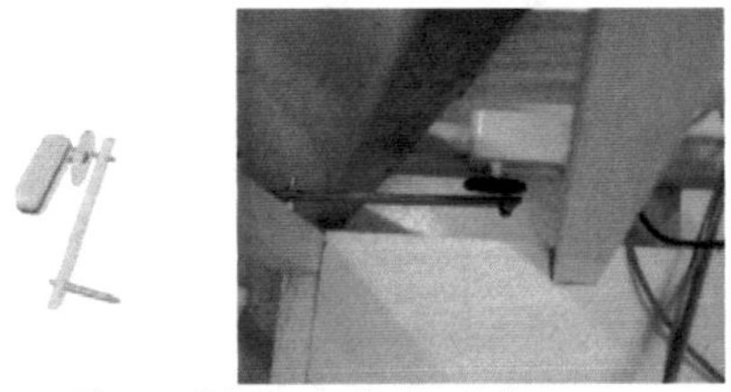

Figure 3 Motor Shake Joint

2.5. Exception warning

When it is detected that the data acquired by the sensor exceeds the set threshold, the data is processed first to check whether the sensor data is fluctuating. If the abnormal data detected by the non-sensor is detected, the main control chip controls the GSM module to send a data abnormality warning. In the mobile phone number preset in advance, if the data does not return to normal after a certain period of time, the call is made until the data is restored.

2.6. Cloud Server

The communication protocol between the system and the cloud server is the MQTT protocol. [7-9] The MQTT is a connection protocol for M2M and the Internet of Things, and adopts a lightweight publishing and subscription message transmission mechanism. Mosquitto is an open source messaging agent that implements the MQTT v3.1 protocol, providing a lightweight, publish/subscribe messaging push mode that makes device-to-device short message communication easy to use. [10-11] After building the ubuntu system on the cloud server, install mqtt mosquitto 1.4.5. [8-9]

3. System test

In order to check whether each part of the sensor can run according to the predetermined program, and check whether the parts can cooperate with the operation, the system will perform the overall simulation data test to check whether the system can run as expected, record the data, and check whether the sensor data is normal.

3.1. Simulated temperature and humidity test

Test content: Verify that the temperature and humidity sensor is working properly, and if the temperature and humidity are abnormal, can you trigger the voice reminder according to the setting.

Test procedure: Cover the DHT22 temperature and humidity sensor with a moist hot towel to simulate the baby bedwetting scene.

Test result: After the towel is put on, it takes about 5 seconds. TFT and cloud can display real-time temperature and humidity. After about 10 seconds, the scheduled mobile phone number receives the abnormal reminder message, and the mobile phone number is scheduled to receive the phone call after about 5 minutes.

3.2. Simulated body temperature test

Test content: Verify that the non-contact infrared temperature measurement module works normally. When the body temperature is abnormal, it can trigger the reminder according to the setting.

Test procedure: Use a cup of hot water and place it in front of the infrared temperature measurement module to simulate the baby's body temperature.

Test result: After the cup is put about 3 seconds, the temperature data displayed on the TFT is refreshed. After about 5 seconds, the cloud refreshes to the body temperature data. After about 10 seconds, the predetermined mobile phone number receives the abnormal reminder message, and the mobile phone is scheduled to be about 5 minutes. Receive a phone call

3.3. Simulated abnormal air test

Test content: Verify that the warm air quality sensor is working properly, and can trigger the voice reminder according to the setting when the air is abnormal.

Test procedure: The lighter releases contaminated gas near the sensor, simulating anomalous air quality.

Test result: When the gas in the lighter is released, the predetermined mobile phone number receives the abnormal reminder message about 10 seconds, and the mobile phone number is scheduled to receive the phone call after about 5 minutes.

3.4. Test Data

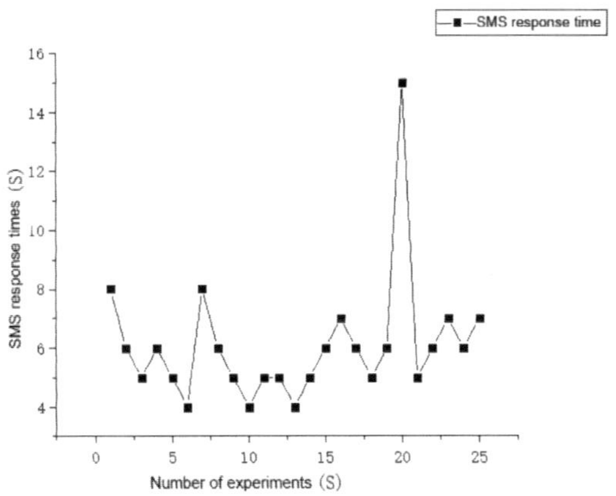

Figure 4 SMS response time

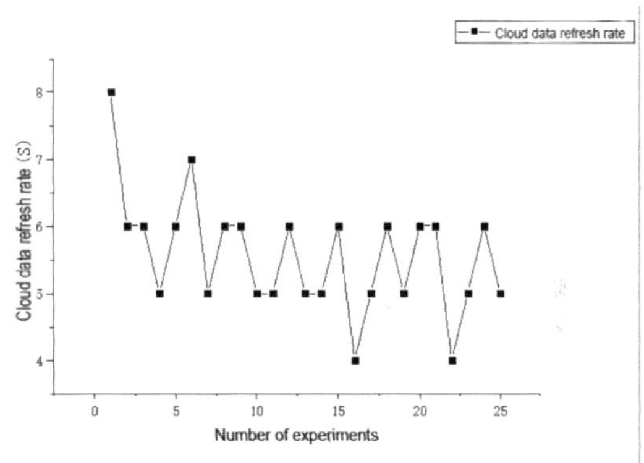

Figure 5 Cloud data refresh rate

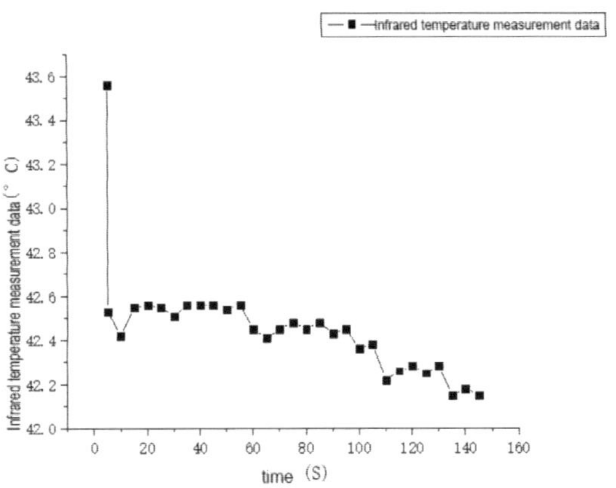

Figure 6 Infrared temperature measurement data

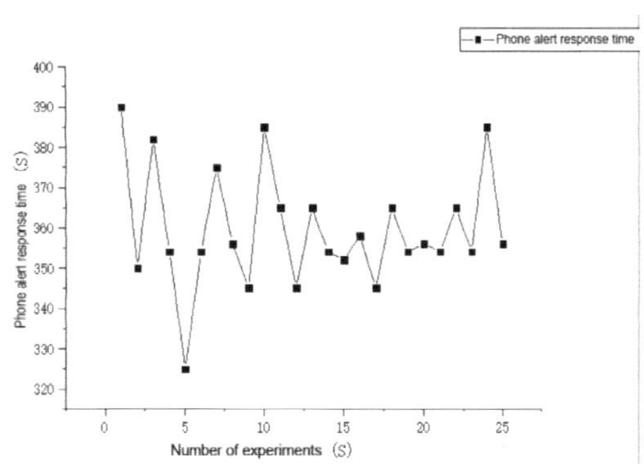

Figure 7 Telephone response time

4. Conclusion

The use of some cost-effective sensors to achieve real-time monitoring of the overall data of the crib at very low cost, the overall cost of the crib is not high, there is a greater price advantage, the use of GSM remote reminder can avoid the voice reminder to the baby At the same time, the non-contact infrared temperature measurement can monitor the body temperature of the baby in real time, which can conveniently view the real-time data of the crib from the cloud, so that parents can better understand the changes of the baby's growth environment and provide a comfortable environment for the baby to grow up.

References:

[1] He Wenhua. Remote smart crib research design [J]. Electronic world, 2016(9):58-58.
[2] Aloÿ A, Yi J, Thomas C, et al. A Study of LoRa: Long Range & Low Power Networks for the Internet of Things:[J]. Sensors, 2016, 16(9):1466.
[3] Dey S , Roy A , Das S . Home automation using Internet of Thing[C]// 2016 IEEE 7th Annual Ubiquitous Computing, Electronics & Mobile Communication Conference (UEMCON). IEEE, 2016.
[4] Zhang Ke, Wang Zhe. Intelligent baby shaker design [J]. Electronic production, 2018(7):31-32.
[5] Fan Xinglong. Application of ESP8266 in smart home monitoring system [J]. MCU and embedded system applications, 2016, 16(9):52-56.
[6] Skraba A , Kolozvari A , Kofjac D , et al. Streaming pulse data to the cloud with bluetooth LE or NODEMCU ESP8266[C]// 2016 5th Mediterranean Conference on Embedded Computing (MECO). IEEE, 2016.
[7] Wang Peiyuan, Du Yusheng. An Internet of Things Intelligent Monitoring System Based on MQTT Protocol [J]. Information technology and information technology, 2018(1):107-110.
[8] Yao Dan, Xie Xuesong, Yang Jianjun, etc.. Research and Implementation of Internet of Things Communication System Based on MQTT Protocol [J]. Information and Communication, 2016(3):33-35.
[9] Kodali R K, Soratkal S R. MQTT based home automation system using ESP8266[C]// Humanitarian Technology Conference. 2017.
[10] Atmoko R A , Riantini R , Hasin M K . IoT real time data acquisition using MQTT protocol[J]. Journal of Physics: Conference Series, 2017, 853:012003.
[11] Hwang H C , Park J S , Shon J G . Design and Implementation of a Reliable Message Transmission System Based on MQTT Protocol in IoT[J]. Wireless Personal Communications, 2016, 91(4):1765-1777.

Acknowledgments

Fund Project: 2019 National College Student Innovation and Entrepreneurship Training Program Fund Project (20191390139), 2019 Guangdong University Student Science and Technology Innovation and Cultivation Special Funding Project (505- pdjh18b0642), 2019 Guangdong University Student Science and Technology Innovation Cultivation Special Funding Project (505- Pdjh18b0643)

A reliability growth projection model based on similar failure mechanisms

Zhongsheng Li*, Jinwei Fan and Dongju Chen

Jidian building, room 304, pingleyuan 100, chaoyang district, 100124, Beijing, China
Beijing key laboratory of advanced manufacturing technology, Beijing University of Technology, Beijing, China

*13910909582@163.com

Abstract. Reliability growth test is a common method to evaluate complex electromechanical systems in mechanical engineering. Reliability growth models are usually used to predict the reliability index of systems after the reliability test. AMSAA (army materials systems analysis activity) projection-model-stein (APMS) model is widely used for delayed corrective strategy, since it only requires limited assumptions but covers all failure data. However, there is a controversy about whether the Stein factor introduced in this model is reasonable. To alleviate the problem, an APMS development model is presented in this paper, because different types of parts have different inherent failure mechanisms. The study showed the projection results of the proposed model are credible. Moreover the Stein shrinkage factors were more reasonable and acceptable, as they were derived from the components with similar inherent failure mechanisms. These provide a technical foundation for the wide application of the proposed model.

1. Introduction

The requirement of high reliability for the complex systems is obvious in mechanical engineering. Reliability test is a common method to evaluate complex electromechanical systems, which involves surfacing failure modes, analyzing them and implementing fixes to the observed modes. Reliability growth models are generally used to quantify realized reliability growth value at the end of the initial test phase and project expected reliability growth value based on assumed, fix effectiveness factors (FEFs) of corrective actions [1, 2]. AMSAA (army materials systems analysis activity) projection-model-stein (APMS) model is widely used for delayed-corrective-strategy testing, because it only requires limited assumptions but covers all failure data [3, 4]. Meanwhile the APMS model is statistically more robust against the effects of fix-effectiveness-factors variability than the AMSAA projection model [5]. Moreover, the Stein-based model can provide a more accurate reliability-projection value than other models, given that minimizing the expected squared error between the assessed and initial mode failure rates [6].

Since the seminal work of stein [7], the 'Stein Estimator', has received enormous attention in the theoretical study and aroused great interest in the practical applications [8]. However, there is controversy about whether the stein factor introduced in this model is reasonable [9], due to inadmissible data involved. To alleviate this problem, an development model of APMS Model, in which the stein estimators are calculated individually by different failure mechanisms, is presented. It is known that different types of components have different failure mechanisms. For example, the failure mechanisms of mechanical products mainly include fatigue and corrosion. The failure mechanisms of electronic products mainly include thermal effect, electrochemistry, electromagnetic compatibility, electro-static discharge, etc. The main objective of the model proposed is to be able to continuously assess the reliability growth of a system across multiple test phases, through dividing the failure data into several sections according to similar failure mechanisms.

2. APMS Model

APMS projection model, estimates the fraction of system failure rate, caused by the not yet occurring modes, after the test through stein shrinkage estimator, utilizing an optimality criterion to minimize the sum of squared errors, between the estimated initial failure ratios and the real rates. The base of the model is that below ratios of occurrence for separate failure modes are a case of a Gamma distribution random sample [3].

It is supposed the system has $k>1$ potential failure modes that have initial failure rates $\lambda_1,...,\lambda_k$. Let n_i denote the number of failures observed for mode $i(i=1,...,k)$ that occur during the test. The maximumlLikelihood estimate of λ_i is $\hat{\lambda}_i = n_i / T$. After mitigation of the failure modes observed during the test period [0, T], the realized system-failure-rate-intensity function is given as

$$\rho_S(T) = \frac{n_A}{T} + \sum_{i\in obs(B)} (1-d_i^*)\cdot\tilde{\lambda}_i + \sum_{i\in \overline{obs(B)}} \tilde{\lambda}_i \tag{1}$$

Where n_A is the observed number of Mode-A failures, $obs(\mathrm{B})$ means the observed Mode-B failures, d_i^* is a realization of d_i (FEF). $\tilde{\lambda}_i$ represents the stein estimation of λ_i, is defined by

$$\tilde{\lambda}_i \equiv \theta_S\cdot\hat{\lambda}_i + (1-\theta_S)\cdot\frac{\sum_{i\in B}\hat{\lambda}_i}{k_B} \tag{2}$$

The shrinkage factor $\theta_S \in [0,1]$ in equation (2), that minimizes the expected sum of squares, $E\left[\sum_{i\in B}\left(\lambda_i - \bar{\lambda}\right)^2\right]$, can be expressed by

$$\theta_S = \frac{\sum_{i\in B}\left(\lambda_i-\bar{\lambda}\right)^2}{\frac{\lambda_B}{T}\cdot(1-\frac{1}{k_B}) + \sum_{i\in B}\left(\lambda_i-\bar{\lambda}\right)^2} \tag{3}$$

Where $\bar{\lambda} = \frac{1}{k_B}\sum_{i\in B}\lambda_i$, and $\lambda_B = n_B / T$, n_B is the observed number of Mode-B failures.

3. APMS development model

It is supposed there are $j(j=1,...,q)$ sections of the complex system, in term of the similar inherent failure mechanism and correction strategy. That is, the stein estimate values are calculated individually as per identified sections. As a result, the stein projection of the system–failure-rate intensity developed by identified sections is given as:

$$\rho_{S,S}(T) = \frac{n_A}{T} + \sum_{j=1}^{q}\sum_{i\in obs(B)}(1-d_{ij}^*)\cdot\tilde{\lambda}_{ij} + \sum_{i\in \overline{obs(B)}}\tilde{\lambda}_i \tag{4}$$

Where d_{ij}^* is a realized d_i of the *j*th section. $\tilde{\lambda}_{ij}$ represents the stein estimation of λ_i for the *j*th divided section, are defined by

$$\tilde{\lambda}_{ij} \equiv \theta_{S,j}\cdot\hat{\lambda}_{ij} + (1-\theta_{S,j})\cdot\frac{\sum_{i\in B(j)}\hat{\lambda}_{ij}}{k_{B,j}} \tag{5}$$

Similarly $$\theta_{S,j}=\frac{\sum_{i\in B(j)}\left(\lambda_{ij}-\bar{\lambda}\right)^2}{\frac{\lambda_{B,j}}{T}\cdot(1-\frac{1}{k_{B,j}})+\sum_{i\in B(j)}\left(\lambda_{ij}-\bar{\lambda}\right)^2} \quad (6)$$

Where $\bar{\lambda}=\frac{1}{k_{B,j}}\sum_{i\in B(j)}\lambda_{ij}$, and $\lambda_{B,j}=n_{B,j}/T$.

Meanwhile $$\sum_{i\in obs(B)}\tilde{\lambda}_i\cong\frac{\hat{\lambda}_B}{1+\beta_{S,B}\cdot T} \quad (7)$$

With $\hat{\lambda}_B$ denoting n_B/T and $0<\beta_{S,B}\le\lambda_B$, where $\beta_{S,B}$ and λ_B are respectively equal to $\frac{1}{\lambda_B}\sum_{i\in B}\lambda_i^2$ and $\sum_{i\in B}\lambda_i$.

The first term in equation (4) is the constant failure intensity contribution due to the Mode-A failures. The second term in equation (4) is the expected failure intensity contribution due to Mode-B failures that have occurred, and completed corrective actions at the end of the test, as per distinguished sections. The last term is the unconditional expected failure intensity contribution due to the Mode-B failures that have not been surfaced by time T .

4. Case Study

A reliability-engineering test of NC machines with three testing phases was conducted in sequence. All corrective actions of surfaced B-modes were implemented at the end of the current test phase, prior to starting a follow-on test phase. The current test phase will be referred to as phase I, and the follow-on test phase as phase II. The failure data are shown in table 1, table 2 and table 3 below. The failure data are divided into five sections according to diffirent failure mechanisms: Base, Feed system, Electronic-control system, Hydraulic and Lubrication system, Pneumatic system.

Table 1. Failure data of NC machines in phase I.

Sections	First-time (h)	Mode	Freq.	d_i	Sections	First-time (h)	Mode	Freq.	d_i
Electronic Control	52.6	B1	1	0.63	Hydraulic & Lubrication	48.0	A	1	---
	80.3	A	1	---		98.7	B8	1	0.59
	158.3	B2	1	0.72		113.0	B9	1	0.63
Feed	57.9	B3	1	0.77					
	73.1	B4	1	0.92	Pneumatic	16.3	B10	1	0.78
	105.9	B5	2	0.75		69.5	B11	1	0.90
	129.4	A	1	---		100.9	B12	1	0.84
Base	25.2	B6	1	0.86		140.7	B13	1	0.88
	102.6	B7	2	0.91					

Regarding phase I, the system is tested for T_1=200 hours. As shown in table 1, there is a total number of 18 failures and all corrective actions will be incorporated at the end of the 200-hour test. There are 3 Mode-A failures and 15 Mode-B failures during the test. There are 13 unique Type-B failure modes seen, which means there are 13 distinct corrective actions incorporated into the system at the end of test.

Table 2. Failure data of NC machines in phase II.

Sections	First-time (h)	Mode	Freq.	d_i	Sections	First-time (h)	Mode	Freq.	d_i
Electronic Control	104.5	A	1	---	Hydraulic & Lubrication	111.6	B6	1	0.69
	165.4	B1	2	0.68		196.5	A	1	---
						339.8	B7	2	0.75
Feed	83.1	B2	2	0.79	Pneumatic				
	125.3	B3	1	0.82		71.0	B8	3	0.74
	384.2	A	2	---		170.6	B9	2	0.91
						308.7	B10	2	0.82

Base	59.3	B4	1	0.88
	236.1	B5	3	0.93

For the data in Phase II, the system is tested for T_2=400 hours. As shown in table 2, there is a total number of 23 failures and all corrective actions will be incorporated at the end of the 400-hour test. There are 4 Mode-A failures and 19 Mode-B failures during the test. There are 10 unique Type-B failure modes seen, which means there are 10 distinct corrective actions incorporated into the system at the end of test.

Table 3. Failure data of NC machines in phase III.

Sections	First-time (h)	Mode	Freq.	d_i	Sections	First-time (h)	Mode	Freq.	d_i
Electronic Control	96.8	A	1	---	Hydraulic & Lubrication	149.5	B7	3	0.70
	366.2	B1	2	0.72		553.1	B8	1	0.76
	433.6	B2	2	0.65					
Feed	86.0	B3	1	0.85					
	158.7	A	2	---	Pneumatic	218.4	B9	3	0.82
	300.4	B4	3	0.74		298.6	B10	1	0.86
	389.7	B5	2	0.81		582.3	B11	2	0.75
Base	187.6	B6	2	0.90					

For the data in Phase III, the system is tested for T_3=800 hours. As shown in table 3, there is a total number of 25 failures and all corrective actions will be incorporated at the end of the 800-hour test. There are 3 Mode-A failures and 22 Mode-B failures during the test. There are 11 unique Type-B failure modes seen, which means there are 11 distinct corrective actions incorporated into the system at the end of test. The predicted results of three phase data are shown in table 4.

Table 4. Calculation result of three phase data.

	Phase I	Phase II	Phase III
$\rho_S(T)$	0.064306	0.034449	0.018313
$\rho_{S,S}(T)$	0.063729	0.0342	0.018353
Observed Value	0.09	0.0575	0.03125

The projection performance are plotted in Figure1. The dashed-dotted lines show the projection value of failure-intensity-function value of APMS projection model. The dashed lines indicate the projection value of the failure-intensity-function value of APMS development-projection model. Finally, the solid lines are the observed value during the test.

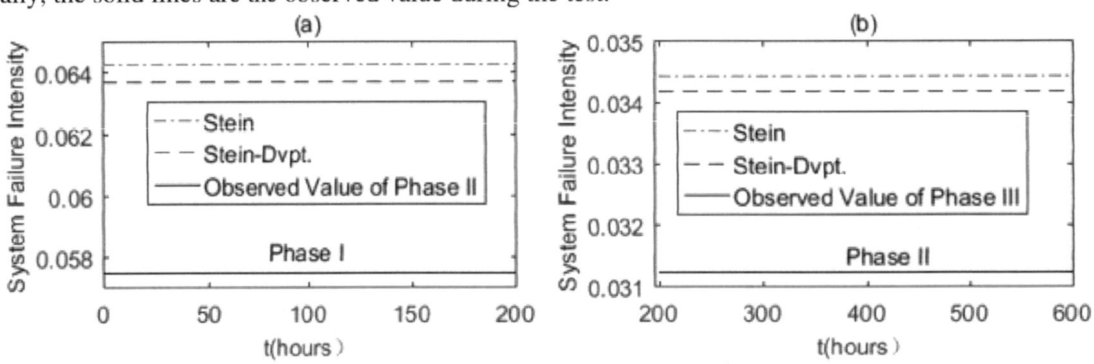

Figure 1. Projection performance of two phases.

As shown in figure 1, the prediction results of the proposed APMS development model are more accurate than APMS Model, by comparing the predicted intensity rate of Phase I with the observed intensity rate of Phase II, similarly Phase II with Phase III. Furthermore, the observed intensity rate at a certain stage is less than the predicted intensity rate during the test. This is probably because the conservative FEFs are used instead of the realized FEFs, for example, the FEF of mode B8 is only 0.59 in table 1. Additionally, the stein shrinkage factors of the proposed model, which derive from five partitioned sections, have the similar failure mechanism. That is, the proposed model is easier to understand and more acceptable in engineering than the APMS Model.

5. Conclusions

With the purpose of evaluating systems reliability under development, the paper presents an APMS development-projection model based on partitioned sections by failure data reconstitution. Then systems failure-intensity function is derived on the basis of reorganized data. Based on the study, the following results are obtained:

(1) The stein shrinkage factors of partitioned sections constructed in term of similar inherent failure mechanisms, are easier to understand and accept in engineering field. This provides a technical foundation for the wide application of this proposed model.

(2) As the stein factors are solved separately, the presented model seems more plausible than the APMS model for reliability prediction of complex systems, which contains many aspects, such as mechanical, hydraulic, pneumatic, electrical, software and integrated systems.

Acknowledgements

This work was supported by the National Science and Technology Major Project of China [grant numbers 2016ZX04003001].

Reference

[1] Honecker S L and Yenal U 2017 Quantifying the effect of a potential corrective action on product life *Proc. of Annual Reliability and Maintainability Symp.* (Orlando FL USA/ IEEE) pp 1-5
[2] Awad M 2015 Reliability growth test planning of repairable systems using subsystem failure distribution data *Int. Conf. on Industrial Engineering & Systems Management* (Seville Spain/ IEEE) pp 155-60
[3] Ellner P M and Hall J B 2005 Amsaa maturity projection model based on stein estimation *Proc. of Annual Reliability and Maintainability Symp.* (Alexandria VA USA/ IEEE) pp 271-7
[4] Department of Defense 2011 *Reliability Growth Management* MIL-HDBK-189C
[5] Wayne M 2009 A reliability growth simulation test bed *Proc. of Annual Reliability and Maintainability Symp.* (Fort Worth TX USA/ IEEE) pp 389-94
[6] Mortin D E and Ellner P M 2005 Reliability growth for military vehicles – emerging methodology and remaining challenges *SAE 2005 World Congress & Exhibition* (Warrendale PA USA/ SAE) **114** pp 28-34
[7] Stein C 1956 Inadmissibility of the usual estimator for the mean of a multivariate normal distribution *Proc. of the 3rd Berkeley Symp. on Mathematical Statistics and Probability,* Vol **1** (University of California Press) pp 197-206
[8] Guo M X and Ghosh M 2012 Mean squared error of james – stein estimators for measurement error models *Stat Probabil Lett* **82** (11) pp 2033-43
[9] Guo J Y, Sun Y Q and Yu X Y 2011 Annotation of reliability growth technology development trends *Journal of Harbin University of Science and Technology* **16** (2) pp 4-14

Adaptive sliding mode control of electro-hydraulic servo system based on RBF network

S H Zhou[1], H Wang[1], J Li[1] and L Y Lan[1]
[1] School of Mechanical Engineering and Automation, Northeastern University, No. 11 Lane 3, Shenyang, Liaoning, China

Email address: wanghfoto@foxmail.com (H Wang)

Abstract. In order to solve the steady-state error in position tracking control for electro-hydraulic servo universal testing machine, caused by uncertain parameters in the system model, an adaptive sliding mode control strategy based on RBF neural network is proposed for this situation. This paper utilizes the adaptive ability of RBF neural network to improve the control quality of the electro-hydraulic position servo system. The strategy has three parts: the equivalent control, the reaching law control and the compensation control based on RBF network. Simulations verify that the control system can track the reference curve well with unknown parameters.

1. Introduction

Electro-hydraulic position servo systems are widely used in testing machines because of their good position capabilities, high power-to-weight ratio, high stiffness, fast response. However, the actual servo systems have highly nonlinear characteristics and many uncertain parameters, such as elastic stiffness of elastic load, hydraulic fluid bulk modulus. Sliding mode control is robust with respect to system uncertainties through the use of switching control. However, the previous sliding mode controller design requires the boundary of the system model uncertainty parameters to be known. In practical engineering, the boundary of the system uncertainty parameters is likely to be unknown, which leads to the control effect of the designed sliding mode controller is not ideal.

In order to solve this issue, a lot of researchers have proposed some advanced algorithms. Paper [1] uses adaptive algorithm and the principle of fuzzy approximation to estimate the unknown parameters of the system. Paper [2] proposes an adaptive output feedback control scheme for the hydraulic position servo system with uncertain parameters and unmeasurable states. Paper [3] constructs a discontinuous projection-based adaptive control law to handle parametric uncertainties. Paper [4] investigates a fast speed adaptive law to estimate the variant perturbation. Neural networks have received much attention because of their adaptability, and many studies have combined them with SMC. In paper [5], an adaptive particle swarm optimization wavelet neural network with double sliding modes controller is proposed to address the complex nonlinearities and uncertainties in the electric load simulator. Paper [6] introduces a recurrent type 2 fuzzy wavelet neural network to approximate the unknown nonlinear functions of the dynamic systems through tuning by the desired adaptive law. In paper [7] and paper [8], RBF neural network is used to approximate the unknown system parameters. Paper [9] and [10] use RBF neural network to compensate the model uncertainties.

This paper focuses on the position control system of electro-hydraulic servo fatigue testing machine and proposes an adaptive sliding mode robust tracking control method which combining the sliding mode control based on the approach law with the RBF neural network. The method is easy to implement, and the simulation research on the position control system model of the fatigue testing machine with elastic load has achieved satisfactory results.

2. System Description

The hydraulic system is shown in Figure 1 comprises a double-rod cylinder, a 4-way servo valve and a load force. In the following, the dynamic model of the system will be given.

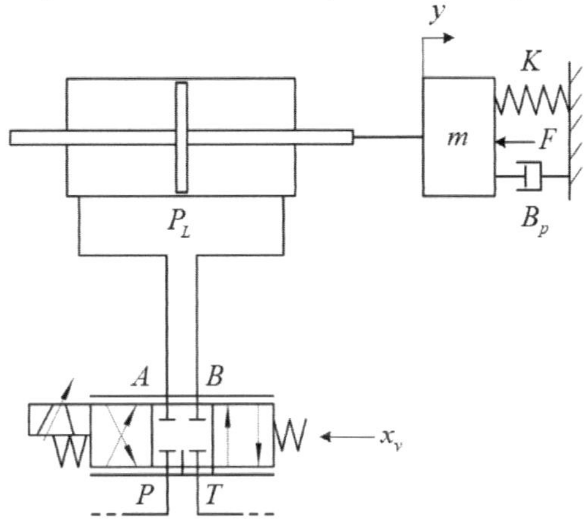

Figure 1. Schematic diagram of the hydraulic system

2.1. Dynamic models

The mathematical model of the electro-hydraulic servo system actuator is as follows:

$$\begin{aligned} Q &= K_q x_v - K_c P_L \\ Q &= A_P Y s + \frac{V_t}{4\beta_e}\cdot P_L s + C_{tc} P_L \\ A_P P_L &= mYs^2 + B_p Ys + KY + F_L \end{aligned} \tag{1}$$

Where Q is flow through the valve orifice, K_q is flow gain, x_v is valve opening, K_c is pressure coefficient, pressure difference between the two cylinder chambers is P_L, A_P is the cylinder effective area, Y is the position of the cylinder measured by displacement sensor, V_t is the effective cylinder volume, β_e is the hydraulic fluid bulk modulus, C_{tc} is leakage, m is total load mass, B_p is damping, K is equivalent spring coefficient of the elastic load, F_L is disturb.

In order to simplify the analysis, the servo valve is considered as a proportional link, which means control valve dynamics is neglected and system input is the valve opening $x_v = u$; the load of the fatigue testing machine is equivalent to the elastic load. When the elastic test piece is clamped, the open-loop transfer function of the system is:

$$G(S) = \frac{U_s}{\Delta U} = \frac{K_{sv} K_q K_s K_a A_p / KK_{ce}}{(s/\omega_r + 1)(s^2/\omega_0^2 + 2\xi_0 s/\omega_0 + 1)} \tag{2}$$

Where K_{sv} is magnification factor for servo valve, K_a is gain of servo amplifier, K_s is gain of displacement sensor. Where:

$$K_{ce} = K_c + C_{tc},\ \omega_0 = \left(\frac{4\beta_e A_p^2}{mV_t} + \frac{K}{m}\right)^{1/2},\ \omega_r = \frac{K_{ce} K}{A_p^2\left(1 + KV_t/4\beta_e A_p^2\right)},\ \xi_0 = \frac{1}{2\omega_0}\left[\frac{4\beta_e K_{ce}}{V_t\left(1 + KV_t/4\beta_e A_p^2\right)} + \frac{B_p}{m}\right].$$

2.2 Converting the transfer function to a controllable standard state space equation

Defining state variables: $\mathbf{x} = \left[x_1\ x_2\ x_3\right]^T = \left[y\ \dot{y}\ \ddot{y}\right]^T$, Converting the transfer function into a controllable standard state space as follows:

$$\begin{cases} \dot{x}_1 = x_2 \\ \dot{x}_2 = x_3 \\ \dot{x}_3 = -a_1 x_1 - a_2 x_2 - a_3 x_3 + bu \\ y = x_1 \end{cases} \tag{3}$$

Where $a_1 = \omega_0^2 \omega_r$, $a_2 = \omega_0^2 + 2\xi_0 \omega_0 \omega_r$, $a_3 = \omega_r + 2\xi_0 \omega_0$, $b = \omega_0^2 \omega_r K_v$, $K_v = K_{sv} K_q K_s K_a A_p / K K_{ce}$.

For an electro-hydraulic position servo system such as in equation (3), when parameters, such as K, m, K_q, K_c, β_e, are uncertain, system parameter $a_i (i = 1,2,3)$ and control gain b will be different from the nominal parameters used when designing the controller. At this point, make assumptions:

- Position reference $[\dddot{r}, \ddot{r}, \dot{r}, r]$ are known as bounded and continuous;
- System state $\mathbf{x}$ is the parameter which can be obtained.

According to the assumptions, system can be described as a linear uncertain system as follows.

$$\begin{aligned} \dot{\mathbf{x}} &= (\mathbf{A}_n + \Delta\mathbf{A})\mathbf{x} + (b_n + \Delta b)u \\ y &= x_1 \end{aligned} \tag{4}$$

Where $\mathbf{A}_n$, b_n are the system's rated parameters, $\Delta\mathbf{A}$, Δb are the uncertain parts of the system.

The goal of this paper is to compensate the disturbance caused by the system in the presence of uncertain parameters, so that the actual displacement of the actuator can follow the reference signal as much as possible within a limited time.

3. Design of sliding mode controller based on RBF neural network compensation

The design of sliding mode control is divided into two parts: the first part is to select the sliding surface, on which the system should meet the required performance; the second part is to choose the ideal sliding mode control law to make the system state can reach the sliding surface in a limited time, ensuring that the system enters the sliding mode.

3.1 Sliding mode control based on the approach law

In this paper, the design of the sliding mode controller adopts the equivalent control based on the approach law and the structure of the control input is:

$$u_{SMC} = u_{eq} + u_{sw} \tag{5}$$

Where u_{SMC} is the output of SMC, u_{eq} is equivalent control output, u_{sw} is switching control output.

Choose linear sliding surface function to design the sliding surface:

$$s = \mathbf{C} \cdot \mathbf{x} \tag{6}$$

Where $\mathbf{C} = [c_1, c_2, \cdots c_{n-1}, 1], c > 0$ and must be Hurwitz.

Concerning about equation (3), design the sliding surface as:

$$\begin{cases} e_1 = r - x_1 \\ e_2 = \dot{r} - x_2 \\ e_3 = \ddot{r} - x_3 \end{cases} \tag{7}$$
$$s = c_1 e_1 + c_2 e_2 + e_3$$

Make $\dot{s} = 0$ to obtain the equivalent control law:

$$\begin{aligned} \dot{s} &= c_1 e_2 + c_2 e_3 + \dddot{r} - \dot{x}_3 \\ &= c_1 e_2 + c_2 e_3 + \dddot{r} - (-a_1 x_1 - a_2 x_2 - a_3 x_3 + bu) \\ &= 0 \end{aligned} \tag{8}$$

The equivalent control law is:

$$u_{eq} = \frac{1}{b}\left[(c_1 \ddot{r} + c_2 \ddot{r} + \dddot{r}) + a_1 x_1 + (a_2 - c_1)x_2 + (a_3 - c_2)x_3\right] \tag{9}$$

Selecting the exponential approach law as the sliding mode controller switching control:

$$\dot{s} = -q\,\mathrm{sgn}(s) - ks, \quad q > 0, k > 0 \tag{10}$$

make $\dot{s} = c_1e_2 + c_2e_3 + \ddot{r} - x_3 = -q\,\mathrm{sgn}(s) - ks$, we can get the control law of sliding mode controller based on the approach law:

$$u_{\mathrm{SMC}} = \frac{1}{b}\left(c_1e_2 + c_2e_3 + \ddot{r} + a_1x_1 + a_2x_2 + a_3x_3 + q\,\mathrm{sgn}(s) + ks\right) \tag{11}$$

3.2 Compensation control based on RBFNN

In this paper, we use RBFNN to compensate the uncertain part of the model. The control structure of SMC based on RBFNN compensation can be expressed as:

$$u = u_{\mathrm{SMC}} + u_{\mathrm{RBF}} \tag{12}$$

The control structure is illustrated in Figure 2:

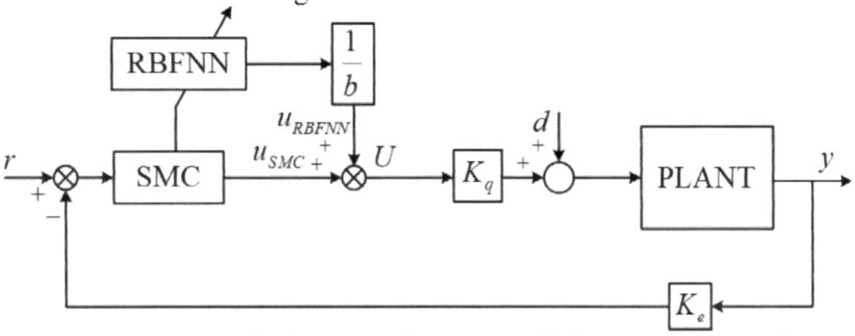

Figure 2. Block diagram of RBFNN sliding mode control

In this paper, the structure of the RBFNN, as illustrated in Figure 3, is 3-2-1, which means the numbers of input-layer neuron, hidden-layer neurons, liner output-layer are respectively to be 3, 2, 1.

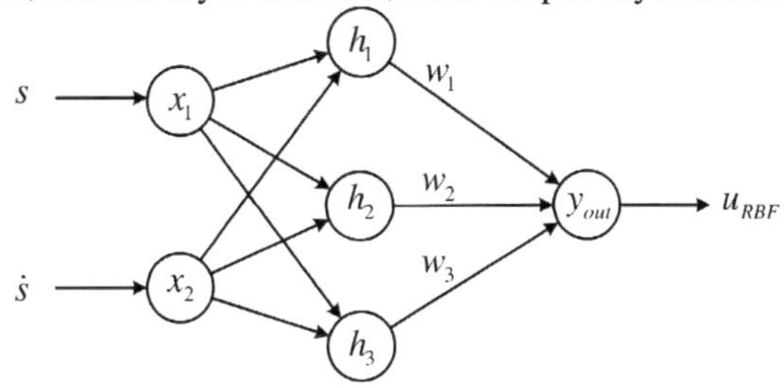

Figure 3. RBFNN Structure in this paper

Where w_j is the hidden-layer-to-outputs interconnection weights, $j = 1,2,3$, h_i is the hidden layer with RBF activation function, , $i = 1,2$, which can be expressed as:

$$h_j = \exp\left(-\frac{\left\|x_i - \theta_j\right\|^2}{2b^2}\right), b > 0 \tag{13}$$

Where x_i is the input of the network, θ_j is the center of RBFNN. The output of the output-layer is the linear weighted sum of the hidden layer radial basis functions output values. The output of output-layer can be defined as:

$$y = \sum_{j=1}^{3} w_j h_j \tag{14}$$

Finally, the output of RBFNN can be expressed as:

$$u_{\mathrm{RBF}} = \frac{1}{b}\mathbf{W}^{\mathrm{T}}\mathbf{h} \tag{15}$$

Use $\mathbf{x} = [s(k), \dot{s}(k)]^T$ as RBF network inputs, take $E = s(k)\dot{s}(k)$ as evaluation metric, adjust the weight of RBF network by using gradient descent method, we can get the adjustment method as:

$$\Delta\omega_j(k) = -\eta_j\frac{\partial E(k)}{\partial\omega_j(k)} = -\eta_j\frac{\partial E(k)}{\partial\dot{s}(k)}\frac{\partial\dot{s}(k)}{\partial u(k)}\frac{\partial(u_{SMC}(k) + u_{RBF}(k))}{\partial\omega_j(k)} \tag{16}$$

Where:

$$\frac{\partial E(k)}{\partial \dot{s}(k)} = s(k), \quad \frac{\partial \dot{s}(k)}{\partial u(k)} = -b, \quad \frac{\partial (u_{\mathrm{SMC}}(k) + u_{\mathrm{RBF}}(k))}{\partial \omega_j(k)} = \frac{\partial u_{\mathrm{RBF}}(k)}{\partial \omega_j(k)} = \frac{1}{b} \cdot h_j(k).$$

Use momentum factor α to accelerate the learning rate of the network, the weight is update as:

$$\omega_j(k+1) = \omega_j(k) + \eta_j s(k) h_j(k) + \alpha[\Delta\omega_j(k-1) - \Delta\omega_j(k-2)] \quad (17)$$

4. Simulations

The simulation environment of this paper is MATLAB R2016a, use the traditional sliding mode controller based on the approach law to compare with the controller designed in this paper. The reference signal of the system is $s_d(t) = 10\sin(2\pi t)$ mm, nominal parameters of fatigue testing machine model for simulation are as follows:

$K = 1\times10^7$ N/m , $B_p = 8.82\times10^{-2}$ N·s/m , $A_P = 1.01\times10^{-2}\,\mathrm{m}^2$, $V_t = 1.05\times10^{-3}\,\mathrm{m}^3$, $\beta_e = 700$Mpa , $m = 150$kg , $K_q = 0.4175\mathrm{m}^3/(\mathrm{A}\cdot\mathrm{s})$, $K_s = 100$V/m , $K_a = 4\times10^{-3}$ A/V , $K_{ce} = 0.5\times10^{-11}(\mathrm{m}^3/\mathrm{s})/\mathrm{Pa}$.

The sliding mode controller parameters are selected as follows:

$C_1 = 5000$, $C_2 = 700$, $q = 100$, $k = 200$.

Parameters of RBFNN are as follows:

$b_i = 50$, $\mathbf{W}(0) = \mathbf{0}$, η=8 , α=0.3 , the centers are set as follows:

$$\boldsymbol{\theta} = \begin{bmatrix} -50 & -25 & 0 & 25 & 50 \\ -50 & -25 & 0 & 25 & 50 \end{bmatrix}^{\mathrm{T}} \quad (18)$$

Model state at the begin of the simulation is $\mathbf{x}(0) = [0\,0\,0]^{\mathrm{T}}$. Through simulation, it can be found that, as shown in Figure 4, when the actual parameters of the system are consistent with the rated parameters of the controller, the sliding mode control based on exponential approach law and the sliding mode control based on RBFNN compensation can track the reference signal fast, but obviously RBFNN slip the mode controller converges faster than the typical sliding mode controller.

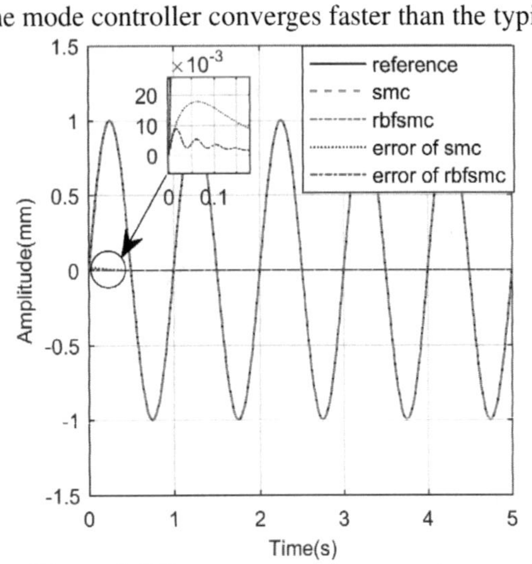

Figure 4. When parameters are consistent

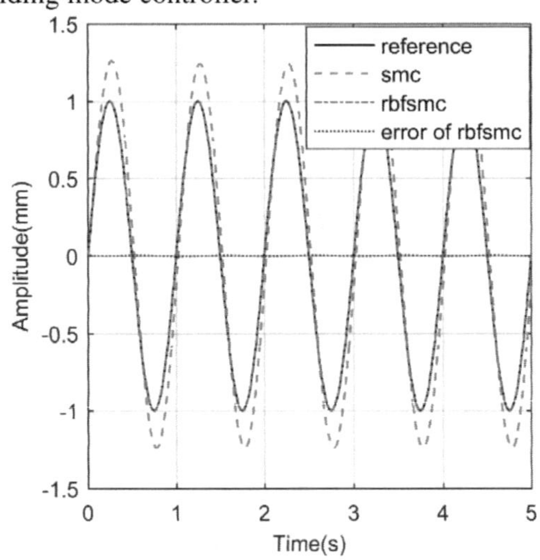

Figure 5. When $K = 0.8\times10^7$ N/m

As shown in Figure 5, Figure 6, Figure 7, when there are unknown parameters in the model, the control results of the sliding mode controller based on the exponential approach law have steady-state error. The sliding mode controller combined with RBF compensation and exponential approach law is insensitive to unknown parameters, its curve and reference signal curve almost coincide, and still have good tracking performance.

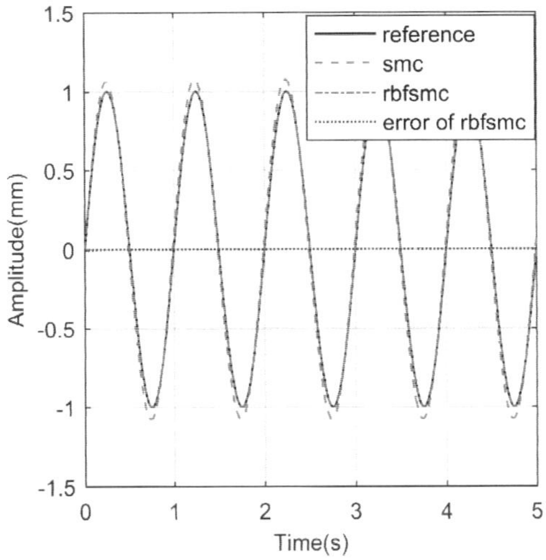

Figure 6. When β_e=1000Mpa

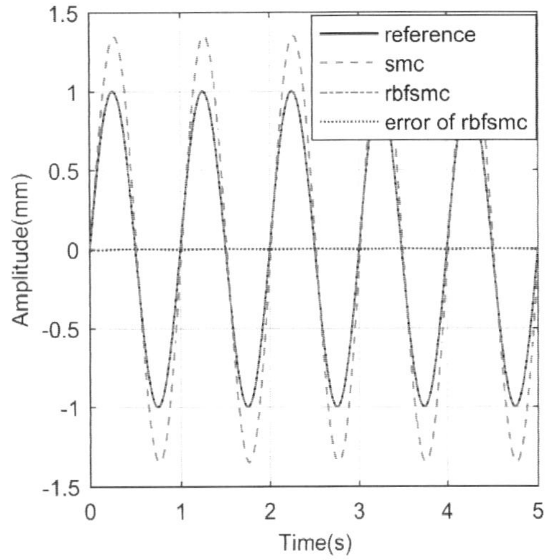

Figure 7. When $K = 0.8\times10^7$ N/m , β_e=800Mpa

5. Conclusion

In this paper, the RBF neural network is used to compensate the interference caused by unknown parameters in the system, thus basically eliminating its influence on the electro-hydraulic servo position control. The connection weight of the RBF neural network which is used in this paper does not need offline training and the structure is simple and easy to implement. The simulation results show that the controller designed in this paper can still have excellent tracking control effect when the controller rated parameters are inconsistent with the actual system parameters.

6. References

[1] Huang X, Wan Y, Sun Y and Hou J 2019 Indirect Adaptive Fuzzy Sliding-Mode Control for Hydraulic Manipulators *Int. Conf. on Mechanical Design (Singapore)* pp 229-42.

[2] Ren H and Gong P 2017 Adaptive control of hydraulic position servo system using output feedback *Proc. Institution Mechanical Engineers* I **231** 527-40.

[3] Yang G, Yao J, Le G and Ma D 2016 Adaptive integral robust control of hydraulic systems with asymptotic tracking *Mechatronics* **40** 78-86.

[4] Tran D T, Jeong K, Jun G, Kiro JS, Kiro MJ and Ahn KK 2017 Adaptive gain back-stepping sliding mode control for electrohydraulic servo system with uncertainties *2017 14th Int. Conf. on Ubiquitous Robots and Ambient Intelligence (Jeju)* pp 534-39.

[5] Wang C, Hou Y, Gao Q, Hou R and Deng T 2016 Electric load simulator system control based on adaptive particle swarm optimization wavelet neural network with double sliding modes *Advances in Mechanical Engineering* **8** 1-10.

[6] Chen X, Li D, Yang X and Yu Y 2018 Identification recurrent type 2 fuzzy wavelet neural network and L2-gain adaptive variable sliding mode robust control of electro-hydraulic servo system (EHSS) *Asian J. of Control* **20** 1480-90.

[7] Moawad N M, Elawady W M and Sarhan A M 2019 Development of an adaptive radial basis function neural network estimator-based continuous sliding mode control for uncertain nonlinear systems *ISA transactions* **87** 200-16.

[8] Li J and Tian H 2018 Position control of SMA actuator based on inverse empirical model and SMC-RBF compensation *Mechanical Systems and Signal Processing* **108** 203-15.

[9] Fei J and Yang Y 2016 Robust neural network control of MEMS gyroscope using adaptive sliding mode compensator *Robotica* **34** 497-512.

[10] Wang Z, Zhu S, Chen Q, Zhang X and Song Y 2015 Sliding mode control of electro-hydraulic servo system for lower-limb exoskeleton based on RBF neural network *2015 IEEE 10th Conf. on Industrial Electronics and Applications (Auckland)* pp 79-83.

Study on the envelope process of the flexspline of harmonic reducer based on the angle of wave generator

Tao Zhang#, Zhifeng Liu, Congbin Yang, Qiushi Hu

Beijing University of Technology, No.100, Pingleyuan, Chaoyang District, Beijing, China

zhangtaoleadership@163.com

Abstract. Harmonic reducer is the core part of robot joint, its tooth shape design theory and meshing characteristics will directly affect the motion control accuracy and vibration characteristics of robot. This paper studies the meshing process of the circular spline and flexsplines of the harmonic reducer based on the double arc profile and the elliptical wave generator. The main innovation of this research is to take the input angle of the wave generator as an independent variable, which more clearly represents the enveloping process and meshing interval distribution of the tooth profile. The results show that the secondary meshing process of the double circular arc tooth profile exists on both sides of the long axis of the wave generator. But the second conjugate contact interval is small, so the optimization of tooth shape should focus on the first conjugate interval. This study provides a theoretical basis for the design of tooth profile parameters of harmonic reducer.

1. Introduction

Since the harmonic gear has the advantage of a high transmission accuracy and compact, it has been widely used in high-precision equipment. In recent years, more and more attention has been paid to circular arc tooth profile, especially double circular arc tooth profile. Compared with the involute profile, it has many advantages, such as more pairs of meshing teeth, larger meshing interval, more uniform stress distribution, etc., which improves the meshing performance of the harmonic reducer and makes the transmission more stable [1]. Xin has done a lot of research work in the tooth profile design and meshing theory of harmonic gear drive. Among them, innovative research results have been achieved in the design and device development of double arc tooth profile harmonic gear[2]. Dong studies on the movement of the tooth profile by the dynamic instantaneous center line of the flexible tooth relative to the rigid one and proposed the kinematic optimization model of backlash under different loads[3,4]. Wang found that the meshing interference can be effectively avoided by optimizing the radial deformation coefficient of the double circular arc tooth profile[5]. Chen et al. studied in detail the relationship between the design parameters of tooth profile and the meshing clearance and interference in the process of single tooth enveloping, and used the finite element simulation to verify[6]. Based on the image recognition algorithm and the laser ranging sensor, Ma et al. extracted the movement speed of the flexible tooth profile and the change rule of the neutral layer of the harmonic reducer under different wave generator speed conditions[7,8].

2. Modeling of transmission mode with circular spline fixed and flexspline as output components

2.1. Deformation function of flexible bearing based on elliptical wave generator

The three core parts of harmonic reducer drive are circular spline, flexspline and wave generator. The wave generator rotates through the large deformation of the outer flexible bearing to drive the meshing transmission. The parameter equation of the wave generator can be expressed as

$$\begin{cases} x = a_T \sin(\varphi) \\ y = b_T \cos(\varphi) \end{cases} \tag{1}$$

The curve equation of the neutral layer after the flexible bearing is installed in the wave generator can be expressed as

$$\omega(\varphi) = \left|\vec{L}\right| - r_m \tag{2}$$

The research shows that it is feasible and convenient to use sine function to approximate the original function from the point of view of fitting results and error analysis. The fitting error is kept in the order of 0.001mm. So the radial deformation function can be expressed as

$$\omega = \omega_0 \cos(2\varphi) \tag{3}$$

Similarly , the tangent displacement and normal angle of the neutral layer curve can be expressed as

$$\begin{cases} v = -\int_0^{\varphi} \omega(\varphi) d\varphi = -\frac{\omega_0}{2}\sin(2\varphi) \\ u = \arctan(\frac{\rho'}{\rho}) \approx \frac{1}{r_m}\frac{d\omega}{d\varphi} = \frac{1}{r_m}(2\omega_0 \sin(2\varphi)) \end{cases} \tag{4}$$

2.2. Basic tooth shape of double circular arc of flexspline

The standard double arc tooth profile of the flexspline is composed of two arcs and a common tangent. The intersection of the center line of the flexible bearing and the center line of the tooth profile is the coordinate origin, and the center line of the tooth profile is the double arc tooth profile of the Y axis as shown in Figure 1. Considering that we study the process of the wave generator rotating anticlockwise, it is necessary to change the equation of the flexspline tooth profile into the left one so as to mesh with the right one of the circular spline.

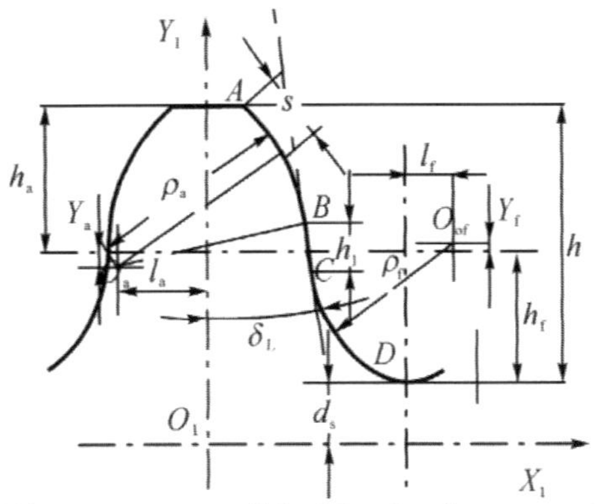

Figure 1. The parameters of double circular arc tooth profile.

The tooth profile equation of AB segment on the left side:

$$\begin{cases} x_{AB} = -(\rho_a \cos(\alpha_a - s/\rho_a) + x_{oa}) \\ y_{AB} = \rho_a \sin(\alpha_a - s/\rho_a) + y_{oa} \\ s \in (0, l_1), l_1 = \rho_a(\alpha_a - \delta_L) \end{cases} \tag{5}$$

The tooth profile equation of BC segment on the left side:

$$\begin{cases} x_{BC} = -(\rho_a \cos\delta_L + x_{oa} + (s-l_1)\sin\delta_L) \\ y_{BC} = \rho_a \sin\delta_L + y_{oa} - (s-l_1)\cos\delta_L \\ s \in (l_1, l_2), l_2 = l_1 + h_l / \cos\delta_L \end{cases} \tag{6}$$

The tooth profile equation of CD segment on the left side:

$$\begin{cases} x_{CD} = -(x_{of} - \rho_f \cos(\delta_L + (s-l_2)/\rho_f)) \\ y_{CD} = y_{of} - \rho_f \sin(\delta_L + (s-l_2)/\rho_f) \\ s \in (l_2, l_3), l_3 = l_2 + \rho_f(\pi/2 - \arcsin((Y_f + h_f)/\rho_f) - \delta_L) \end{cases} \tag{7}$$

The meaning and calculation formula of tooth profile parameters can refer to reference [5]

2.3. Kinematic frame and tooth profile equation of circular spline

Suppose the tooth profile equation of the flexspline is:

$$\begin{cases} x_r = x(s) \\ y_r = y(s) \end{cases} \tag{8}$$

Transformation matrix from flexspline coordinate to circular spline coordinate is

$$M_{gr} = \begin{bmatrix} \cos(\beta) & \sin(\beta) & \rho\sin(\Delta\varphi) \\ -\sin(\beta) & \cos(\beta) & \rho\cos(\Delta\varphi) \\ 0 & 0 & 1 \end{bmatrix} \tag{9}$$

Enveloping tooth profile of flexspline:

$$\begin{cases} x_g = x_r\cos(\beta) + y_r\sin(\beta) + \rho\sin(\Delta\varphi) \\ y_g = -x_r\sin(\beta) + y_r\cos(\beta) + \rho\sin(\Delta\varphi) \end{cases} \tag{10}$$

According to envelope theory, the conjugate profile of a tooth of circular spline can be expressed as

$$\begin{cases} x_g = x_g(s, \varphi_H), y_g = y_g(s, \varphi_H) \\ \dfrac{\partial x_g}{\partial s} \cdot \dfrac{\partial y_g}{\partial \varphi_H} - \dfrac{\partial y_g}{\partial s} \cdot \dfrac{\partial x_g}{\partial \varphi_H} = 0 \end{cases} \tag{11}$$

3. Analysis of meshing process based on the angle of wave generator

In order to better understand the engagement process, we will conduct research through detailed case analysis, and the design parameters of harmonic reducer are shown in Table 1.

Table 1. The design parameters of harmonic reducer

Parameters	Symbol	Value	Unit
Module	m	0.25	mm
Teeth number of the FS	z_1	200	--
Teeth number of the CS	z_2	202	--
Modification coefficient	w_0	1.12m	--
Radius of the FS neutral line	r_m	24.3	mm

The single tooth envelope process is shown in Figure 2 and Figure 3. When the wave generator rotates counterclockwise, the tooth profile of the flexspline rotates clockwise and then conjugate with the tooth top and tooth root arc of the circular spline. Considering that the direction of the external torque is counterclockwise, the left profile of the flexspline and the right profile of the circular spline contact in the whole enveloping process.

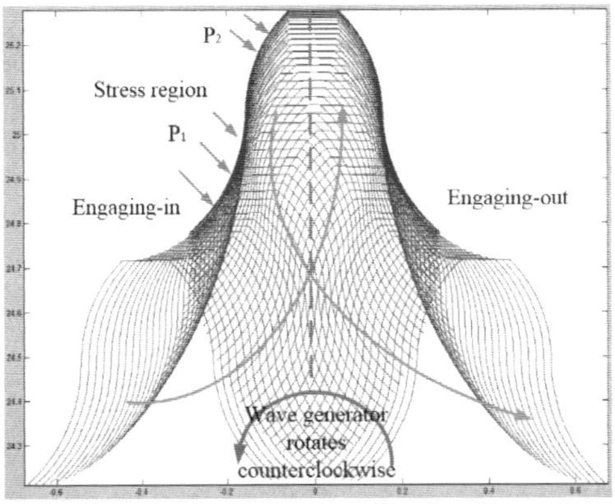

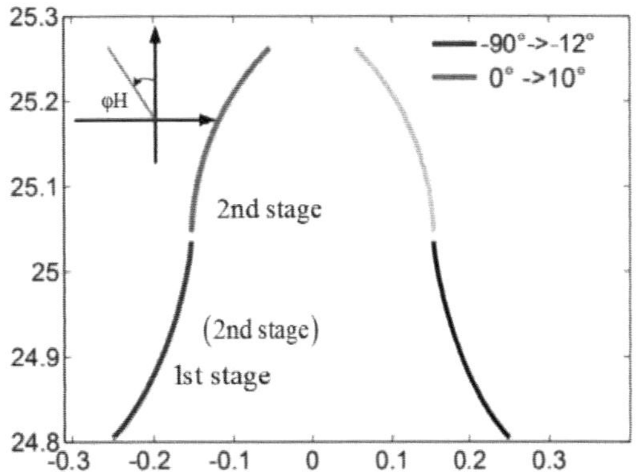

Figure 2. Enveloping process of a single tooth profile. **Figure 3.** Tooth profile of circular spline.

As shown in Figure 2, although the double arc tooth profile is composed of three curves, the conjugate contact between circular spline and flexspline only involves concave arc and convex arc. In the first stage of engagement, conjugate contact occurs in the convex tooth profile of flexspline and circular spline. At this time, the corresponding rotation angle interval of the wave generator is [-60°,-10°]. The angle of wave generator corresponding to the second stage of engagement is [0°,10°], the convex arc of the flexspline is conjugate with the concave arc of the circular spline, and the concave arc of the flexspline is conjugate with the convex arc of the circular spline in the meantime.

From the sign of the rotation angle of the wave generator, we can see that the first conjugate interval and the second conjugate interval are located on both sides of the long axis of the wave generator as shown in Figure 4. The secondary meshing process of the double circular arc tooth profile does exist, and exists on both sides of the long axis of the wave generator. But the second conjugate contact interval is small, so the optimization of tooth shape should focus on the first conjugate interval.

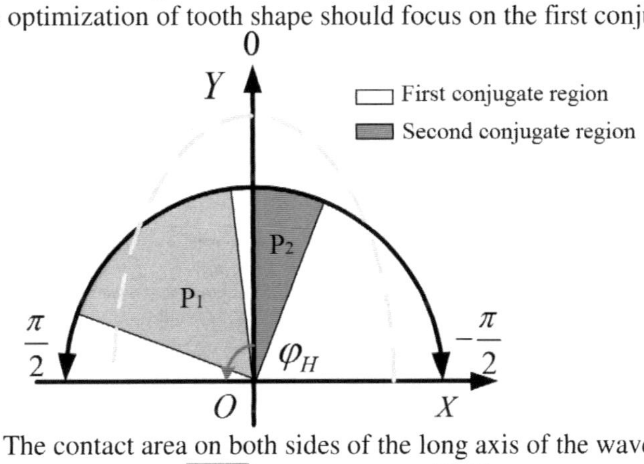

Figure 4. The contact area on both sides of the long axis of the wave generator.

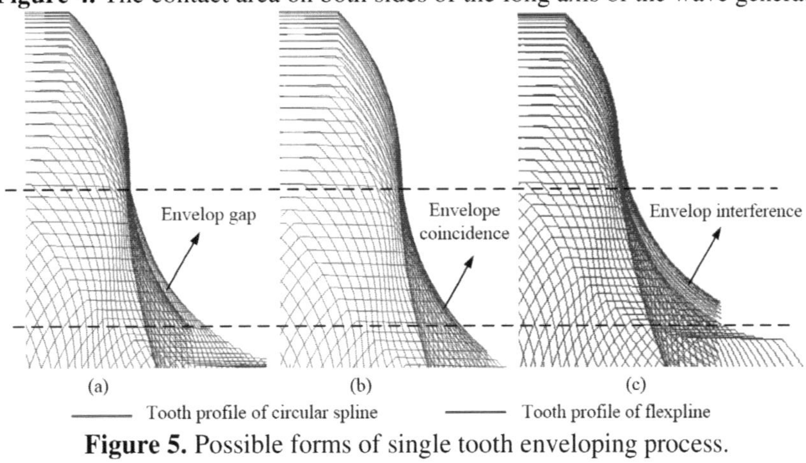

Figure 5. Possible forms of single tooth enveloping process.

Three different types of envelope results are described in Figure 5. Figure 5 (b) is the most appropriate envelope result, where the conjugate arc at the top of the flexspline tooth coincides with the conjugate arc at the root. If we can't realize the complete coincidence through optimization, then we must ensure that top circular envelope tooth profile of the flexspline is above the root circular envelope tooth profile as shown in Figure 5 (a), which is one of the most important conditions for proper transmission. When the enveloping results as shown in Figure 5(c) appear, it shows that the double arc tooth profile parameters of the flexspline need to be optimized to avoid enveloping interference.

4. Conclusion

In this paper, we establish the contact model of single tooth and multi tooth on the left side of the flexspline and the right side of the circular spline when the wave generator rotates anticlockwise. The meshing clearance, meshing interval and secondary meshing process have been studied. The conclusions obtained from the study are listed below:

(1) For the double arc tooth profile of flexspline, there are two conjugate intervals located on both sides of the long axis of the wave generator. One interval is called single-conjugate-contact region, which is 10 to 60 degrees ahead of the rotation angle of the wave generator. The other interval should be called single-conjugate-contact region, which is 0 to 10 degrees behand of the rotation angle of the wave generator.

(2) In the envelope process, if we can't realize the complete coincidence of the top circular envelope tooth profile and the root circular envelope tooth profile through optimization, then we must ensure that top circular envelope tooth profile of the flexspline is above the root circular envelope tooth profile, which is one of the most important conditions for proper transmission.

5. References

[1] Hongbing Xin, Huiyang He and Ruijin Xie 1997 *J. Changchun Inst. Opt* **23** 47
[2] Hongbing Xin 2001 *J. China Mech. Eng* **22** 30
[3] Huimin, Dong J 2001 *J. Mech. Tran* **35** 7
[4] Huimin, Dong and Shuhai Liu J 2003 *J. Mech. Tran* **27** 23
[5] Jiaxu Wang, Xiangxiang Zhou, Junyang Li, Ke Xiao and Qi Li J 2016 *J. HN. U* **43** 56
[6] Xiaoxia Chen, Yusheng Liu, Jingzhong Shuzhong Lin and Wei Xu 2014 *J. MECH. MACH. THEORY* **73** 1
[7] Donghui Ma, Jianing Wu, Tao Liu and Shaoze Yan 2017 *J. Sci. China* **08** 57
[8] Donghui Ma, Jianing Wu and Shaoze Yan 2016 *J. Sci. Chi. Tech. Sci* **59** 1305

Acknowledgments

The authors would like to thank the National Natural Science Foundation of China No. 51805012, for supporting the research

Researches on suppliers selection based on improved AHP-TOPSIS method

Xinchen Wang, Yanhua Ma, Pengfei Gao*, Lu zhai, Bing Gu

Department of Industrial Engineering, University of Jilin, Changchun, China

Corresponding author's e-mail: gaopf19@mails.jlu.edu.cn

Abstract. Selection of supplier can be considered as the significant strategic decision for company's competitive advantage. Generally, the scales of criteria for selection were used to evaluate the performance of suppliers. But the traditional criteria are limited to reflect the comprehensive ability of the suppliers. Meanwhile, the quantitative scale, which is used to evaluate the suppliers, is too subjective to describe the real station of suppliers. Therefore, the improved analytic hierarchy process (AHP) is conducted to reanalyze the criteria and distribute the weight for each criterion. Technical for order preference by similarity to ideal solution (TOPSIS) used to calculate the relative proximity for each supplier and rank them based on the corresponding information. Finally, the calculate process is programmed by Excel VBA and MATLAB respectively. The two results show the same ranking order, which verify the improved AHP-TOPSIS method can effectively guide company to select supplier.

1. Introduction

Suppliers can be considered as inevitable sources of external risks in modern supply chains [1]. Supplier evaluation and selection is a significant strategic decision for reducing operating costs and improving organizational competitiveness to develop business opportunities [2]. However, the traditional criteria are limit to evaluate the comprehensive ability for the suppliers, which can not offer a reasonable selection for the company to survive in fierce competition in the market. For example, the culture, flexibility, delivery are seldom considered in the criteria system [3]. Meanwhile, the criteria, which used in AHP, is distributed the weight by experts opinions. This process can bring the subjective factor to calculation result which fails to obtain the reasonable solution. Therefore, the new criteria system and objective methods are widely study in the supplier selection problem. A decision model for selection of suppliers is propose and fuzzy TOPSIS method is used to reduce the subjectivity of experts' suggestions [4]. The same approaches such as the grey-based method, best worst method, integrated framework method [5], grey system [6] and rough set methodologies [7] and other methods can effectively reduce the subjectivity degree. Some new methods, which avoid the criteria scale, also be proposed, such as chance-constrained data envelopment analysis [8], integrated QFD-MCDM framework [9] and other methods can solve the ranking of supplier. With the development of cities, the environment and green supply are emphasized in the selection process. Fuzzy axiomatic design approach, which consider the environmental protection and sustainable development, is proposed to evaluate the suppliers [10]. The strength of the fuzzy NGT-VIKOR based solution approach is its practical applicability and evaluation of green supplier development programs under limited or no quantitative information [11]. AHP-Entropy-TOPSIS framework also be adopted

IOP Conf. Series: Materials Science and Engineering **784** (2020) 012031 doi:10.1088/1757-899X/784/1/012031

to solve the green suppliers selection problem [12]. Furthermore, the multi-criteria decision making approaches are concluded to solve the selection of green supplier and normal suppliers [13,14].
In this paper, the criteria of supplier selection are reanalyzed to fit the comprehensive requirements from the market competition. Then, the improved AHP method distributes the weight for each criterion. Finally, the TOPSIS is adopted to calculate the quantitative result and rank suppliers. EXCEL VAB and MATLAB are used to program the calculation process and verify the effectiveness of method.

2. Using AHP to allocate index weight

2.1. Reanalyzing the criteria in AHP model

The traditional criteria which were used to evaluate the suppliers can be summarized as QSTC (Q: quality, S: serve, T: time, C: cost) in figure 1. However, it only focuses on the static performance for the company rather than the dynamic performance which cannot help the leader of company determine the selection of suppliers. Meanwhile, the traditional criteria fail to reflect the comprehensive competitive ability for the suppliers in the future market competition. Therefore, the sub-criteria and AHP model are modified based on the Delphi method to add more criteria to complete the evaluate system and formulate the improved AHP model.
Comparing with original evaluate system and summarizing the questionnaire result, the experts group give the suggestion that adds the enterprise culture to the criteria system and redistribute the weight for each sub-criteria. The improved AHP model and criteria system is shown in figure 2.

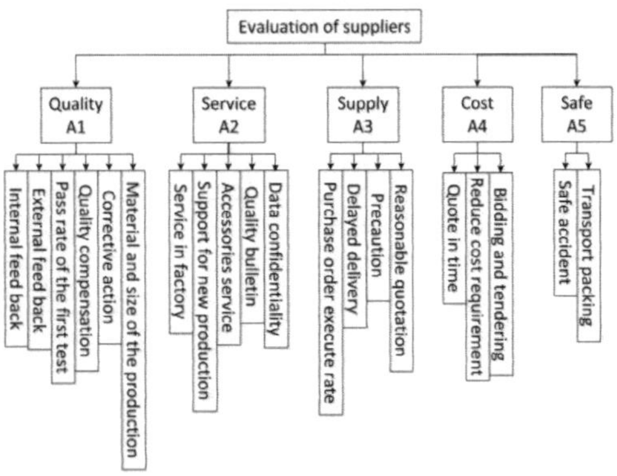

Figure 1. Original AHP model and criteria

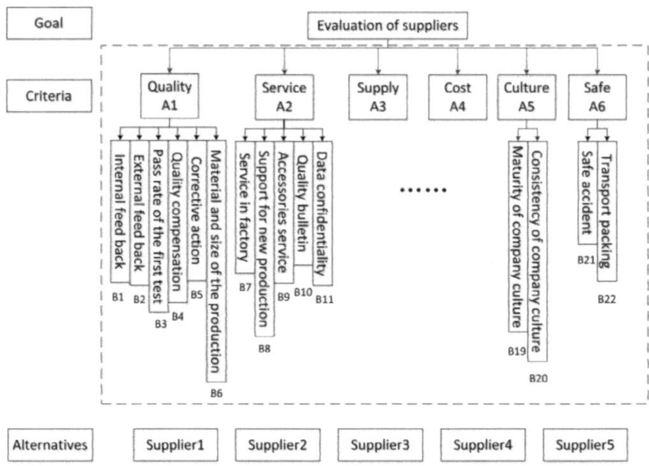

Figure 2. Improved AHP model and criteria

2.2. Constructing judgment matrix

AHP uses a pair-wise comparison of the criteria importance with respect to the goal. This pair-wise comparison allows finding the relative weight of the criteria with respect to the main goal. Therefore, the defined scale or ratio should be introduced for the judgment matrix. If the quantitative data can be used to consist of the matrix, the comparisons can be easily expressed and the calculation work of AHP will be considerably reduced.
In this paper, index scale is introduced for the pair-wise comparison. In the general scale table, number 9 indicates the extremely important for pair-wise comparison of the criteria. However, in the index scale, the equation $a^8 = 9$ is used to express the same degree and the value a can be accurately solved that $a = 1.316$. Before the compare between the two criteria, the important scale a should be given. The other corresponding index scale numbers are shown in the table 1. Furthermore, the quantitative expert opinion is shown in table 2.

Based on the information in the table 2, the judgment matrix **M** of goal hierarchy M with respect to first criteria hierarchy (**A-M** matrix) can be expressed as

$$\mathbf{M}=\begin{array}{c} \quad A1 \quad A2 \quad A3 \quad A4 \quad A5 \quad A6 \\ \left[\begin{array}{cccccc} 1 & 2.618 & 1.618 & 2.058 & 3.330 & 5.388 \\ 0.382 & 1 & 2.058 & 2.058 & 2.618 & 4.236 \\ 0.468 & 0.486 & 1 & 1 & 3.330 & 4.236 \\ 0.468 & 1.618 & 1 & 1 & 3.330 & 5.388 \\ 0.300 & 0.382 & 0.300 & 0.300 & 1 & 2.618 \\ 0.186 & 0.236 & 0.236 & 0.186 & 0.382 & 1 \end{array}\right] \end{array}\begin{array}{c} \\ A1 \\ A2 \\ A3 \\ A4 \\ A5 \\ A6 \end{array} \qquad (1)$$

2.3. The check of consistency

Next step is to calculate the hierarchical single sorting and consistency check which are the normal AHP calculation processes. The result of the consistency check for A-M matrix can be shown on the table 3. The same calculate processes can be used to solve the consistency check of A1—B, A2—B, A3—B, A4—B, A5—B, A6—B. All of judgment matrixes pass the test of consistency check. The final weight of criteria is shown on the table 4.

Table 1 Index scale and meaning for judgment matrix

Subjective feeling Equally important	Slightly important	Obviously important	Strongly important	Extremely important	Form
Index form $a^0=1$	$a^2=1.618$	$a^4=2.618$	$a^6=4.236$	$a^8=6.854$	a^{k-1}
Reciprocal form $a^0=1$	$a^{-2}=0.618$	$a^{-4}=0.382$	$a^{-6}=0.236$	$a^{-8}=0.146$	a^{1-k}

Table 2 Quantitative scale for human feeling

Subjective feeling	Equally important	Slightly important	Slightly obviously	Obviously strongly	Strongly important	Extremely important
m:n	0.9-1.1	1.1-1.3	1.3-1.5	2.5-4	4-6	8-10

Table 3 Matrix A-M consistency check calculation process

Matrix	n(dimension)	λ max	CI=(λ max-n)/(n-1)	RI	CR=CI/RI
M	6	6.191822	0.038364	1.24	0.030939

Table 4 The weight of criteria for suppliers selection

Criteria	A_1	A_2	A_3	A_4	A_5	A_6
Weight of criteria	0.3193	0.1821	0.1681	0.2138	0.0754	0.0413

3. Ranking suppliers based on the TOPSIS

TOPSIS is further used to calculate the decision matrix. The suppliers which need be assessed by company is indicated as C_i (i = 1, 2, 3, …, m), m is the number of suppliers, the components which are offered for the company from each suppliers should be the same one. The assess criteria is B_j (j=1, 2, 3, …,n). The assess value of the alternatives set C = {C_1, C_2, C_3, …, C_m} with respect to criteria set can be express as $x_{ij}(i=1,2,3\cdots,m; j=1,2,3\cdots,n)$. Furthermore, the decision matrix is unified to obtain the normative decision matrix.

After standardizing the relevant data of the evaluation criteria for scheme, the positive and negative ideal solution of the data is calculated to obtain the reference scale. The corresponding scales are used to evaluate the best supplier and worst supplier. The weighted decision matrix and positive and negative ideal solutions are shown in table 5.

The Euclidean distance between each supplier with the best supplier and the worst supplier is calculated. Furthermore, the relative proximity η_i between each supplier and the best supplier is solved to find which result is close to the positive ideal solution. More close to the positive ideal solution, the better the supplier is; on the contrary, the worse it is. The calculation results are shown in table 6.

Table 5 Weighted decision matrix and positive and negative ideal solutions

	C_1	C_2	C_3	C_4	C_5	Positive solution S_j^+	Negative solution S_j^-
B_1	0	0.020956	0.041912	0.020956	0.020956	0.041912	0
B_2	0	0.020956	0.041912	0.041912	0	0.041912	0
…	…	…	…	…	…	…	…
B_{22}	0.031766	0	0.031766	0.015883	0	0.031766	0

Table 6 Euclidean distance and relative proximity results

Supplier	Sd_j^+ Euclidean distance value	Sd_j^- Euclidean distance value	η_i
1	0.180124632	0.140122322	0.43754459
2	0.157910418	0.132233352	0.455751133
3	0.045845924	0.231561669	0.834734431
4	0.091509891	0.205018689	0.691396051
5	0.135056632	0.135056632	0.5

4. Results

The improved AHP is used to identify the weight for each criteria, furthermore, the TOPSIS is used to rank the suppliers. The calculation process is shown on the above chapter, the final solution is calculated by the EXCEL VBA and MATLAB which is shown in table 7.

Based on the value of relative proximity, supplier 3 obtains the largest score 0.28, which means supplier 3 is the best supplier among the five selections. The supplier 4 obtains 0.23 that ranks second, so the supplier 3 and supplier 4 should be considered as the corporate supplier.

For the criteria weight, quality, cost, technology service and supply are account the mainly weight in the first hierarchy criteria which account 31.93%, 21.38%, 18.21% and 16.81% respectively.

Table 7 The result of excel and Matlab

Result	Supplier 1	Supplier 2	Supplier 3	Supplier 4	Supplier 5
Excel VBA	0.1461	0.1524	0.2782	0.2304	0.1929
MATLAB	0.1460	0.1521	0.2786	0.2307	0.1926
Relative error	0.068%	0.192%	0.143%	0.130%	0.156%

5. Conclusion

In this paper, the improved AHP-TOPSIS method is proposed to solve the multi-goal selection of supplier. Firstly, the traditional AHP model is modified to complete the criteria system of suppliers, the company culture is added to evaluate the performance of the five suppliers. Secondly, the weight of each criterion is calculated by the AHP, the index scale is adapted to express the quantitative experts' opinion. Thirdly, TOPSIS uses the weight of criteria which are obtained from the AHP calculation to rank the supplier. The top two suppliers are finally selected to establish the corporation relationship. Comparing with the experts ranking, the ranking which solved by AHP-TOPSIS method is same to the experts ranking. Therefore, the above result can verify that AHP-TOPSIS can effectively solve the selection of suppliers.

References

[1] Rezaei J., T. Nispeling, J. Sarkis and L. Tavasszy, A supplier selection life cycle approach integrating traditional and environmental criteria using the best worst method, 2016 *Journal of Cleaner Production* **135** 577-588.
[2] Dweiri F., S. Kumar, S. A. Khan and V. Jain, Designing an integrated AHP based decision support system for supplier selection in automotive industry, 2016 *Expert Systems with Applications* **62** 273-283.
[3] Chan F. T. and H. K. Chan, An AHP model for selection of suppliers in the fast changing fashion market, 2010 *The International Journal of Advanced Manufacturing Technology* **51** 1195-1207.
[4] Fallahpour A., E. U. Olugu, S. N. Musa, K. Y. Wong and S. Noori, A decision support model for sustainable supplier selection in sustainable supply chain management, 2017 *Computers & Industrial Engineering* **105** 391-410.
[5] Luthra S., K. Govindan, D. Kannan, S. K. Mangla and C. P. Garg, An integrated framework for sustainable supplier selection and evaluation in supply chains, 2017 *Journal of Cleaner Production* **140** 1686-1698.
[6] Rajesh R. and V. Ravi, Supplier selection in resilient supply chains: a grey relational analysis approach, 2015 *Journal of Cleaner Production* **86** 343-359.
[7] Bai C. and J. Sarkis, Integrating sustainability into supplier selection with grey system and rough set methodologies, 2010 *International Journal of Production Economics* **124** 252-264.
[8] Azadi M. and R. F. Saen, Developing a new chance-constrained DEA model for suppliers selection in the presence of undesirable outputs, 2012 *International Journal of Operational Research* **13** 44-66.
[9] Yazdani M., P. Chatterjee, E. K. Zavadskas and S. H. Zolfani, Integrated QFD-MCDM framework for green supplier selection, 2017 *Journal of Cleaner Production* **142** 3728-3740.
[10] Kannan D., K. Govindan and S. Rajendran, Fuzzy Axiomatic Design approach based green supplier selection: a case study from Singapore, 2015 *Journal of Cleaner Production* **96** 194-208.
[11] Awasthi A. and G. Kannan, Green supplier development program selection using NGT and VIKOR under fuzzy environment, 2016 *Computers & Industrial Engineering* **91** 100-108.
[12] Freeman J. and T. Chen, Green supplier selection using an AHP-Entropy-TOPSIS framework, 2015 *Supply Chain Management: An International Journal* **20** 327-340.
[13] Govindan K., S. Rajendran, J. Sarkis and P. Murugesan, Multi criteria decision making approaches for green supplier evaluation and selection: a literature review, 2015 *Journal of Cleaner Production* **98** 66-83.
[14] Ho W., X. Xu and P. K. Dey, Multi-criteria decision making approaches for supplier evaluation and selection: A literature review, 2010 *European Journal of operational research* **202** 16-24.

User requirement and usability testing framework for information system development: case study of financial institution

P O-thatawong[1] and A Jiamsanguanwong[1]
[1]Department of Industrial Engineering, Faculty of Engineering, Chulalongkorn University THAILAND

ponglada.o@student.chula.ac.th and arisara.j@chula.ac.th

Abstract. Due to the fact that financial institution has its limited competency in developing the information system, outsourcing is often used to support the information system development which same as the case study of the issuing documents system in financial institute. The financial institution has received customer's complaints regarding the mistake of issuing documents caused by the system functionality and human error in operations. Thus, the purpose of this study is to design framework of user requirement and usability testing for information system development to prevent the error in system functionality and set some quality checking process of documents before delivery to customer to prevent mistakes from human error in work operation. System Usability Scale showed that after proposed the framework of user requirement and usability testing to employees who related with the system development, their level of satisfaction toward the framework and non-framework of user requirement and usability testing raise up to 30.93% and 20%, respectively. After six months of implemented quality checking process of documents revealed zero complaints from customers regarding to the mistake of issuing documents.

1. Introduction

It cannot be denied that the progression of technology has been an essential in current period. Information technology has been the crucial business resource and has assisted industries in terms of their opportunities, preferably bank and financial industries where the information technology plays an important role to reform financial institution to raise the speed and trustworthiness [1, 2].

Financial institutions have developed a system to issue documents for customers regarding statement, invoice, and receipt, using a third-party company to support the process of system development. The process of document system development can be seen in figure 1. Most of the issues from the system development are often caused by the process of gathering the needs and testing [3]. After launching the document system, it was found that there were complaints regarding customer's financial statements, **the first complain** was incorrect financial balance statement that occurred on July and October accounted for 133,370 customers and 3,700 customers, respectively. **The second complain** was wrong document delivery occurred on August for 333,000 customers. **The third complain**, on September, was incorrect information in document accounted for 27,512 customers.

The root cause finding and analysis was conducted using 5 Gen [4], and why-why analysis. It was found that the issue of the first complain was mainly caused by the process of gathering user requirement and testing in system development process (shows in dot square in figure 1). The process of gathering user requirement was performed by conducting focus group interview which merely rely on employee's experience, resulting in providing imprecise information and the system itself was not

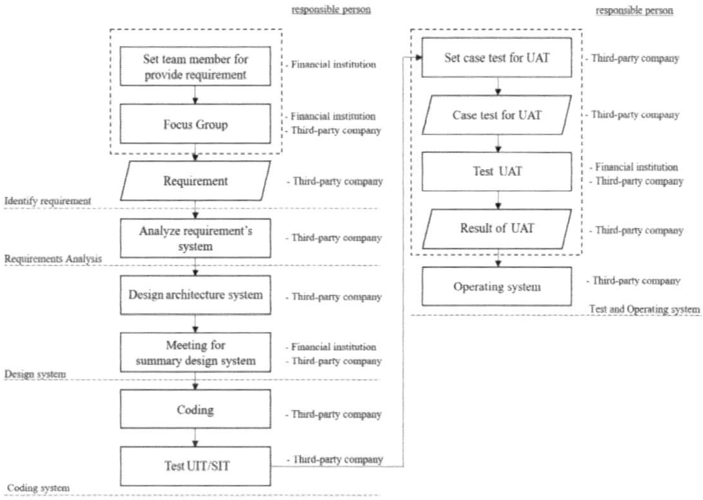

Figure 1: The process of document system development

perfectly function. In system testing, user acceptance test is currently used in which the financial institution has no standard procedure in determining test scenario. This may not literally cover all working functions so that the system error may reach the customer. The issue of the second and third complaints have been found that both issues caused by human error of the staff from the operation of improving the system but improper set the function back to the normal condition before operate the system. Unfortunately, the financial institution has no policy to check these documents before sending out to customer. Thus, wrong documents were sent out.

Collecting requirements are acquiring and collecting process to ensure stakeholder's expectations and employ process modeling. Due to the vagueness of formulated requirements, many system development projects were fail due to poor collecting requirement process in the early process. The successful software system is based on required functionalities under particular limitations required from the system [5, 6]. Mattmann et al. [6]. Also pointed out the need of systematically work during requirements collection and provided critical evaluation.

Thus, the purpose of present study is to design framework for user requirement and usability testing in system development to prevent the future issue of improper functions in system development. This study also provides quality check policy for current document system to prevent any error reach to the customer which may cause by human error in work operation.

2. Methodology

2.1. User Requirement and Usability Testing Framework

The user requirement and usability testing framework is proposed based on literature review regarding to current waterfall model [7, 8], as seen in figure 1. The details of framework is shown in table 1. This framework has divided the process of gathering user requirement into 5 working steps according to Metacognition [9-13] and work process tool for systematic collecting user requirement. Each step was clearly identified the responsible persons, activity, input, and output. The first step is to specify the business need in order to set goal of developing system and team member. The second step is to assess the current working condition in order to form the current work process, work constraints, and relevant parties with the system. The third step is to identify problems/ inefficient workflow with current work process in order to propose solution and set "to-be" work process. The fourth step is to study the feasibility of the "to be" process, modified, and set system requirement. The final fifth step is to summarize system requirement in standard form (REQ-1, REQ-2) which were designed according to software requirement specification (SRS) [14]. This documents will be used for communication and documentation.

The usability testing Framework has 4 working steps which applied from usability testing process [15, 16], as shown in table 2. The first step is to set the test goal refer to the requirement documents from user requirement framework (REQ-1, REQ-2). The second step is to set the test scenario according to the requirement of the system in REQ-2, the team has to identify all scenario of using and all type of input to the system and output from the system. The third step is to identify the users and number of users for testing the system. The fourth step is to identify the test metrics such as success rate, error rate, efficiency in term of time on task, and user satisfaction etc. These metrics is selected based on the goal and objective of system testing. The result from usability could be used to revise the system and help detect some error in system functionality.

2.2. Quality check policy

Quality check policy is proposed regarding to the human error from the third-party company who responsible for improving and updating system that caused error in system issuing the documents to customer such as wrong period and form. The quality checking policy is to check the documents after

Table 1: User requirement framework

Phase	Responsible person	Activity	Input	Output	Remark
1. Specifying the business need	- Project approver - Project manager	1. Small group meeting to study goals / Define the team for system development of the project.	Current problems (e.g. Customer complaints)	Summary of goals / scope and project responsible person	All output of each activities are recorded in documents REQ - 1 and REQ - 2 that are designed from SRS
2. Assessing stakeholder requirements	- Project manager - Section Manager who wants to develop - Staff performing and who is involved.	2. Study the former working process (As-Is) (e.g. Interview / Questionnaire/ Visit the workplace/ Observe the working environment)	Current work	Summary of current work procedures (As-Is) and list of relevant parties	
3. Identifying problems	- Project manager - Section Manager who wants to develop - Staff performing and who is involved.	3. Study current problems and solutions.	Current problems (e.g. Customer complaints) Current work procedures (As-Is)	A solution to the current and long-term solutions.	
		4. Specify the solution along with the work process.	A solution to the current and long-term solutions. Current work procedures (As-Is)	Work procedures that need to be modified (To-Be)	
4. Developing requirements	- Project manager - Section Manager who wants to develop -Third-party company	5. Study the feasibility of the process required (To-Be).	Work procedures that need to be modified (To-Be)	Summary of possible system requirements	
5. Setting requirements	- Project manager - Section Manager who wants to develop	6. Summary requirement	Summary of possible system requirements	Document report of requirement (REQ-2)	

Table 2: Usability testing framework

Activity	Responsible person	Input	Output	Remark
1. Determination of test system / project goals	- Project approver - Project manager	Document report of requirement (REQ-1, REQ-2)	Test Goal	
2. Determine the test case according to the work that is occurring in the system.	- Project manager - Section Manager who wants to develop - Staff performing and who is involved.	Test goal Work procedures that need to be modified (To-Be) Document report of requirement (REQ-1, REQ-2)	Test case	
3. Specify Participant, who must be the actual use and number of participant.	- Project manager - Section Manager who wants to develop	Document report of requirement (REQ-1, REQ-2)	user and number of user	
4. Choose indicators for testing • **Efficiency (Task time/Lostness)** • **Effectiveness (Success rate/Error rate)** • **Customer's satisfaction (System Usability Scale)**	- Project manager - Section Manager who wants to develop	Test goal	Indicators index	

it releases from system, including period checking, working function, form checking, balancing, date of financial transactions. The checking frequency is only one sample for each of 6 type of documents for checking the condition of function whether it is in proper working function. Beside from manual checking policy, the system was also added self-check function for the balancing by having the system take the initial balance plus with total deposits and subtracting total withdrawals, must equal the final transaction balance. If not, the system will cancel working flow automatically and report those who control the system.

2.3. Validation
The user requirement and usability testing framework is validated by using system usability scales (SUS) comparing between current work process and the new framework with 30 employees from the financial institute who have experienced in the process of developing document system with third party company. The SUS is 10 questions on 5 points-Likert scale from (1) totally disagree to (5) totally agree. The score ranges from 0 to 40, the higher the better.
For quality check policy, the validation is conducted by piloting the quality check policy for 6 months and record all the complaints and error regarding the documents issuing from the system.

3. Results
The result from SUS revealed that employees rated the proposed user requirement framework as significantly higher SUS scores than their current work process of user requirement, as shown in table 3. As well as, the result of the proposed usability testing framework as significantly higher in SUS scores than the current user acceptance testing.
After piloting the quality check policy for 6 months (January to June), the results of error detected before reaching to customer is shown in table 4. During these 6 months, the financial institute received no complaint from customer regarding wrong balance, wrong issue of document date, and its form. This showed that the quality check policy could help the financial institute prevent the quality issues from their documents system to customer.

4. Discussion
The designed framework would help the financial institute to shape their process for system development to be more precisely developed toward the need of the system, as well as other industries that may have the same issue. In terms of the framework, it literally helped the institution the better work procedures of documents' requirement and system testing. Additionally, the framework of developing information system in requirement and testing system will assist the institution work and collect documents systematically and detailly.

Table 3: Average and standard deviation of SUS Score of current work process and proposed framework

	Framework for collect requirement				
	N	Average	SD	t	Sig.
Current work process	30	17.20	3.80	-14.57	.000
Proposed framework	30	29.57	1.94		
	Framework for Test system				
	N	Average	SD	t	Sig.
Current work process	30	20.60	2.40	-15.63	.000
Proposed framework	30	28.60	2.03		

Table 4: The number error documents detected from January to June 2019

Month	*Number of customer documents (Account)*	*Number of error documents detected (Account)*
Jan	406,096	9
Feb	358,777	14
Mar	359,878	7
Apr	406,867	5
May	365,987	12
Jun	368,255	8

5. Conclusion

The purpose of present study is to design framework for user requirement and usability testing in system development to prevent the future failures of improper functions in system development and provides quality check policy for current document system to prevent any error reach to the customer which may cause by human error in work operation. The results revealed that employees had positive attitude toward the new framework and believed that it is better in term of usability of work process. For limitations, as the research is merely emphasizing on the system of issuing documents and studying only in the financial institution, as such, there would possibly be other factors that should be concerned about other requirements and testing system, which could be needed to conduct more research. However, there is inevitable to improve the work process quality to be more effective and efficient and reduce repetitive manual tasks.

6. References

[1] Dangolani, S.K., The Impact of information technology in banking system (A case study in Bank Keshavarzi IRAN). Procedia-Social and Behavioral Sciences, 2011. 30: p. 13-16.
[2] Berger, A.N., The economic effects of technological progress: Evidence from the banking industry. Journal of Money, credit and Banking, 2003: p. 141-176.
[3] Georgiadou, E., Software process and product improvement: a historical perspective. Cybernetics and Systems Analysis, 2003. 39(1): p. 125-142.
[4] Chiarini, A., C. Baccarani, and V. Mascherpa, Lean production, Toyota Production System and Kaizen philosophy: A conceptual analysis from the perspective of Zen Buddhism. The TQM Journal, 2018. 30(4): p. 425-438.
[5] Engineering Design: A Systematic Approach. Third Edition. Mechanical Engineering, 2007. 129(7): p. 58-58.
[6] Mattmann, I., S. Gramlich, and H. Kloberdanz, Getting Requirements Fit for Purpose-Improvement of Requirement Quality for Requirement Standardization. Procedia CIRP, 2016. 50: p. 466-471.
[7] Alshamrani, A. and A. Bahattab, A comparison between three SDLC models waterfall model, spiral model, and Incremental/Iterative model. International Journal of Computer Science Issues (IJCSI), 2015. 12(1): p. 106.
[8] Balaji, S. and M.S. Murugaiyan, Waterfall vs. V-Model vs. Agile: A comparative study on SDLC. International Journal of Information Technology and Business Management, 2012. 2(1): p. 26-30.
[9] Kramarski, B. and Z.R. Mevarech, Enhancing mathematical reasoning in the classroom: The effects of cooperative learning and metacognitive training. American Educational Research Journal, 2003. 40(1): p. 281-310.
[10] Kapa, E., Transfer from structured to open-ended problem solving in a computerized metacognitive environment. Learning and Instruction, 2007. 17(6): p. 688-707.
[11] Lawanto, O. Metacognition changes during an engineering design project. in 2009 39th IEEE Frontiers in Education Conference. 2009. IEEE.
[12] Pennequin, V., et al., Metacognition and low achievement in mathematics: The effect of training in the use of metacognitive skills to solve mathematical word problems. Thinking & Reasoning, 2010. 16(3): p. 198-220.
[13] Huang, F. and L. Bin, Software defect prevention based on human error theories. Chinese Journal of Aeronautics, 2017. 30(3): p. 1054-1070.
[14] Davis, A., et al. Identifying and measuring quality in a software requirements specification. in [1993] Proceedings First International Software Metrics Symposium. 1993. IEEE.
[15] Geisen, E. and J. Romano Bergstrom, Chapter 1 - Usability and Usability Testing, in Usability Testing for Survey Research, E. Geisen and J. Romano Bergstrom, Editors. 2017, Morgan Kaufmann: Boston. p. 1-19.
[16] Battleson, B., A. Booth, and J. Weintrop, Usability testing of an academic library web site: a case study. The journal of academic librarianship, 2001. 27(3): p. 188-198.

IOP Conf. Series: Materials Science and Engineering **784** (2020) 012033 doi:10.1088/1757-899X/784/1/012033

Head gimbal assembly automated visual inspection by using digital image processing

Phonlawut Jaturawichanan[1] and Somkiat Tangjitsitcharoen[2]
Department of Industrial Engineering, Faculty of Engineering, Chulalongkorn University, Thailand

phonlawut.jaturawichanan@gmail.com[1], somkiat.ta@eng.chula.ac.th[2], +66865377322[1], +66022186853[2]

Abstract – This paper presented an approach of Digital Image Processing (DIP) and measurement properties of image region for inspecting the part in Head Gimbal Assembly (HGA) process. Current, the inspection process is performed by the skilled employee, which depends on employee experiences. The inspection method proposed by using digital image processing to do an image enhancement in order to detect the abnormalities and defects on part. The measurement properties of image region is proposed to identify the abnormality occurred on HGA part after image enhancement, in production process. This research aimed to develop the algorithm to inspected HGA part by considered the number of pads, convex area value, and eccentricity value. Then used measurement properties of image region to enhance the algorithm for classifying part in production process. This approach has been developed and proved with the real part in HGA production process. It's proved that the algorithm and method proposed can be classified the part in HGA production process.

1. Introduction

The Hard Disk Drive (HDD) industry in Thailand is a large electronics industry. Thailand is the main production based of HDD industry. In addition, investment has been expanded as well as technologies transferred in Thailand. Hard Disk Drive has 2 major components, which are Head Gimbal Assembly (HGA) and accessory equipment for HDD. In this research will focus on the *"Head Gimbal Assembly"* or HGA. Head Gimbal Assembly process will consist of several processes, which are assembly, cleaning and inspection. Within HGA process, inspection process is the main process for capturing the defects occurring during HGA process. Inspection process required skilled employee to perform, which depends on the employee experiences. To train the new employee for inspection process, it take at least 1 month before working with a real part in production process. Hence, this research aim to develop the algorithm to help and reduce the operation cost in HGA production. It is leading to HGA manufacturing and inspection improvement.

For developed an algorithm for inspecting the HGA part, *"Digital Image Processing"* or DIP and *"Measurement properties image of region"* are the proposed methodologies to perform an inspection process. As many searches show the results that *"Image Thresholding"* can be used to detected or inspected an abnormality, which appear on an object, by converting grayscale image to binary image (black & white image). Hence, this technique is the most effective in image with high level of contrast [1] – [4]. Figure 1 is shown the result of input image after applied an image thresholding.

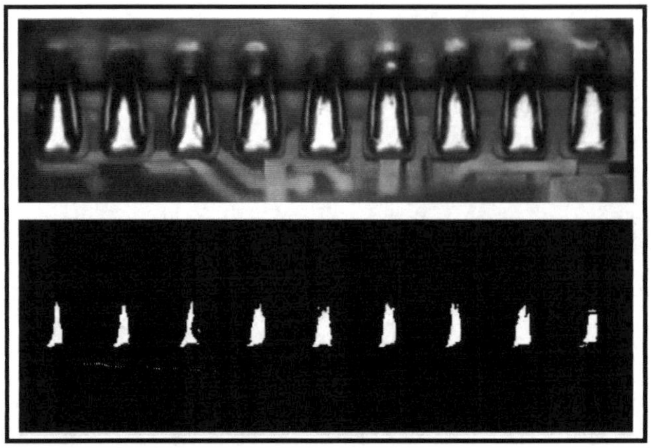

Figure 1: Image thresholding process

After applied image thresholding, *"Morphology"* is one of the DIP technique. It can be classified into 2 sub-techniques, which are *"Dilation"* and *"Erosion"*. Dilation is an expansion technique. It is used for fulfilled or expanded an interested area in a picture. This research also applied this technique after image thresholding in order to make an image suitable to process in the next step [5]. Erosion is a technique, which used for remove unnecessary area (size reduction) or collapse an object. Figure 2 is shown the result after applied dilation and erosion. Next step is about the justification process, whether this processed part is accepted or rejected? To justify the result, this research proposed *"Measurement properties image of region"* to identify the number of pad appeared in an input image, convex area value, and eccentricity value. These referred researches are evident that digital image processing and measurement properties of image region can be applied to classified part characteristics, part abnormality (appearance), and defects classification in HGA production process by using image thresholding, Morphology (dilation) techniques and measurement properties of image region.

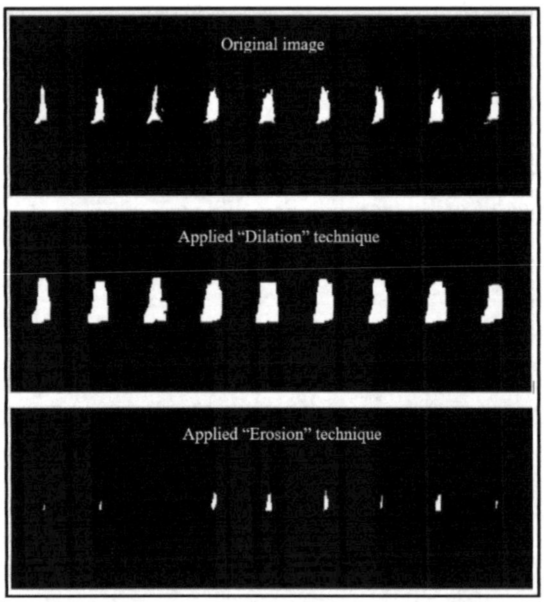

Figure 2: Result after applied "Dilation" and "Erosion"

2. Digital Image Processing (DIP) and Measurement properties of image region

In the research, DIP and measurement properties of image region are the techniques, which selected for performing an experiment. DIP is designed for image enhancement by using *"Image*

Thresholding" and *"Morphology"* in order to get a desired image. Image thresholding is one of the important technique in image segmentation, when T is the *"Thresholding value"*. This technique will replace each pixel in an image with a black color pixel, if the image density $I_{i,j}$ is less than some fixed constant T (That is, $I_{i,j} < T$), or a white color pixel, if the image density is greater than that constant [6]. Measurement properties of image region is the technique that use to measure a variety of image quantities ad features a black and white image. In this research, *"Convex area"* and *"Eccentricity"* are the properties, which using to determine the difference of an image after applied image thresholding as shown in the Figure 3-5.

Accepted

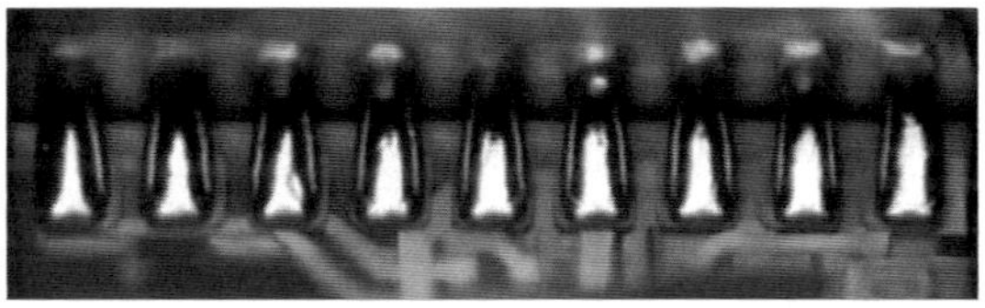

# of Pad_1	Convex Area_1	Eccentricity_1
9	5571	8.1

Figure 3: Accepted HGA part with convex area (5571) and eccentricity (8.10) value

Rejected

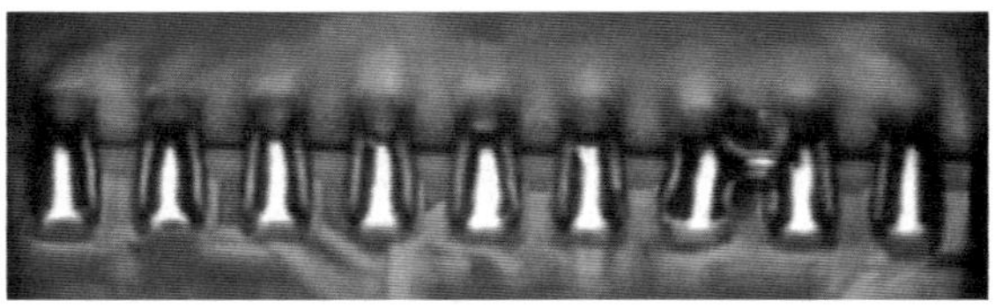

# of Pad_2	Convex Area_2	Eccentricity_2
9	4721	8.47

Figure 4: Rejected HGA part with convex area (4721) and eccentricity (8.47) value

Rejected

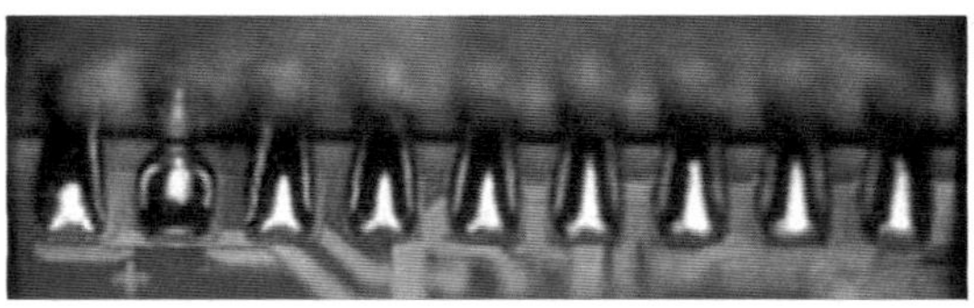

# of Pad_3	Convex Area_3	Eccentricity_3
9	2789	7.69

Figure 5: Rejected HGA part with convex area (2789) and eccentricity (7.69) value

Convex area can be calculated by Number of pixels in *"Convex Image"*, returned as a scalar. Eccentricity of the ellipse that has the same second-moments as the region, returned as a scalar. The eccentricity is the ratio of the distance between the foci of the ellipse and its major axis length. The value is between 0 and 1. (0 and 1 are degenerate cases. An ellipse whose eccentricity is 0 is actually a circle, while an ellipse whose eccentricity is 1 is a line segment.) [7] – [8].

IOP Conf. Series: Materials Science and Engineering **784** (2020) 012033 doi:10.1088/1757-899X/784/1/012033

3. Experimental equipment and condition

This experiment performed by collected the sampling of the HGA in the production process. Then used the digital camera to capture HGA images, which stored in memory retrieved to computer and using digital image processing and measurement properties of image region to analyse an input image. Developing an algorithm, then load an image data in an algorithm. After input the digital image data in an algorithm, it will do the digital image processing, which will enhance an image, feature extraction, and calculated convex area value, and eccentricity value. Then used an enhanced images and calculated data to classify the HGA part. Figure 6 is demonstrated the algorithm of digital image processing and measurement properties of image region.

3.1 Equipment setup and condition

Digital Image Processing (DIP) is performed by digital camera with hi-power microscope (with illuminate controlled) and computer in order to take a picture of solder area on HGA part, then input into algorithm to process. Illuminate nearby inspection process has more influence to quality of a picture taken. Hence, during this experiment, the pictures data has collected under controlled environment.

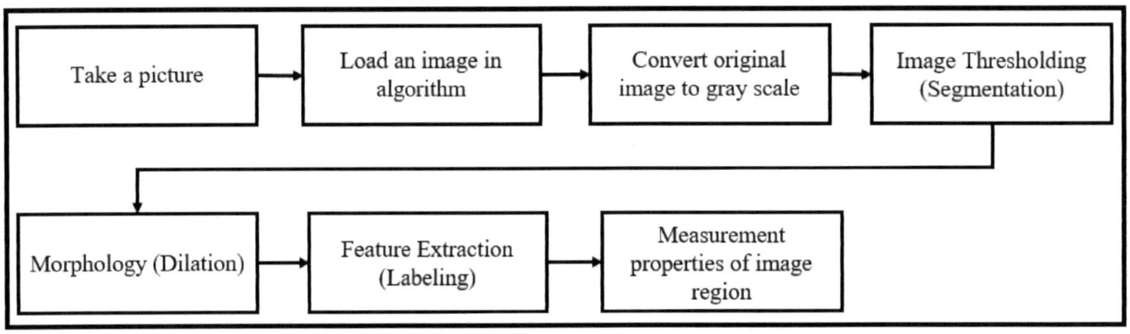

Figure 6: Digital Image Processing algorithm

3.2 Experiment performing

This experiment will be separate into 2 phases, which are digital image processing and measurement properties of image region. Digital Image Processing (DIP) is designed to process at the first phase of this experiment, in order to enhancement and prepare an image, then do a feature extraction in order to labeling a number on segmented objects. Next step is measurement properties of image region. This step will calculated the convex area value, and eccentricity value, which will use to identified the HGA part. The steps of experiment algorithm are consist of 7 steps (Figure 3). First step, take a picture of sample part, which collected in the real production process. Second step, load a picture in an algorithm. Third step, an algorithm will convert an original picture to grayscale picture. Fourth step, after getting a grayscale picture. The algorithm will do segmentation by checking each pixel, then compare to the target, which set at 250. If the value of pixel is less than 250, it will turn that pixel to 0, otherwise equal 255 (Figure 1). Fifth step, to removed noise after image thresholding by fulfilled an interested area with "dilation" 2 times (Horizontal axis and vertical axis). Sixth step, labeling each object and count on a labelled object in picture to check how many object(s) that algorithm can be detected? Last step, calculate the necessary parameters, which are convex area value, and eccentricity value.

4. Result and discussion

According to the DIP and measurement properties of image region, the result show that DIP can enhanced an image by using image thresholding technique in order to detect the number of HGA pads appeared in an image, then compare the result to the target, which is equal to 9 pads. After applied image thresholding, we observed that some image has an output of HGA pads equal to 9 pads as accepted condition. Hence, measurement properties of image region will be the second condition that will make an algorithm more accuracy. Measurement properties of image region algorithm will be implemented after image thresholding. Figure 7 is illustrated the measurement properties of image region algorithm.

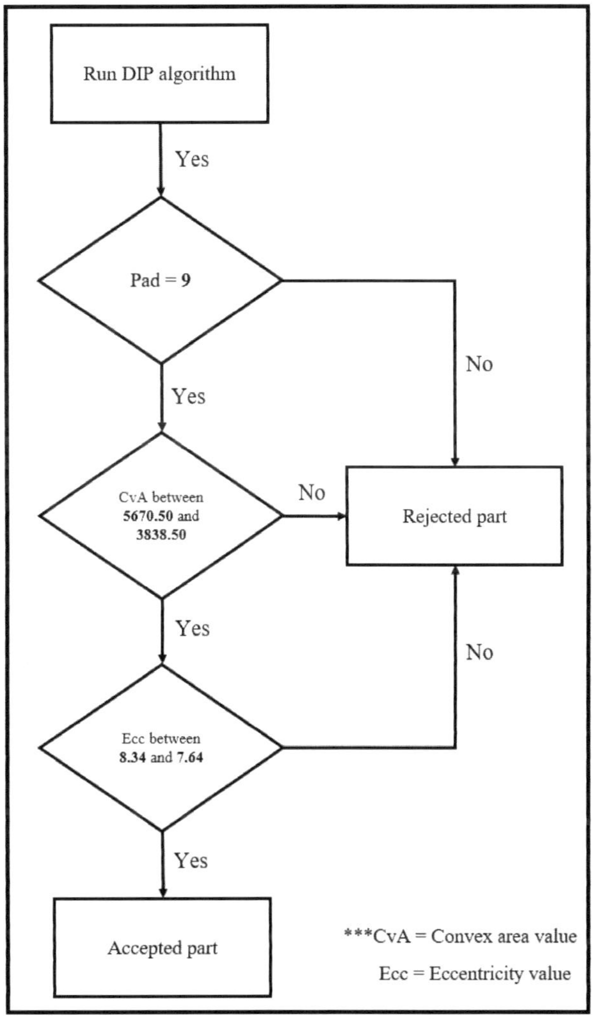

Figure 7: Justification algorithm by using "Measurement properties of image region"

This algorithm will be set number of HGA pads, convex area and eccentricity value as target in order to distinguish HGA part. All conditions must be satisfy, then the result will be accepted. Otherwise, this algorithm will be rejected that HGA part.

4.1 Target value

To set the target values, which are number of HGA pads, convex area and eccentricity value. For number of HGA pads, it set by physical count the acceptable HGA part. To set the convex area and eccentricity value, we applied *"Statistical Process Control"* or SPC to establish the control limit, which included *"Upper Control Limit"* or UCL and *"Lower Control Limit"* or LCL [9]. Hence, after applied SPC to convex area and eccentricity value. The result show that convex area value has UCL equal to 5671 and LCL equal to 3838. Eccentricity value has UCL equal to 8.34 and LCL equal to 7.64. Figure 8 – 9 is demonstrated that convex area and eccentricity value can distinguish the rejected HGA part.

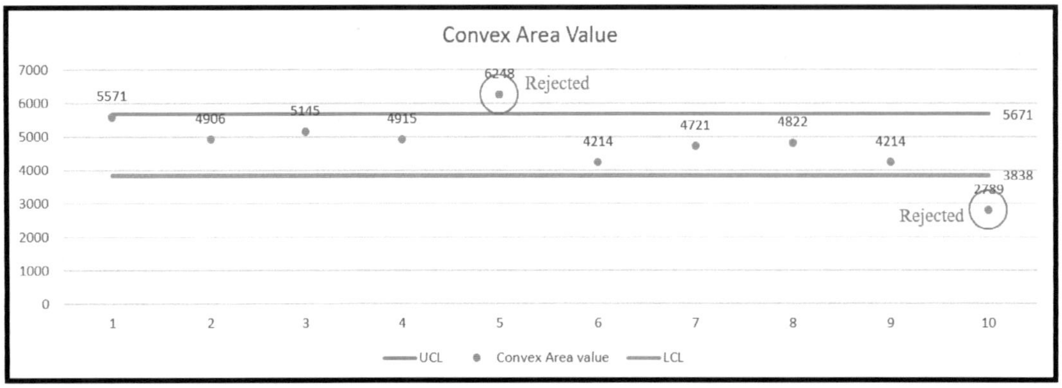

Figure 8: Convex area value with control limit

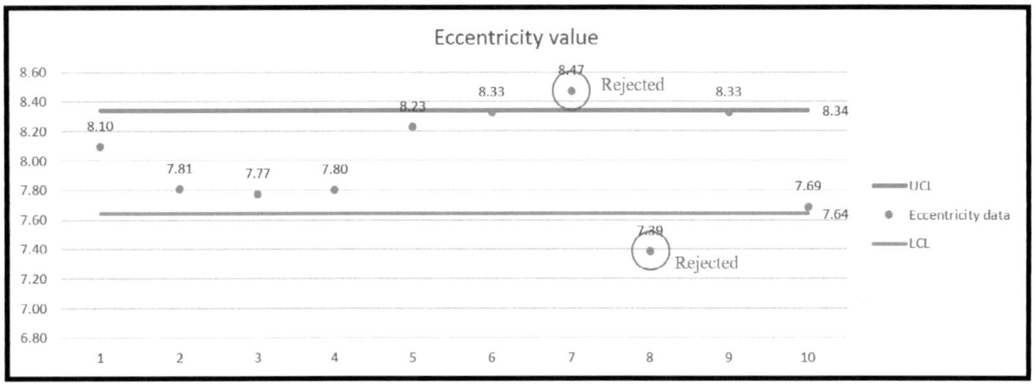

Figure 9: Eccentricity value with control limit

4.2 Algorithm verification

To verify the measurement properties of image region, we repeated an experiment by using the new set of data input in an algorithm, then see the result. Figure 10 is some of the result of the new data set.

1

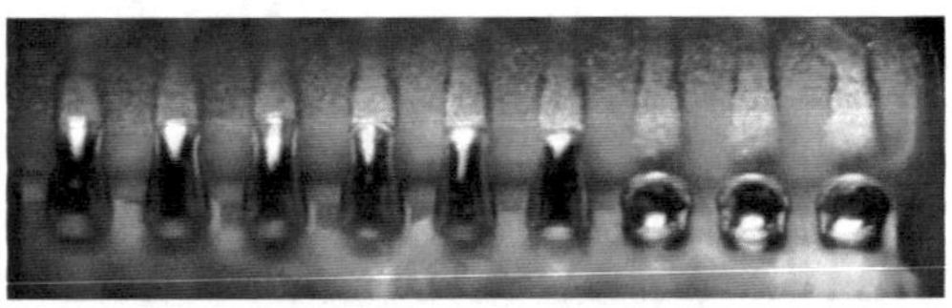

# of Pad_1	Convex Area_1	Eccentricity_1
9	3952	11.81

2

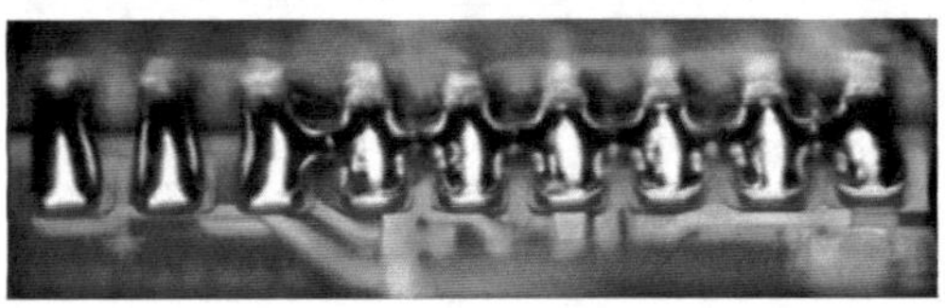

# of Pad_2	Convex Area_2	Eccentricity_2
9	10750	16.52

Figure 10: Result from new data set

5. Summary

As the result shown, Digital Image Processing (DIP) and measurement properties of image region can be utilized to detect a part characteristic, part abnormality (Appearance), and defect occurred during HGA production process by using image thresholding (Segmentation) in order to obtained the number of pad on HGA, then calculated the convex area value, and eccentricity value of each pad. The result interpreted by compared between the calculated results obtained from algorithm and target result. In the future, we can develop this algorithm by using a "Neural Network". The neural network will help this algorithm re-calculate the target values, which are convex area value and eccentricity value. And also detect the change in variation of HGA production process.

References

[1] Krishna.S, V. S., S.Gopalakrishanan (2015). "Image Processing of Metal Parts using Quality Inspection Method in Comparison with MATLAB." IJSRD - International Journal for Scientific Research & Development 3(04).

[2] Chantasitiporn, J. (2548). Medicine tablet indentification using feature extraction based on Neural Network. Department of computing, Silapakorn University.

[3] Aunkaew, S. (2550). A Study of Inspection of White Mould on Surface Rubber Sheets Using Image Processing. Thaksin. J., 10(2).

[4] Bruno, A., Gugliuzza, F., Ardizzone, E., Giunta, C. C., & Pirrone, R. (2019). Image Content Enhancement Through Salient Regions Segmentation for People With Color Vision Deficiencies. *Iperception, 10*(3), 2041669519841073. doi:10.1177/2041669519841073

[5] HENDRIKS, C. L. L. (2004). *Structure Characterization Using Mathematical Morphology.*

[6] N., S., & S., V. (2016). IMAGE SEGMENTATION BY USING THRESHOLDING TECHNIQUES FOR MEDICAL IMAGES. *An International Journal (CSEIJ), Vol.6.* doi:10.5121/cseij.2016.6101

[7] Information on https://www.mathworks.com/help/images/ref/regionprops.html

[8] A.Hamdan, M., A.Alqadi, P. Z., & M.Subaih, B. (2016). A Methodology to Analyze Objects in Digital Image using Matlab. *International Journal of Computer Science & Mobile Computing, Vol. 5*(Issue. 11), pg.21 – 28.

[9] WOODALL, W. H. (2000). Controversies and Contradictions in Statistical Process Control. *Journal of Quality Technology, Vol. 32.*

IOP Conf. Series: Materials Science and Engineering **784** (2020) 012034 doi:10.1088/1757-899X/784/1/012034

Research of electric heating control based on model predictive control algorithm

Zhang Chao[1] , Liu Bo[2]

[1]Institute of Mechanical and Electrical Engineering, Central South University Changsha 410000, China

[2]Institute of Mechanical and Electrical Engineering, Central South University Changsha 410000, China

[1]zchaos123@foxmail.com [2]liuboyh@126.com

Abstract. In order to improve the heating efficiency and control precision of electric heater, the method that applying the model predictive control algorithm to electric heating control was proposed. According to the analysis of electric heating characteristics, it was proposed that taking PID controller and the electric heater as the generalized object of the control to improve the dynamic response and the linearity. For realizing model predictive control, the system framework of this algorithm and the operating steps of controller were designed. The result of simulation and experiment proved this method effective and superior over the traditional control. Finally, the analysis on some parameters of controller provides a reference for others who interested in this study.

1. Introduction

As a common industrially equipment, electric heater is mostly used for heating gas and liquid to a setting temperature . Due to the complicated heating process, heater was often a nonlinear object with a large time delay and inertia, and it is difficult to construct its accurate mathematical model. Therefore, the traditional PID control often can not provide a good control effect,[1] and it is necessary to establish a more advanced control method. To improve the electric heating control ,there were many scholars made a lot of contributions. Liu Hao used fuzzy PID control to optimize the electric heating system on line,which obviously improved the control precision of electric heating. [2] Hang Bolin used the Least squares to establish the mathematical model of oil and gas electric heater based on his experimental data, and provided others a great reference for advanced control research on it.[3] Jia Li proposed that the cascade control of Neuro-fuzzy and PID on temperature control of thermal power units and verified the effectiveness of this method.[4]

Until now, advanced control research on the field of electric heating is rare, and the most are improvements based on PID control. As an advanced control method with the predictive capability on controlled objects, model predictive control is good at handling those objects with large time delay and inertia,[5] and also usually used on boiler heating control. This study hopes that extending this control method to the field of electric heating to improve the efficiency of industrial production and the control precision. The analysis of electric heater and its realizing process of model predictive control provide a reference for others who want to research this method further.

2. The principle of model predictive control [6]

The model predictive control consists of three steps as follows:

(1). Predictive model

The future response of system can be predicted by the current information (Original output: y_0 and model coefficients: a) of the controlled object and the future inputting variation Δu :

$$\hat{y}_1(k+i) = \hat{y}_0(k+i) + a_i \Delta u(k) \quad i = 1,2,\ldots,N \qquad (1)$$

For multiple future input, the predictive output of system in the future as follows:

$$\hat{y}_1(k+i) = \hat{y}_0(k+i) + \sum_{j=1}^{M} a_{i-j+1} \Delta u(k+j-1),\ i = 1,2,\ldots N \qquad (2)$$

(2). Rolling optimization

The target of optimization is the square of the inputting variation Δu and deviation between the predictive output $\hat{y}_1$ and the setting value y_r :

$$J(k) = \sum_{i=1}^{P} q_i [y_r(k+i) - y_M(k+i)]^2 + \sum_{j=1}^{M} r_j \Delta u^2(k+j-1) \qquad (3)$$

Setting the controlling steps to M, the predictive steps to P:

$$\begin{bmatrix} \hat{y}_M(k+1) \\ \cdots \\ \hat{y}_M(k+M) \\ \cdots \\ \hat{y}_M(k+P) \end{bmatrix} = \begin{bmatrix} \hat{y}_0(k+1) \\ \cdots \\ \hat{y}_0(k+M) \\ \cdots \\ \hat{y}_0(k+P) \end{bmatrix} + \begin{bmatrix} a_1 & & & \\ a_2 & a_1 & & \\ a_M & a_{M-1} & \cdots\ a_2 & a_1 \\ a_{M+1} & \cdots & \cdots & a_2 \\ a_P & \cdots & \cdots & a_{P+M-1} \end{bmatrix} \cdot \begin{bmatrix} \Delta u(k) \\ \cdots \\ \Delta u(k+M-1) \end{bmatrix} \qquad (4)$$

The optimizing target expressed in the form of matrix (**Q** is error weighting matrix, **R** is control weighting matrix):

$$J(k) = \left\| y_{rp}(k) - \hat{y}_{PM}(k) \right\|_Q^2 + \left\| \Delta u_M(k) \right\|_R^2 \qquad (5)$$

The matrix: $C^T = [10\cdots0]$ for first element, the operation parameter group $C^T(A^TQA+R)^{-1}A^TQ$ simplified as the control coefficient matrix $[d_1 \cdots d_P]$. The optimized variation of input as follows:

$$\Delta u_M(k) = C^T(A^TQA+R)^{-1}A^TQ\left[y_{rp} - \hat{y}_{r0}\right] \qquad (6)$$

(3). Dynamic correction

Weighted correction of the error between the predicted value and the actual output value:

$$e(k+1) = y(k+1) - \hat{y}_1(k+1) \qquad (7)$$

$$\hat{y}_{cor}(k+1) = \hat{y}_1(k+1) + he(k+1) \qquad (8)$$

3. Realization of the electric heating model predictive control

3.1 The characteristic analysis of electric heater

The electric heater, as a device that transforms electrical energy to heat energy and heats the medium in the industry, its heating temperature is usually controlled by monitoring of output or interior of equipment, as shown in Figure1. However, the electric heater possesses two obvious characteristics that non-linearity and a large time inertia. In addition, its accurate mathematical model is hard to constructed that due to the influence of variables in heating.

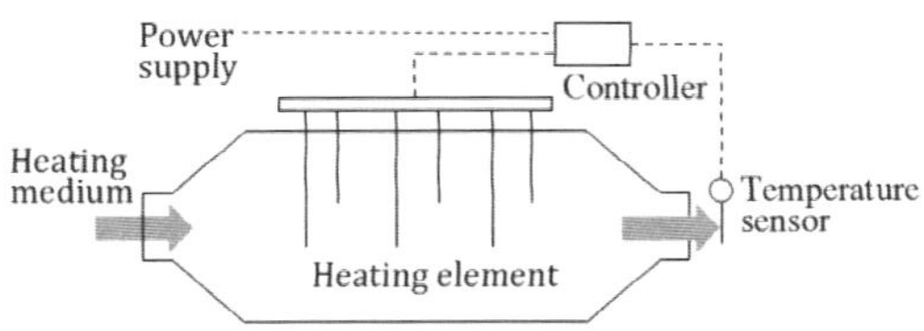

Figure1. Control of electric heater

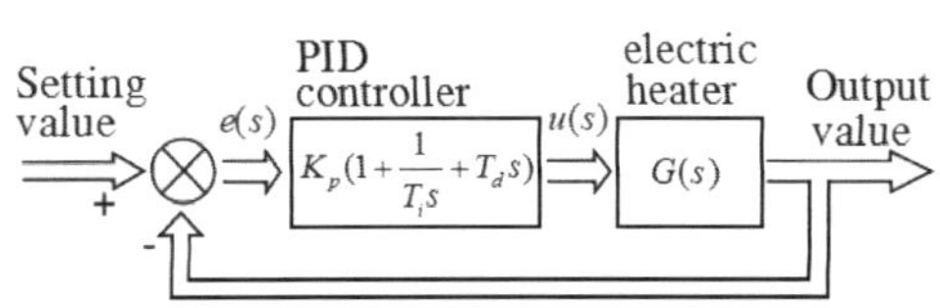

Figure2. Traditional PID method

The existence of non-linearity is because of the heating process that only the part of heat conduction is linear, both the thermal convection and radiation are non-linear, so that the characteristics of the electric heating process is non-linear. The large time inertia comes from the inertia of the representation of temperature in heat transfer process due to the influence of the heat capacity and the thermal resistance.[7] Different from the boiler system, the electric heater has the smaller time delay on the representation of characteristics. And the time delay is mainly originated form the travel distance between the central position of heating effect and the monitoring position of sensor. In a summary, the control on electric heater should consider the characteristics of non-linearity and large time inertia for better effect.

However, the current electric heating control was often the traditional PID control, as shown in Figure 3, which adjusts according to the deviation between the setting value and the actual output. The efficient electric heating process should adjust the set value according to the change of the heating rate and the traditional PID obviously can not do it. In addition, in the control of electric heating, the positive overshoot of input should be avoided as much as possible. Because the temperature rising is unidirectional in the heating actuator, and temperature can increase by increasing of electric current, but reduce only rely on heat transfer between the medium and itself. Therefore, the heat transfer may not keep up with the adjustment of controller and cause the control fails temporarily in case the excessive overshoot occurs.

As an optimal controlling method, model predictive control not only utilizes the deviation of actual output and expected output, but also uses the predictive model to predict the output in future, and uses rolling optimization to determine the future optimal input, so that the deviation between the actual output and the expected output is the smallest in the future. The implementation of this method also need some requirements on the control object:

(1). Only for linear model

It can be known from the principle that the prediction of the output is obtained by overlying the original output and the effect of input variation. Therefore, it is necessary to ensure that the control object is a linear model and conform to the linear superposition principle.For some non-linear object, it must be converted to linear.

(2). The almost dynamic information of the control object should be covered

The prediction of model predictive control is based on the Characteristics of the object's step response, therefore the modeling scope should cover all or the main dynamic information of the object. In the control of electric heating, the controlling period will become very short while the control precision is too high. Since the calculation range of the model predictive control should cover the most of the dynamic response, the amount of calculation will become very large when the control period is too short, and the performance requirements of the controller will become very high, it does not correspond to the industrial design requirements. Therefore, the responding speed of the controlled object should be appropriately increased.

From the above conditions, the controlled object should be designed to ensure that it conforms to the requirements of the model predictive control . According to the theory on degenerative feedback, the PID control can accelerate the dynamic response of the control object. Therefore, the electric heating system with the PID controller were designed as the control object, as the generalized object, which can effectively accelerate the dynamic response of the system and improve the controlling effectiveness by the PID.[8] In addition, the generalized object that composed of PID controller and electric heater also ensures that the non-linear electric heating model converted into a linear object with good linear characteristics.

3.2 Realization of model predictive control

According to the principle of model predictive control algorithm, the system includes three parts: predictive model, rolling optimization and dynamic correction. The predicted future output calculated by the controlling input and model parameters to realize the predictive model, and the real output is monitored and realize the dynamic correction by the deviation between predictive output and real. The optimized input is obtained by rolling optimization with the corrected predicted output. And the three are closed loop structures.

For better use of the model predictive control,the system structure of this algorithm was build. The calculation of the system is mostly matrix operation, the deviation between the setting value r and the predictive output $\hat{y}_P(k)$ is calculated with the control coefficient matrix $[d_1 \cdots d_P]$ to obtain the

IOP Conf. Series: Materials Science and Engineering **784** (2020) 012034 doi:10.1088/1757-899X/784/1/012034

optimized controlling variation $\Delta u(k)$. Then this current predicted output will become the Original initial output in next step, and calculated by the controlling variation $\Delta u(k)$ and the model coefficient matrix $[\alpha_1 \cdots \alpha_N]$ to obtain the next predicted output $\hat{y}_P(k+1)$. Thus, the continuous rolling optimization in model predictive control was formed as shown in Figure 3.

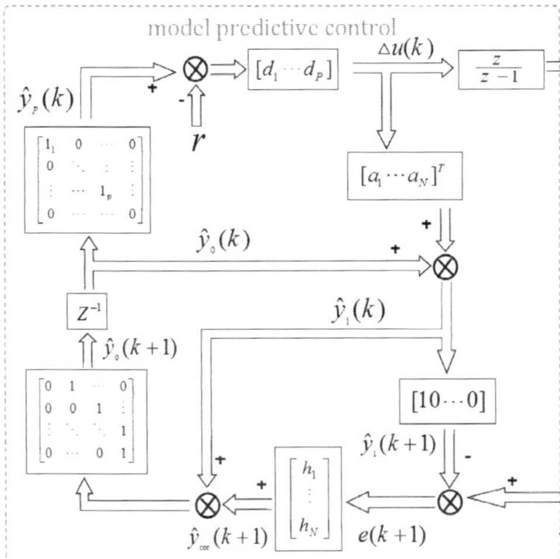

Figure3. System framework

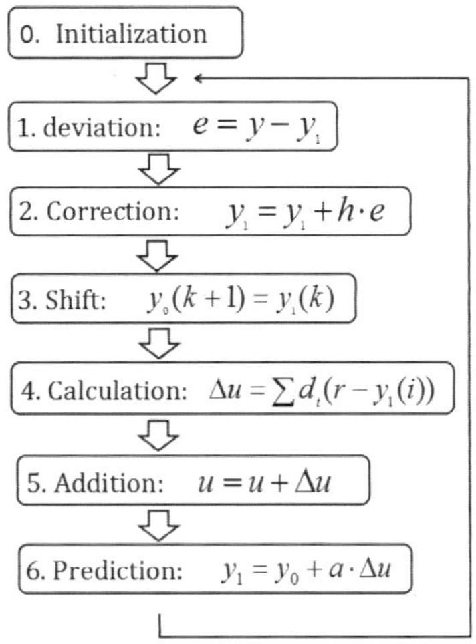

Figure4. Operating steps

From the system framework, some important parameters deeply affect the effect of controller. For example, the model coefficient matrix $[\alpha_1 \cdots \alpha_N]$ and its length *N*, the control coefficient matrix $[d_1 \cdots d_P]$ and its length *P*. These are some parameters must be established before the online operation of control. The operating steps as shown in figure 4 will realize the model predictive control after all the following parameters be set.

(1). model coefficient matrix ***a***

The model coefficient matrix should be established according to the characteristics of the controlled object's step response, which can be get from the mathematical model of the object or identified from the experimental results.

(2). control coefficient matrix ***d***

The control coefficient is used in the rolling optimization section, as the most critical parameter in the control method, consist of the model coefficient matrix, the error weighting matrix and the control weighting matrix, determines the controlling strategy and optimization.

(3). Correction coefficient matrix ***h***

The correction coefficient matrix ***h*** is composed of multiple weighting coefficients which correct the deviation between the predicted output and the predicted output to prevent the system mismatch.

(4). Sampling period: *T* and model time step: *N*

The sampling period *T* should conform to the Shannon sampling theorem. And the time range calculated by $T * N$ need to cover the responding time of the entire controlled process. Therefore, the sampling period *T* is related inversely to the length of the model time step *N*. The smaller *T*, the larger *N*, the resistance of the system is more better and the larger also the amount of calculation at this time. It should be moderate when selecting the sampling period and the model time steps.

(5). Predictive time step: *P* and control time step: *M*

The essence of the optimization is optimizing the output errors in *P* steps of the model and determining to complete in the future controlling variation in *M* steps. These two values greatly affect the optimization effect. The selecting principle is ensuring that the range of $T * P$ contain the main dynamic response information of the object. And the calculating amount in controller should be considered in the selection of *M* and *P*.

(6). Error weighting matrix ***Q*** and control weighting matrix ***R***

The error weighting matrix ***Q*** and the control weighting matrix ***R*** are related to the optimizing performance of the control. The former involves the proportion of the response error term, and the latter limits the incremental rate of input changing. In general, the two can be simply set to unit matrix.

4. The analysis on simulation and result of experiment

In order to verify the effectiveness and superiority of this method in the control of electric heating objects, the 75kw explosion-proof resistive gas electric heater was used as the object in the simulation and experiment. The model characteristics can be approximated by the transfer function as follows:

$$G(s) = \frac{5.3}{120.3s+1} e^{-0.85s} \quad (9)$$

The model coefficient matrix and control coefficient matrix came from the transfer function in formula(9), the setting temperature was 200, according to the Shannon sampling theorem, the sampling time of the main controller set to 1s, and the prediction time steps: 10, the control time steps to 2, the error weighting matrix, the control weighting matrix and the correction coefficient matrix used the normal default value. The control results are as follows:

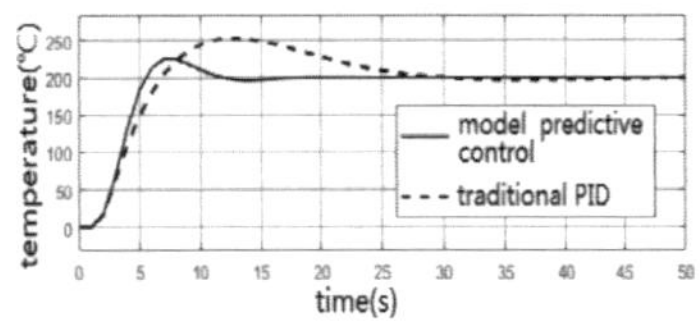

Figure5. Output in simulation

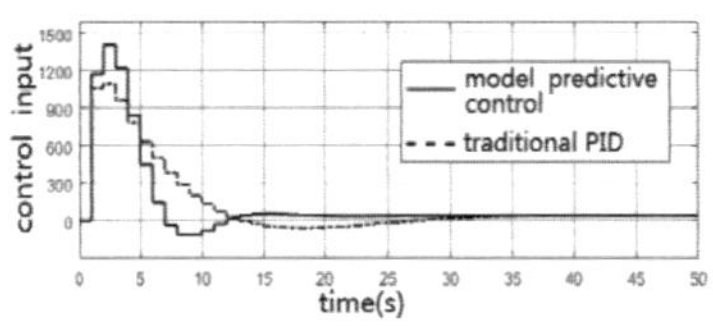

Figure6. Input in simulation

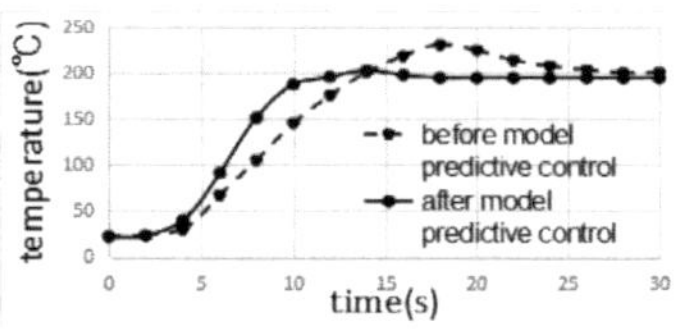

Figure7. Output in experiment

It can be seen from figure5 that compared with the conventional PID control, the overheating amount of the electric heating system on the model predictive control is smaller, and the adjusting time of the whole system is obviously shorter, the system is more stable and the anti-disturbance is stronger. Obviously, model predictive control has stronger adaptability and corrective ability for the system with electric heating characteristics, and is more fit on industrial application requirements. From the figure7, the results from actual experiment was almost the same as simulation. Although the actual responding speed not as fast as the expectation, the model predictive control was still more efficient than traditional methods.

The reason can be known from figure6, the change of control input in the model predictive control is more flexible than the traditional PID control. The former can advance the prediction of the future multi-steps, and track to obtain the optimal current input according to the predicted trajectory, the latter obviously can not do this, it deeply reflects the superiority of the predictive control on algorithm.

In order to better use of this control method in the future, the main parameters of model predictive controller was studied further. From the output of system on the different value of P and M, the influence of the predictive time step and the control time step on control can be seen. This following result came from the simulation because the maximum or minimum of setting on *P* or *M* may cause the damage to electric heater. It can be seen from figures8 and figures9 that the influence of the value selection of these two parameters is very obvious. When the predictive time step *P* was too small, the system will be unstable, and the large *P* will cause the system to react slowly. To ensure the stability of the system and sufficient responding speed, for the heating control object, the value selection of *P* can be a number between 10-20.

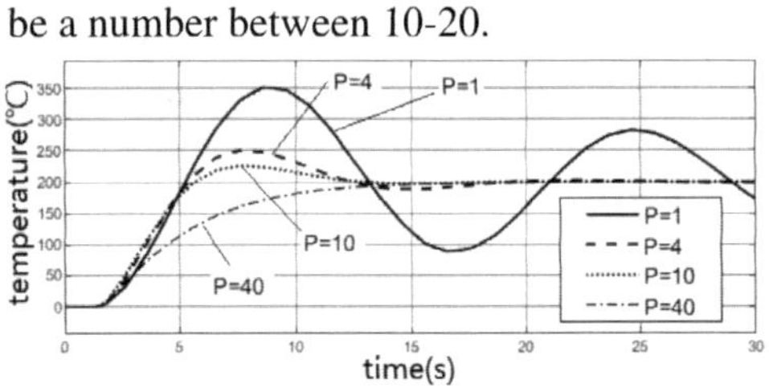

Figure8. The affect of value:P

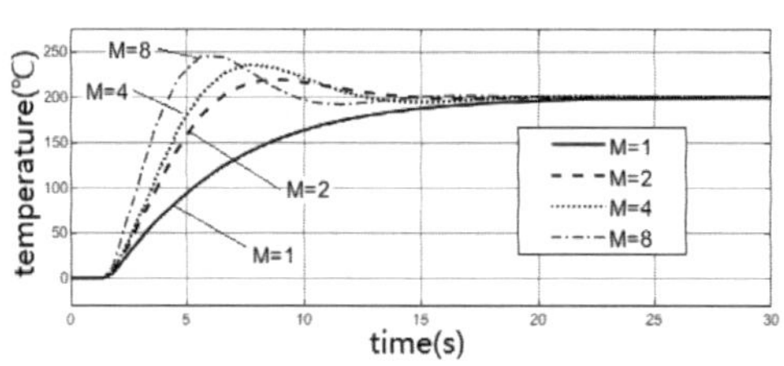

Figure9. The affects of value:M

For the control time step *M*, when the *M* value was too small, it was obvious that the system response is extremely slow. As the value of *M* increasing, the respond of system became more quickly, but the excessive *M* causes the system to be unstable, and the overshoot starts to increase obviously, the robustness of system was reduced. Since the *M* also acted as an order in the rolling optimization process, its influence on the calculating amount is very great, so it is possible to select a small M in an appropriate condition if the control requirement is ensured, for the electric heating system, the M setting to 2-4 can meet the needs of control well.

5. Conclusion

This study was devoted to realize the model predictive control on electric heater, and proposed that treating the electric heater and PID controller as the generalized object to improve the linearity and dynamic characteristics of the control object. For the realization of model predictive control, the system framework of the algorithm and the operating steps of the controller were established. The simulation and experimental results showed that this method is effective and superior to the traditional PID on the electric heating control. Finally, through the analysis on simulation, it is considered that the prediction time step setting as 10-20 and the control time step setting as 2-4, can achieve a better control effect with a small amount of calculation.

6. Reference

[1] Jiang Xiaoping, Tian Yuan, Liu Jian, Wang Lijun. Design of electric heating temperature controller based on nonlinear model [J]. Automation and instrumentation, 2010,25 (03): 29-32
[2] Liu Hao. Design and experimental research of electric heating temperature control system [D]. China Institute of Metrology, 2013
[3] Hang Berlin, Gao Guowei, Sun long, Zhang Junmei. Establishment of regression model for oil gas electric heater [J]. Petrochemical equipment, 2011,40 (06): 9-12
[4] Jia Li, Chai zongjun. Neural fuzzy PID cascade control of main steam temperature of thermal power unit [J]. Control engineering, 2013,20 (05): 877-881
[5] Xi Yugeng, Li Dewei, Lin Shu. Model predictive control: current situation and challenges [J]. Journal of automation, 2013,39 (03): 222-236
[6] Zou Tao, Ding Baocang, Zhang Duan. Introduction to application of model predictive control engineering [M]. Beijing. Chemical Industry Press, 2010
[7] Sisi Liu ; Runnan Dong ; Xi Chen. Application of the Main Steam Temperature Control Based on Sliding Multi-Level Multi-Model Predictive Control[C]. 2017 International Conference on Computer Systems, Electronics and Control (ICCSEC). 25-27 Dec. 2017
[8] Sun Huanyu. Application of predictive control algorithm in boiler temperature control system [D]. Changchun University of technology, 2013

AUTHOR INDEX

Institute of Physics
Dirac House, Temple Back
Bristol BS1 6BE UK

ISSN: 1757-8981
ISBN 978-1-7138-2668-2

CURRAN ASSOCIATES INC.
proceedings.com